Magnesium Batteries
Research and Applications

Energy and Environment Series

Editor-in-chief:
Heinz Frei, *Lawrence Berkeley National Laboratory, USA*

Series editors:
Nigel Brandon OBE FREng, *Imperial College London, UK*
Roberto Rinaldi, *Imperial College London, UK*
Vivian Wing-Wah Yam, *University of Hong Kong, Hong Kong*

Titles in the series:
1: Thermochemical Conversion of Biomass to Liquid Fuels and Chemicals
2: Innovations in Fuel Cell Technologies
3: Energy Crops
4: Chemical and Biochemical Catalysis for Next Generation Biofuels
5: Molecular Solar Fuels
6: Catalysts for Alcohol-Fuelled Direct Oxidation Fuel Cells
7: Solid Oxide Fuel Cells: From Materials to System Modeling
8: Solar Energy Conversion: Dynamics of Interfacial Electron and Excitation Transfer
9: Photoelectrochemical Water Splitting: Materials, Processes and Architectures
10: Biological Conversion of Biomass for Fuels and Chemicals: Explorations from Natural Utilization Systems
11: Advanced Concepts in Photovoltaics
12: Materials Challenges: Inorganic Photovoltaic Solar Energy
13: Catalytic Hydrogenation for Biomass Valorization
14: Photocatalysis: Fundamentals and Perspectives
15: Photocatalysis: Applications
16: Unconventional Thin Film Photovoltaics
17: Thermoelectric Materials and Devices
18: X-Ray Free Electron Lasers: Applications in Materials, Chemistry and Biology
19: Lignin Valorization: Emerging Approaches
20: Advances in Photoelectrochemical Water Splitting: Theory, Experiment and Systems Analysis
21: Electrochemical Reduction of Carbon Dioxide: Overcoming the Limitations of Photosynthesis
22: Integrated Solar Fuel Generators
23: Magnesium Batteries: Research and Applications

How to obtain future titles on publication:
A standing order plan is available for this series. A standing order will bring delivery of each new volume immediately on publication.

For further information please contact:
Book Sales Department, Royal Society of Chemistry, Thomas Graham House, Science Park, Milton Road, Cambridge, CB4 0WF, UK
Telephone: +44 (0)1223 420066, Fax: +44 (0)1223 420247,
Email: booksales@rsc.org
Visit our website at www.rsc.org/books

Magnesium Batteries
Research and Applications

Edited by

Maximilian Fichtner
Helmholtz Institute Ulm, Germany
Email: m.fichtner@kit.edu

ROYAL SOCIETY
OF **CHEMISTRY**

Energy and Environment Series No. 23

Print ISBN: 978-1-78801-434-2
PDF ISBN: 978-1-78801-640-7
EPUB ISBN: 978-1-78801-896-8
Print ISSN: 2044-0774
Electronic ISSN: 2044-0782

A catalogue record for this book is available from the British Library

The Royal Society of Chemistry is a charity, registered in England and Wales, Number 207890, and a company incorporated in England by Royal Charter (Registered No. RC000524), registered office: Burlington House, Piccadilly, London W1J 0BA, UK, Telephone: +44 (0) 20 7437 8656.

For further information see our web site at www.rsc.org

Printed in the United Kingdom by CPI Group (UK) Ltd, Croydon, CR0 4YY, UK

Preface

Magnesium is one of the few elements on Earth that offers real long-term potential for technical mass applications. One reason for this is that reserves of Mg minerals are widespread and easy accessible and these reserves should last approximately 430 000 years (seawater not included), if the consumption remains constant. Another reason is that magnesium is recyclable and virtually non-toxic. Rather, it is an essential element in nature and central in photosynthesis and for metabolic functions in animals and humans. In batteries, the Mg ion can transport and store two charges per ion and, thus, offers the potential for higher storage capacities, which is complemented by the fact that a Mg metal anode can work safely in a cell with a liquid electrolyte. In contrast, Li ions are stored in graphite in the anode of current Li ion batteries, for safety reasons – and are thereby diluted by a factor of eight.

A good, working and rechargeable Mg battery would not only be technical progress, which could potentially improve the performance of battery-driven applications, it would also be a relief, because it opens up the possibility for the mass production of big batteries, which is, in the mid- to long-term, more difficult to achieve with current Li ion technology due to the limitation of certain raw materials and the expensive recycling process for lithium.

The work on Mg batteries started with some scattered publications a few decades ago. The feasibility was demonstrated in principle but it became clear that the classical concepts for electrolytes and electrodes cannot just be transferred or applied to Mg battery technology and more questions were raised than answered. Then, the field moved more in the focus of battery research in the 2000s, after a breakthrough had been achieved with the

Energy and Environment Series No. 23
Magnesium Batteries: Research and Applications
Edited by Maximilian Fichtner

Published by the Royal Society of Chemistry, www.rsc.org

development of a new and effective electrolyte. However, there were still no more than 10 to 15 publications per year. This changed considerably in the 2010s when the number of publications steadily increased so that at the end of the decade, between 120 and 150 publications per year had been made on various topics of Mg batteries.

After this initial phase, with its quest for the first electrolytes, cathodes and anodes and after both progress and frustration, it is our aim to present here a first extensive summary of the principles, state-of-the art and prospects of Mg batteries.

The book will start with a general overview and motivation for a Mg battery by M. Fichtner, followed by an electrolyte part with a chapter on the development of non-aqueous systems by R. Mohtadi and O. Tutusaus and a chapter on solid-state magnesium ion conductors by C. Battaglia *et al.* Theoretical modelling of multivalent ions in inorganic hosts is critical for an understanding of the dynamics and storage process of Mg in materials and will be introduced by G. Gautam and P. Canepa. The anode side seems simple and straightforward because pure Mg metal has been claimed to be suitable as an anode material. However, the chapter by E.M. Sheridan *et al.* shows that there are many more options, which all have their prospects and drawbacks. Complementing work is presented by M. Matsui in a chapter where the electrochemical properties of magnesium metal and intermetallic anodes are discussed.

One of the most critical issues in the field is the development of effective insertion cathodes for magnesium batteries. Intercalation and conversion materials and related mechanisms are introduced by H.D. Yoo and S.H. Oh. B. Ingram will further expand upon this discussion with a contribution on high voltage cathodes, which includes a techno-economical evaluation based on a methodology developed in their laboratory. The part on cathode materials will be complemented by an overview on organic electrodes by J. Bitenc and R. Dominko and a further contribution from Z. Zhao-Karger, who gives an overview of the properties and status of Mg sulfur batteries.

An unusual but interesting concept is the dual-ion battery, which relies on the combination of the Mg ion plus a second ion, both reversibly stored in different electrodes. The concept, first results and interesting effects of the co-intercalation of cations and anions will be presented by H. Li *et al.* As aqueous (primary) Mg batteries are still one of the most widespread applications, D. Hoeche will present an overview on the status of this technology.

Finally, although the development of Mg batteries is still at an early stage and the performance of Mg batteries is not yet competitive, data already available from laboratory cells has been collected by C. Tomasini Montenegro *et al.* The data was evaluated for the first life-cycle analysis of Mg batteries and gives clear indications for further directions in the development of Mg battery cells.

At this point, I would like to cordially thank all of the authors and co-authors for their commitment and dedication and their contributions to all of the relevant topics of Mg battery technology, making "Magnesium Batteries – Research and Applications" a comprehensive reflection of the state-of-the art in the field.

Maximilian Fichtner
Helmholtz-Institute Ulm (HIU), Helmholtzstr. 11, Ulm, 89081, Germany

Contents

Energy and Environment Series No. 23
Magnesium Batteries: Research and Applications
Edited by Maximilian Fichtner

Published by the Royal Society of Chemistry, www.rsc.org

CHAPTER 1

Motivation for a Magnesium Battery

MAXIMILIAN FICHTNER*

Helmholtz Institute Ulm (HIU), Helmholtzstr. 11, Ulm, 89081, Germany
*E-mail: m.fichtner@kit.edu

1.1 Introduction

Batteries, as one of the most efficient and versatile energy storage technologies, play a central role in the ongoing global transition from fossil fuels to renewable energy. They are key enablers for the decarbonisation of both the transport sector (electric mobility) and the power sector (stationary storage of intermittent, and decentralized renewable energy sources), and are essential in a broad range of strategic industries (including automotive, power grids, aerospace, portable electronics, medical devices, and robotics).[1,2]

The increasing demand for more and better batteries is currently starting to put pressure on established technologies such as the Li-ion battery and researchers worldwide have started to search for alternatives that have the potential to be "better" in terms of storage capacity, power, safety, and costs. For a few years, sustainable solutions have started to move into focus, as supply risks are foreseen in the mid- to long-term for certain elements that are used in current Li-ion batteries. Magnesium batteries have been regarded as one possible option and preliminary work already started

Energy and Environment Series No. 23
Magnesium Batteries: Research and Applications
Edited by Maximilian Fichtner

Published by the Royal Society of Chemistry, www.rsc.org

in the late 1980s on the study of the properties of such systems. Overall, the work on this new type of batteries has been motivated mostly by three factors:

1. the scientific curiosity of researchers who are entering a *new and less explored field* with numerous scientific and technical challenges – with the benefit that the gained knowledge will lead to a more comprehensive understanding of electrochemical energy storage
2. the ongoing *need for "better" batteries* in order to satisfy the demands of future applications, and
3. the need for *sustainable storage solutions*, based on globally abundant resources that are available in the long term, materials which are inherently safe, non-toxic, recyclable and less costly.

The first item offers a series of new opportunities for doing research on topics which have been less explored so far. As will be shown later, the bivalent nature of the magnesium ion not only offers the possibility of carrying and storing more charge per ion, it also poses serious scientific challenges when it comes to explaining and improving the mobility of the hard cation in liquids and, in particular, in solid host structures.

The second item deals with the expectation that the physical properties of the bivalent Mg should allow the building of batteries that can outperform current Li-ion batteries, in particular by offering higher energy densities. These assumptions have been triggered by the sheer materials data of the magnesium anode, which is highly promising, see also Table 1.2. In that respect, it is regarded as highly promising that the Mg anode can be used in a metallic, *i.e.* undiluted form, rather than the Li anode, which uses a graphite host to intercalate Li ions, in order to prevent the formation of metallic dendrites, see below. This safety measure dilutes the lithium concentration by a factor of ten.

The need for sustainable storage solutions, mentioned under item 3, moves more and more in the focus of battery development, in particular because more and more applications based on large batteries are considered in practice. Electrochemical storage in large batteries for stationary storage systems but also in sports utility vehicles (SUVs) and trucks will consume a large amount of raw materials. The materials demand to realize large applications exhaustively would by far exceed the demand of portable and mobile applications. As an example, the estimated global energy–storage demand of cell phones and tablets is in the order of 40–150 GWh, while the estimated demand of home storage is up to 3000 GWh; for trucks it would be up to 62 500 GWh.[3] In the same study, the demand for Li and Co for the battery industry has been predicted based on various scenarios. In accordance with other studies in the field, the results conclude that, under realistic growth scenarios, there may be a shortage of cobalt in the next 10 years and lithium resources from salt lakes and mines might dwindle in the

2050s. Recycling of cobalt and other heavy metals in batteries is possible and is being done already. However, only a few percent of Li-ion batteries are recycled at the moment and strong efforts will be needed to ramp up the global recycling capacities, in order to deal with the enormous returns of spent batteries from battery vehicles that are expected from the 2030s onwards. The current industrial recycling processes depend on recovering cobalt and nickel metals or their alloys. Lithium is typically not recycled at the moment, mostly because there is limited or no economic benefit to recycle lithium in today's commercialized technologies. In effect, there are large energy costs and the chemical clean-up procedures needed for the variety of cathode chemistries and sortation based on chemistry is challenging and difficult.[4] It is therefore expected that federal regulations will have to be established in order to initiate Li recycling. This may, however, have a negative impact on future battery costs, which are actually expected to drop.

In the following, an outline will be given on major research directions which are dealing with the aims for pushing research to new frontiers and developing technologically viable solutions. Moreover, the need for "better" batteries will be illustrated and the growing demand for sustainable storage solutions will be discussed.

1.2 Overview on Research Topics

At its still early stage of development, two major issues in the work on Mg batteries have been the *development of electrolytes* for transport and reversible stripping/plating of the doubly charged Mg ion and the *development of cathodes* which can either reversibly insert and de-insert Mg ions or convert the Mg into a certain compound by a chemical reaction, see also Chapters 5–10. In both fields it became apparent that the concepts that have been established in other battery types cannot just be transferred to Mg technology. Rather, it has turned out to be essential to considerably expand upon the existing knowledge on Mg chemistry and electrochemistry by doing in-depth research on the new systems in order to generate a knowledge base which may eventually help to enable viable technical solutions. Also *anodes* are a matter of research, in particular composite anodes or alloys with Mg. A new field is the use of new compounds that can insert Mg ions at low potentials.

1.2.1 Electrolytes

Electrolytes for secondary magnesium batteries have been a particular challenge from the beginning after it became clear that the classical approach for making electrolytes, which is for example pursued in Li-ion, NiMH or NiCd technology, may not work: dissolving an electrolyte salt in a solvent and using the solution as an electrolyte was simply not successful for the Mg battery in the first phase of development. There, the anions of the tested salts either reacted with the magnesium anode, or the anode was covered by a passivation layer. In both cases, the electrochemical reaction stops. A similar passivation layer also forms at the anode of Li-ion battery cells (the so-called SEI, solid electrolyte interface), but there, the singly charged and softer Li ion can easily penetrate the layer, while the hard and doubly charged Mg ion can obviously not pass the rigid interphase layers on the Mg surfaces.[5]

Preliminary work in the context of the later magnesium battery dates back to more than 100 years ago when Grignard developed Mg organometallic reagents,[6] which were later tested as electrolytes as they are capable of reversibly stripping and plating magnesium.[7] In this work by Gregory *et al.*, Mg electrolytes were synthesized in ethereal solvents *via* the reaction of an organomagnesium compound with aluminium chloride ($AlCl_3$) or trialkylborane as a Lewis acid. This publication inspired numerous follow-up studies because they not only demonstrated the first feasible electrolyte (even if the performance was not yet convincing), they also published the first table with suggestions for potential insertion materials, investigated some of them and published the first data, see Table 1.1.

With a similar synthetic concept, Aurbach *et al.* prepared a second generation of electrolytes through the reaction between Bu_2Mg and $EtAlCl_2$ in tetrahydrofuran (THF) and demonstrated the first prototype of a good

Table 1.1 Mg^{2+} intercalation materials. Adapted from ref. 1 with permission from the Royal Society of Chemistry.

		Capacity	
Material	$E_{\text{Mg-cathode}}$, V	Moles Mg/mole host	Ah g^{-1}
Co_3O_4	2.28	0.80	0.222
Mn_2O_3	2.40	0.66	0.224
Mn_3O_4	2.40	0.66	0.154
MoO_3	2.28	0.50	0.143
PbO_2	3.10	0.25	0.056
Pb_3O_4	3.10	0.25	0.020
RuO_2	2.55	0.66	0.266
V_3O_5	2.66	0.66	0.194
WO_3	2.16	0.50	0.116
TiS_2	1.63	0.15	0.157
VS_2	1.71	0.34	0.154
ZrS_2	2.60	0.66	0.228
MoB_2	1.15	0.66	0.301
TiB_2	1.25	0.42	0.324
ZrB_2	1.20	0.66	0.313

rechargeable Mg battery.[8] Unlike the former chemically very aggressive and dangerous Grignard reagents the electrolyte was made *via* a Lewis acid–base reaction of an organometal compound as a base and $AlCl_3$ as an acid. The active component in this electrolyte is a binuclear Mg complex where two Mg atoms are bridged by three Cl atoms. The residual six coordination sites are occupied by coordinated solvent molecules. With a Chevrel phase as the cathode, several thousand cycles were demonstrated. Later studies had the goal of further simplifying the synthesis, to reduce the strongly reducing character of the organometallic components in the electrolyte, and, recently, to develop electrolytes that no longer contain corrosive constituents.

A first approach in that direction was realized by Mohtadi *et al.*, who demonstrated reversible stripping and plating of Mg from solutions of $Mg(BH_4)_2$ in THF.[9] The electrolyte did not contain corrosive Cl but had a narrow stability window and only a low salt concentration was possible, leading to low ionic conductivity. A few years later, the same group presented an electrolyte based on Mg closoborate. This anion is much more stable due to the aromatic system of the boron cage. It also offers a higher ionic conductivity. However, the synthesis of such compounds is expensive and time-consuming.[10] Further progress was made when an electrolyte based on a similar rationale, namely a Mg salt with a big and weakly coordinating, very stable anion was presented by Zhao-Karger *et al.*[11] The salt can be synthesized by a simple and straightforward one-step reaction from $Mg(BH_4)_2$ and hexafluoro-isopropanol, rendering $Mg[B(OR^F)_4]_2$ in a quantitative yield, with H_2 gas as the only by-product. The electrolyte can then be made from solutions of the salt in a variety of solvents. The ionic conductivity of such systems is in the order of 10 mS cm^{-1}, which is in the range of current electrolytes for Li-ion batteries. The electrochemical stability window is 4.3 V *vs.* Mg, the electrolyte is stable in air and is non-nucleophilic, making it compatible with a sulfur cathode so that Mg–S cells can also be built. In effect, this work and the pioneering studies before have paved the way back to realizing the "classical" electrolyte concept which was mentioned above.

In spite of the technical advances, there are a lot of issues which need to be understood and overcome. As an example, the control of the interaction of the electrolyte and potential side products with the battery electrodes will be crucial for the development of viable systems with high energy efficiencies. Most publications so far have reported extensive overpotentials which hint at serious kinetic barriers in the transition of the Mg ion from the electrolyte to the surface of the metal anode. In this process, the Mg ion needs to strip off its solvent shell, migrate to the interface, potentially penetrate surface layers and receive electrons so that it becomes metallic and plates at the anode when charging the battery.[12–14] It is the aim of an increasing number of studies to understand these limitations and to find ways how these barriers can be removed or mitigated. It is obvious that further studies at the electrolyte–electrode interface will reveal more details of the structure and composition of the adlayers, their origin and their interference with the plating process. This will be essential for achieving further progress in the field.

1.2.2 Cathodes

The development of viable insertion cathodes is one of the most challenging endeavours in research and development. So far, only a few effective Mg cathode materials have been reported. Gregory *et al.* started to evaluate the ability of transition metal oxides, sulphides, and borides to reversibly insert Mg^{2+} ions, see Table 1.1. Some of these hosts were investigated by Novak *et al.* in the 1990s, with $Mg(ClO_4)$ in tetrahydrofuran (THF) as an electrolyte. The conclusion was that only V_2O_5 showed a reasonable capacity.[15] Later, Aurbach *et al.* achieved the first breakthrough by introducing a new electrolyte using Mo_6S_8 (*i.e.* the Chevrel phase) as a cathode and Mg metal as an anode. This cell is still a good working reference for Mg-ion batteries, with a volumetric energy density comparable to that of a Li-ion battery. In recent years, the experimental work on Mg insertion hosts was still concerned with synthesizing and testing structures that were suggested in Gregory's early work. In addition, conceptually new approaches to identifying suitable materials have been explored, *e.g.* the development of metastable materials.[16] This approach offers new opportunities in identifying a new functional multivalent cathode material. It may also be of general interest as a general concept because metastable compounds can obviously offer sufficiently low migration energies to support reversible cycling with multivalent ions.

In addition to this more conceptual and experimental work, computer-aided design and *ab initio* calculations have been used to reveal suitable atomic configurations and structures for the diffusion of Mg in host materials such as spinels or layered compounds.[17] Recent work has also shown that co-intercalation of solvent or shielding of the hard Mg^{2+} ion by coordinated solvent can lead to a softening of the ion so that faster insertion and de-insertion is enabled in certain matrices.[18]

Moreover, it was found by Matsubara *et al.* (see Chapter 11) that the motion of Mg ions in oxidic hosts can be greatly facilitated by the presence and interaction of other inserted and mobile ions, such as Li^+.

An alternative to cathode insertion materials are conversion electrodes, sulfur electrodes in particular, which offer high capacities at reasonable voltages. As shown in Table 1.2, an effective and efficient Mg–S battery would have quite interesting features when compared to a Li–S battery. The first work on a Mg–S system[19,20] already indicated that Mg–S batteries are feasible, but besides the already mentioned issue of overpotential, there is a capacity loss upon cycling. So far, the origin has mostly been attributed to polysulfide intermediates that form in both Li–S and Mg–S batteries[21] and are partly soluble in organic electrolyte.

Table 1.2 Comparison of theoretical properties of the Mg–S system with Li–S (metal anode) and Li–S (graphite anode), based on materials values.

	Mg/S	Li/S
Theor. grav. cap. anode [mAh g^{-1}]	2205	3861 (LiC_6: 372)
Theor. vol. cap. anode [mAh cm^{-3}]	3832	2062 (LiC_6: 833)
Theor. energy density [Wh L^{-1}]	3200	2800 (LiC_6: 1100)
volumetric change S → M_xS (molar volume S ~16 cm^3 mol^{-1})	molar volume MgS: 19.9 → 24% expansion	molar volume Li2S: 27.7 → 72% expansion

1.2.3 Anodes

Four types of anodes have been investigated for Mg batteries so far:

- sheets of pure Mg metal
- finely dispersed Mg particles in carbon composites
- sheets of Mg-based alloys
- Insertion compounds for Mg at low potential

Using plain Mg metal appears to be the most simple and straightforward approach. Mg metal is not expensive and can be machined by various techniques – although the mechanical properties of Mg are not favourable. A practical advantage is, that the metal sheet can be used both as an electrode and a collector. This has been demonstrated in a collaborative project[22] where the first pouch cells with sulphur cathode and Mg metal anode were fabricated and tested. However, although the concept appears simple and convincing, the power density of such systems is limited and the abovementioned adlayers on the Mg surface obviously provide kinetic barriers for Mg transfer which lead to considerable overpotentials during charging. Therefore, the formation mechanism, the composition, and the prevention or removal of such layers is an important topic of research. This is closely associated with the work on electrolytes, as such layers typically form when impurities are present in the electrolyte and/or decomposition products of the electrolyte salt or the solvent can chemically interact and form a solid film.

Finely dispersed Mg particles in carbon have already been used in early studies on Mg–S batteries[21] as they offer the advantage of a larger surface for the electrochemical reaction.[23] This enhances the rate capability of the system and reduces the overpotentials. However, there are also trade-offs due to the reduced capacity of the electrode due to the addition of carbon. Moreover, finely dispersed magnesium may strongly react with oxygen in air so that passivation may occur when handling the material in an uncontrolled atmosphere.

Compared to the pure metal, magnesium alloys can improve the ductility of the metal anode so that fabrication is easier and more safe. In addition, as Mg metal may not be compatible with electrolytes that are not made from ethereal solvents, a suitable alloy may mitigate or prevent such effect due to its altered electrochemical properties compared to those of the pure element. Alloying elements dissolve and redeposit during galvanostatic cycling without having a negative effect on the electrochemical properties of Mg.

Recently, the search for other anode materials has started and there will be a review on alternative anodes in Chapter 5. Various types of compounds have been tested as hosts for the insertion of Mg ions, including candidates that are known from Li ion technology, such as graphite or titanates. There, it may be beneficial to co-intercalate either solvent molecules (graphite) or other ions such as Li^+ or Na^+ (titanates) in order to achieve reversible behaviour. Other compounds such as phosphorenes and borophenes have been proposed as new candidates and mostly studied by theoretical modelling, but little is known about the behaviour of various structural configurations when Mg ions are inserted and de-inserted.

1.2.4 Mg Deposition and the Lack of Dendrite Formation

The formation of dendrites in batteries with metal anodes and liquid electrodes poses a major safety risk as the fine metallic needles can grow through the separator and cause short-circuits inside the battery, which may trigger the release of hazardous substances, fire, and explosions.

As was confirmed in many studies, the use of Mg metal as an anode does obviously not pose the risk of the formation of dangerous dendrites. This is a particular feature and advantage of the Mg metal anode as most other metals such as Li, Na, and Zn, do form dendritic structures in liquid electrolytes. There are several models in the literature to explain the formation of dendrites in Li-ion batteries, but only recently was a first attempt made towards an explanation of the different trends for dendrite formation of the different metals. Jäckle *et al.*[24] investigated different scenarios for self-diffusion of Li, Na, Mg, Al, and Zn using density functional theory calculations. They studied self-diffusion barriers including diffusion barriers that are relevant for three-dimensional growth such as barriers for diffusion across steps in order to find out whether there are "simple" relationships which would explain the trends of dendrite formation. Their results suggest that Li dendrite growth is an inherent property of the metal, which is also in agreement with experimental observations.

Overall, the following challenges are key and need to be addressed for the development of next generation electrode materials for Mg:

- *Increasing the ionic mobility of the bivalent Mg ion* within host matrices through a better understanding of structure–property relationships. Multivalent ions pose a particular challenge for intercalation electrodes, as their mobility is usually low in host materials. Therefore, strategies need to be developed for designing an "ideal" geometrical and chemical environment of the mobile ion in order to optimize both ion packing density and mobility.

- *Improving the stability of electrode materials*: Current insertion hosts often suffer from structural instability. The challenge here is to design and synthesize tailored positive and negative electrode materials for Mg batteries with excellent cycle life at a high rate.
- Identification of principles governing the storage of magnesium ions in host materials: the structures of the materials need to be clarified to reveal and comprehend *the magnesium storage* mechanisms.
- Synthesizing highly reactive nanocomposites for multivalent conversion electrodes with *high reversibility on the microscale in the solid*. The understanding of the materials' conversion reactions and a quantification of the energy barriers for the fundamental steps is key for faster kinetics and improved reversibility.

1.3 Need for Better Batteries

The term "better batteries" is often used in the literature, mostly to express the fact that higher energy densities are desirable of batteries. In that respect, Mg technology offers a – so far only theoretical – perspective to be better or at least similar to the current Li-ion technology. The term "magnesium battery" rather than "magnesium-ion battery" (similar to "lithium-ion battery") already displays one of the major differences between the lithium and the magnesium technology: in the current Li-ion battery, Li is stored as an ion at the anode of the battery cell, in an insertion material such as graphite, for example. This is done as a safety measure because dendrites may form at a Li metal anode, as already mentioned above. However, they are not formed when magnesium metal or alloy is used as an anode. Therefore, it is possible to utilize an "undiluted" magnesium metal anode, thus offering very high gravimetric and volumetric capacity. This striking advantage is currently compromised by the lack of suitable cathode materials that show good capacity.

Another feature of a "better battery" is a higher rate capability. This is particularly important for applications where fast charging is needed, such as in automotive applications. However, from the current state of knowledge, the data from Mg metal anodes together with the studied cathode materials does show a moderate to slow charging and discharging behaviour so that a battery would have to be optimized in terms of its components in order to allow fast charging.

A key feature of a battery is its lifetime, particularly if it is to be used in large stationary applications for thousands of charge and discharge cycles. Lifetime has a critical impact because big batteries are expensive and both calendar and cycle life have a direct impact on the cost of the stored kWh, and on the cost of the storage itself. The longer the depreciation period of a battery, the lower the cost per charge and discharge cycle. It is difficult to foresee the potential of Mg technology at the moment as it is still in an early stage of development. The numbers that are currently elaborated for one or the other battery configurations are most likely prone to change in the next few years, when progress will be made in the different components of the battery.

Fabrication cost is another important factor and also here it is not clear at the moment what the cost of a Mg battery will be. The low price of Mg metal is often used as an argument when the technology is compared with Li-ion technology, for example. In fact, a sheet of Mg metal as both collector and active material of an anode would be simpler and probably less costly in the case that the currently high price for Mg sheets from laboratory suppliers would be lowered by mass production. What is still unclear however, is the price of the electrolyte, because there is as of yet no standard electrolyte. There is a wide span in the costs of currently published systems, depending on how many synthesis steps are required, in which yield the electrolyte is obtained and whether costly or cheap reagents and solvents are involved.

Also the cathode side remains open. Current configurations do not seem to be competitive and there are as of yet no viable technical solutions. In the case in which the current concepts can be further developed, insertion cathode materials will most likely be based on transition metals as in Li-ion technology. If no viable insertion materials can be identified, conversion electrodes may be an alternative, in particular the magnesium–sulphur battery, where carbon–sulphur composites would be employed as a cathode.

An inherent disadvantage which may lead to higher costs is certainly the low discharge voltage of Mg batteries. Compared to cells that deliver a higher voltage, more cells are necessary in order to store the same amount of energy, which has an impact on the fabrication cost and size and volume of a battery pack.

1.4 Need for Sustainable Solutions

As indicated above, future battery generations will have to take sustainability aspects into account. In that respect there are issues with the current Li-ion technology.[25] One reason is that many of the materials used in Li-ion batteries are on the Critical Raw Materials (CRM) lists, see the EU list,[26] for example. In particular, lithium, graphite, cobalt, nickel and copper are among a list of 33 materials that are used in significant amounts in low carbon technologies, including battery technology. Interestingly, magnesium has been a member of such lists, not because of its abundance but rather due to the fact that most Mg is currently produced in China, which is regarded as a rather political risk. If Mg could also be produced in significant amounts in other regions of the world, the material could be removed from this list. More information on the resource situation of magnesium will be given in the next Chapter.

1.4.1 Cathode

There are strong efforts being undertaken worldwide to reducing the amount of cobalt in NCM materials ($LiNi_xCo_yMn_zO_2$), which are used in power tools, e-bikes and electric drivetrains. The increase in the relative amount of nickel is possible and commercial batteries with NCM compositions of 8,1,1 have been announced for 2019. Further reduction seems possible, but the structural stability of the material decreases with increasing Ni content and thermal runaway starts at lower temperatures compared to the NCM 1,1,1 composition. The high reactivity of Ni^{4+} at the end of charge leads to low thermal stabilities and side reactions with the electrolyte. Thus, decomposition already starts below 150 °C and Ni oxide decomposes under the release of pure oxygen, which is a major safety concern in a battery with an organic electrolyte.[27] In the case that all of these issues can be resolved, Ni is still a critical raw material and the approach may be viable only if substantial amounts of Ni are recycled from used batteries in the mid- to long-term. Hence, large automotive companies such as Volkswagen have recently pointed out that they have the goal to not use Co and Ni in future battery generations. As alternatives to NCM materials, other compositions have been considered, such as manganese spinel ($Li_2Mn_2O_4$) or olivine ($LiFePO_4$), which offer high power but lower capacities. The Li–S battery has been announced as an alternative several times already, but it is still under development. This is mostly because there are fundamental issues that have not yet been solved, for example, the capacity decay due to the formation of soluble polysulfides and side reactions of certain electrolyte solvents with polysulfide intermediates.

It is difficult to make a statement about the sustainability of cathode materials for magnesium batteries at present. As already mentioned, the work is in progress and so far, the materials investigated have been made from elements other than those in the cathodes of Li-ion batteries. Vanadium-based materials have been proposed, but it is too early to regard them as a viable and commercially attractive option. If yes, it may be interesting to know that the abundance of vanadium is similar to that of lithium in the Earth's crust.

1.4.2 Anode

The anode of current Li-ion batteries is made from graphite, which is again rated as a critical raw material. Here, it seems that there are alternatives with long-term options such as synthetic graphite, which can eventually be made from processed biomass. At the moment this is not a commercially attractive option because the synthetic route is more expensive than the mining and clean-up of natural graphite. There is a perspective that graphite may not be needed any more when solid state batteries have overcome their technical and economic obstacles because Li metal is the anode of choice in combination with a solid electrolyte.

In comparison, a magnesium battery would not need natural graphite as the anode may be made directly from a metal sheet or from a composite with carbon. The carbon, however, can be porous or have other particular properties, which may need dedicated synthesis efforts.

1.4.3 Electrolyte

The current electrolyte in Li-ion batteries is typically based on $LiPF_6$ as a conducting salt which poses a potential risk in case the housing of the battery is damaged. When released in humid air, the hexafluorophosphate reacts with H_2O to form HF, which is highly toxic and can cause severe injuries and even death in persons that come into contact with it.

The current electrolytes of magnesium batteries are quite different from their Li counterparts due to their completely different chemical compositions. The early electrolytes based on Lewis acid–base complexes contain the binuclear $Mg_2Cl_3^+$ complex, which is co-ordinated by solvent molecules. Depending on the synthesis method and on the so-called Schlenk equilibrium of the system, there may be different anions, including organometallic compounds, which can be aggressive and toxic. Recently published electrolytes based on salts with large and weakly co-ordinating anions provide a softer chemistry with a more viable perspective. They are not aggressive or very toxic and are even stable in air.

1.5 Magnesium as a Resource

Magnesium is one of the most abundant elements in the Earth's crust. As dolomite, a mixed carbonate of magnesium and calcium, it forms entire mountain chains with a content of 13 mass% Mg. The mineral magnesite, which is found as a white sand in stone pits, is pure $MgCO_3$, with a Mg content of 28.5 mass%. In addition, sea water contains 1270 ppm of Mg (Li: 0.1 ppm) and the Mg content in the brine of salt lakes is also high. According to the assessments of the US Geological Survey (USGS) the current economically viable reserves amount to 8.5 Mio tons, which corresponds to a lifetime of 306 years, given the global annual production of 27 700 tons remains constant. The resources are much larger, however, and amount to 12 billion tons, which would last for 433 000 years, sea water and salt lakes not included. While most of the global resources are found in Russia (27%) and China (25%) and two thirds of the currently produced Mg comes from China, followed by Turkey (10%), Russia (5%) and Austria (3%). Although there seems to be no issue with the abundance of Mg worldwide, it has nevertheless been regarded as a critical raw material, due to the current dominating market power of China. It is certainly an advantage that magnesium is easily recyclable without experiencing any loss in quality. Worldwide, the market share of recycled Mg is in the order of 15 to 20%. An overview of the features of magnesium as a resource is given in Table 1.3.

Table 1.3 Factsheet of magnesium as a resource. (Data: US Geological Survey[28]).

Properties:	• Chemical element "Mg", with atomic number 12 • Silvery light metal with a density of only 1.7 g cm^{-3} • The lightest metallic material • Mg is corrosion resistant, resistant against acids, tough and ductile • The metal burns at 650 °C with a white, very light flame. It also burns under water
Use:	• Aluminium–magnesium alloys (43%) • Die casts (37%) • Desulfurization- and deoxidation-additive in steel- and cast iron production (10%) • Production of fire resistant materials (10%)
Largest producers of magnesium:	• China (68.6%) • Turkey (10.1%) • Russia (4.9%) • Austria (2.7%) • Slovakia (2.2%) • Spain (2.2%)
Available reserves:	8.5 Mio. Tons
Available resources:	>12 Bio. Tons
Statistical lifetime of reserves:	306 years
Statistical lifetime of resources:	433 212 years
Recycling rate:	In Germany 50% of the magnesium is recycled, worldwide 15–20%.
Annual global production of magnesium:	27 700 Mio. Tons

1.6 Conclusion

Mg batteries are one of the few options to complement or even replace Li ion batteries in the future. Although the basic properties of the Mg anode are promising, such as the bivalency of the Mg ion, the high volumetric storage capacity of the anode, the non-toxicity of Mg, and the lack of dendrite formation which allows the building of batteries with a metal anode, battery cells with competitive properties have not yet been reported. This is mostly due to the lack of suitable cathodes allowing high capacities and voltages and the presence of overpotentials, especially during plating and capacity fading due to unstable electrolytes and/or unstable electrodes. The research to address all of these issues in order to realize viable solutions is exciting and so far, no dead-ends are in sight, so the success of this technology will largely depend on the skills and effort that are put into this fascinating field of energy research and technology.

Acknowledgement

The support from the BMBF projects "Mag-S" (grant # 03XP0032A) from the Federal Ministry of Education and Research (BMBF) of Germany and the European FET-PROACTIVE project "E-Magic" (grant #824066) in HORIZON2020 is gratefully acknowledged. This work contributes to the research performed at CELEST (Center for Electrochemical Energy Storage Ulm-Karlsruhe).

References

1. M. M. Thackeray, C. Wolverton and E. D. Isaacs, Electrical energy storage for transportation—approaching the limits of, and going beyond, lithium-ion batteries, *Energy Environ. Sci.*, 2012, **5**, 7854.
2. D. Larcher and J.-M. Tarascon, Towards greener and more sustainable batteries for electrical energy storage, *Nat. Chem.*, 2015, **7**, 19.
3. C. Vaalma, D. Buchholz, M. Weil and S. Passerini, A cost and resource analysis of sodium-ion batteries, *Nat. Rev. Mater.*, 2018, **3**, 18013.
4. J. Heelan, E. Graz, Z. Zheng, Q. Wang, M. Chen, D. Apelian and Y. Wang, Current and Prospective Li-Ion Battery Recycling and Recovery Processes, *JOM*, 2016, **68**(10), 2632.
5. T. Gao, S. Hou, K. Huynh, F. Wang, N. Eidson, X. Fan, F. Han, C. Luo, M. Mao, X. Li and C. Wang, Existence of Solid Electrolyte Interphase in Mg Batteries: Mg/S Chemistry as an Example, *ACS Appl. Mater. Interfaces*, 2018, **10**, 14767.
6. V. Grignard, Some new organometallic combinations of magnesium and their application to the synthesis of alcohols and hydrocarbons, *C. R. Acad. Sci.*, 1900, **130**, 1322.
7. T. D. Gregory, R. J. Hoffman and R. C. Winterton, Nonaqueous Electrochemistry of Magnesium, *J. Electrochem. Soc.*, 1990, **137**(3), 775.
8. D. Aurbach, Z. Lu, A. Schechter, Y. Gofer, H. Gizbar, R. Turgeman, Y. Cohen, M. Moshkovich and E. Levi, Prototype systems for rechargeable magnesium batteries, *Nature*, 2000, **407**, 724.
9. R. Mohtadi, M. Matsui, T. Arthur and S. Hwang, Magnesium borohydride: From hydrogen storage to magnesium battery, *Angew. Chem., Int. Ed.*, 2012, **51**, 9780.
10. O. Titusaus, R. Mohtadi, T. Arthur, F. Mizuno, F. Nelson and Y. Sevryugina, An Efficient Halogen-Free Electrolyte for Use in Rechargeable Magnesium Batteries, *Angew. Chem., Int. Ed.*, 2015, **54**, 7900.
11. Z. Zhao-Karger, M. E. Gil Bardaji, O. Fuhr and M. Fichtner, A new class of non-corrosive, highly efficient electrolytes for rechargeable magnesium batteries, *J. Mater. Chem. A*, 2017, **5**, 10815.
12. Y. Shao, T. Liu, G. Li, M. Gu, Z. E. M. Nie, J. Xiao, D. Lv, C. Wang, J.-G. Zhang and J. Liu, Coordination Chemistry in magnesium battery electrolytes: how ligands affect their performance, *Sci. Rep.*, 2013, **3**, 3130.

13. P. Canepa, G. S. Gautam, R. Malik, S. Jayaraman, Z. Rong, K. Zavadil, K. Persson and G. Ceder, Understanding the Initial Stages of Reversible Mg Deposition and Stripping in Inorganic Nonaqueous Electrolytes, *Chem. Mater.*, 2015, **27**, 3317.
14. O. Tutusaus, M. Rana, N. Singh, T. S. Arthir and F. Mizuno, Study of Electrochemical Phenomena Observed at the Mg Metal/Electrolyte Interface, *ACS Energy Lett.*, 2017, **2**, 224.
15. P. Novak and J. Desilvestro, Electrochemical Insertion of Magnesium in Metal Oxides and Sulfides from Aprotic Electrolytes, *J. Electrochem. Soc.*, 1993, **140**(1), 140.
16. J. L. Andrews, A. Mukherjee, H. D. Yoo, A. Parija and P. M. M. Marley, Reversible Mg-Ion Insertion in a Metastable One-Dimensional Polymorph of V2O5, *Chem*, 2018, **4**, 1.
17. M. Liu, Z. Rong, R. Malik, P. Canepa, A. Jain, G. Ceder and K. A. Persson, Spinel compounds as multivalent battery cathodes: a systematic evaluation based on *ab initio* calculations, *Energy Environ. Sci.*, 2015, **8**, 964.
18. Z. Li, X. Mu, Z. Zhao-Karger, T. Diemant, R. J. Behm, C. Kübel and M. Fichtner, Fast multivalent intercalation enabled by solvated Mg^{2+} ions into self-established metallic layered materials, *Nat. Commun.*, 2018, **9**, 5115.
19. H. S. Kim, T. S. Arthur, G. D. Allred, J. Zajicek, J. G. Newman, A. E. Rodnyansky, A. G. Oliver, W. C. Boggess and J. Muldoon, Structure and compatibility of a magnesium electrolyte with a sulphur cathode, *Nat. Commun.*, 2011, **2**, 427.
20. Z. Zhao-Karger, X. Zhao, O. Fuhr and M. Fichtner, Bisamide based non-nucleophilic electrolytes for rechargeable magnesium batteries, *RSC Adv.*, 2013, **3**, 16330.
21. Z. Zhao-Karger, X. Zhao, D. Wang, T. Diemant, J. Behm and M. Fichtner, Performance Improvement of Magnesium Sulfur Batteries with Modified Non-Nucleophilic Electrolytes, *Adv. Energy Mater.*, 2014, **5**(3), 1401155.
22. "MagS" project (grant no. 03XP0032A) from the Federal Ministry of Education and Research (BMBF) of Germany.
23. B. Sievert, J. Häcker, F. Bienen, N. Wagner and A. Friedrich, Magnesium Sulfur Battery with a New Magnesium Powder Anode, *ECS Trans.*, 2017, **77**(11), 413.
24. M. Jäckle, K. Helmbrecht, M. Smits, D. Stottmeister and A. Groß, Self-diffusion barriers: possible descriptors for dendrite growth in batteries? *Energy Environ. Sci.*, 2018, **11**, 3400.
25. J. F. Peters, M. Baumann, B. Zimmermann, J. Braun and M. Weil, The environmental impact of Li-Ion batteries and the role of key parameters – A review, *Renewable Sustainable Energy Rev.*, 2017, **67**, 491.
26. E. Tzimas, https://setis.ec.europa.eu/setis-reports/setis-magazine/materials-energy/critical-materials-energy-technologies-evangelos, 2015, Online.

27. J. Kim, H. Lee, H. Cha, M. Yoon and M. Park, Prospect and Reality of Ni-Rich Cathode for Commercialization, *Adv. Energy Mater.*, 2018, **8**, 1702028.
28. https://minerals.usgs.gov/minerals/pubs/commodity/magnesium/, 2018, Online.

CHAPTER 2

Non-aqueous Electrolytes for Mg Batteries

R. MOHTADI†* AND O. TUTUSAUS†

Materials Research Department, Toyota Research Institute of North America, Ann Arbor, MI 48105, USA
*E-mail: rana.mohtadi@toyota.com

2.1 Introduction

Owing to its high volumetric capacity (3832 mA h cm^3) and abundance in the Earth's crust, Mg metal has recently attracted increased attention as a battery anode candidate. Another important motivation to investigate this metal is that the electrochemical processes related to Mg reversible plating/stripping have demonstrated the absence of dendrite formation, alleviating the safety concerns associated with lithium metal anodes. One key challenge with a Mg metal anode is its propensity to be easily passivated by most common solvents and salts through metal–electrolyte chemical/electrochemical interactions, which are detrimental for reversible magnesium deposition/stripping. As such, the development of Mg electrolytes is strongly tied to salts and solvents compatible with Mg metal, which, for decades, limited electrolyte designs to a handful of reagents that are highly reactive and/or strongly reducing, as well as being halide-based. Over the past decade, there have been important developments in the electrolyte field that have opened up

†Authors contributed equally to this work.

Energy and Environment Series No. 23
Magnesium Batteries: Research and Applications
Edited by Maximilian Fichtner

Published by the Royal Society of Chemistry, www.rsc.org

new avenues for generating novel designs beyond those previously known. Those exciting discoveries have greatly widened the number of available systems compatible with Mg metal that overcome many of the issues that plagued older systems. In this chapter, we will discuss these new electrolytes, explain their properties and highlight existing challenges.

2.2 Halide-ion Containing Electrolytes

Halide anions, especially chloride, have been historically strongly embedded in Mg electrolyte development mainly because Grignard reagents were readily available precursors in the early days of Mg electrolyte development. Despite the current trend to shift away from Grignard-type precursors, the vast majority of known Mg electrolytes still contain chloride anions.

Halide-containing electrolytes can be divided into two main groups, based on the structure of the magnesium precursor used (Figure 2.1).[1] The first class encompasses electrolytes derived from heteroleptic magnesium compounds (RMgX), and the second class, from homoleptic magnesium compounds (MgR_2 or MgX_2); in both cases, R can be an alkyl, aryl, amido, or alkoxide group. While many of these magnesium precursors alone display moderate electrochemical performance, Lewis acids such as $AlCl_3$ or BR_3, and/or $MgCl_2$ are commonly added as performance enhancers. These react with nucleophilic R groups or halides, shifting ionic dissociation equilibria or transforming the low oxidative stability R groups into more stable complex anions. The overall effect is a noticeable increase in the electrolyte ionic conductivity, anodic stability and coulombic efficiency. It is important to note that ligand scrambling occurs between magnesium precursors and performance enhancer due to many equilibria processes occurring in solution. As a result, solution composition is an often unknown complex mixture of charged and neutral species. Further, it is possible to obtain the same electrolyte system from two different combinations of magnesium precursor and performance enhancer (*e.g.* $Ph_2Mg + AlCl_3$ or $PhMgCl + PhAlCl_2$ or $MgCl_2 + Ph_2AlCl$).

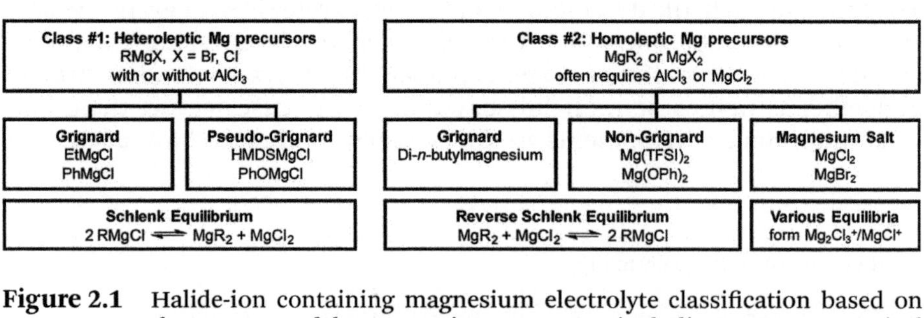

Figure 2.1 Halide-ion containing magnesium electrolyte classification based on the structure of the magnesium precursor, including some prototypical examples, and the existing solution equilibria. In all cases, R is chosen from alkyl/aryl, amido, or alkoxy/phenolate anions, and X is a halide (usually Cl). Reproduced from ref. 1 with permission from Elsevier, Copyright 2017.

The use of chloride-containing Mg electrolytes offers several advantages: (1) assistance in the activation of the native MgO passivating layer on the Mg surface; (2) enhancement of electrolyte performance in terms of high cycling efficiencies (close to 100%) and low overpotentials for both the deposition and stripping processes; (3) allows for reduced cost of precursors, especially when involving $AlCl_3$ or $MgCl_2$. However, there are a number of challenges that have proven to be difficult to overcome so far: (1) limited anodic stability due to chloride oxidation at ~3.3 V *vs.* Mg/Mg^{2+}; (2) corrosiveness to non-noble metal parts typically at voltages above 2.5 V *vs.* Mg/Mg^{2+}; (3) the presence of strong Mg–Cl bonds, difficult to break at the cathode interface, leading to lower energy-density $[MgCl]^+$-storage chemistry and; (4) low cationic transference number of the relatively large complex $[Mg_xCl_y]^{(2x-y)+}$ cations.

2.2.1 Carbon-based Anions

Until recently, Grignard reagents were among the most preferred precursors for preparing Mg electrolytes.[2] However, the number of new systems based on Grignard reagents has shown a steady decrease recently due to their reduced anodic stability. As seen below, the most recent work involving carbon-based anions focuses on Grignard reagents precursors based on aromatic structures, such as phenyl and carboranyl anions, given their increased oxidative stability compared to their alkyl congeners.[3,4]

Nelson *et al.* reported a modification to $PhMgCl/AlCl_3$ (APC) electrolyte to reduce its corrosion properties to non-noble current collectors.[4] By substituting the Lewis acid $AlCl_3$ with chloride-free $Al(OPh)_3$, the chloride content was greatly reduced. Extended electrolysis of a stainless steel sample at 4.5 V *vs.* Mg/Mg^{2+} showed minimal pitting compared to a high density of pitting observed on the APC electrolyte. While the electrolyte showed an impressive anodic stability of ~5 V *vs.* Mg/Mg^{2+} on stainless steel, the same group later demonstrated that such effect was due to the adsorption of electron-insulating aromatic species resulting from the oxidation of anionic phenyl constituents.[5] Based on the results of their study, the authors concluded that such phenomenon could be present in other phenyl-containing solutions and suggested moving away from this type of electrolyte systems.

Carboranyl anions are another type of less known carbanions that are based on 3-dimensional aromatic boron clusters and display high oxidative stability. Carter *et al.* synthesized 1-(1,7-carboranyl) magnesium chloride and isolated a salt composed of a chloride-bridged Mg dimer and a carboranyl magnesium chloride complex anion upon crystallization (Figure 2.2a).[3] A THF solution of the crystallized material showed high coulombic efficiency (>98%) and relatively high oxidative stability (3.2 V *vs.* Mg/Mg^{2+} on Pt). Interestingly, the same anodic stability was measured on 316-stainless steel and aluminium (Figure 2.2b), indicating a lower corrosivity compared to other chloride-containing electrolytes, although a similar adsorption process was also found for $PhMgCl/AlPh_3$ solutions.

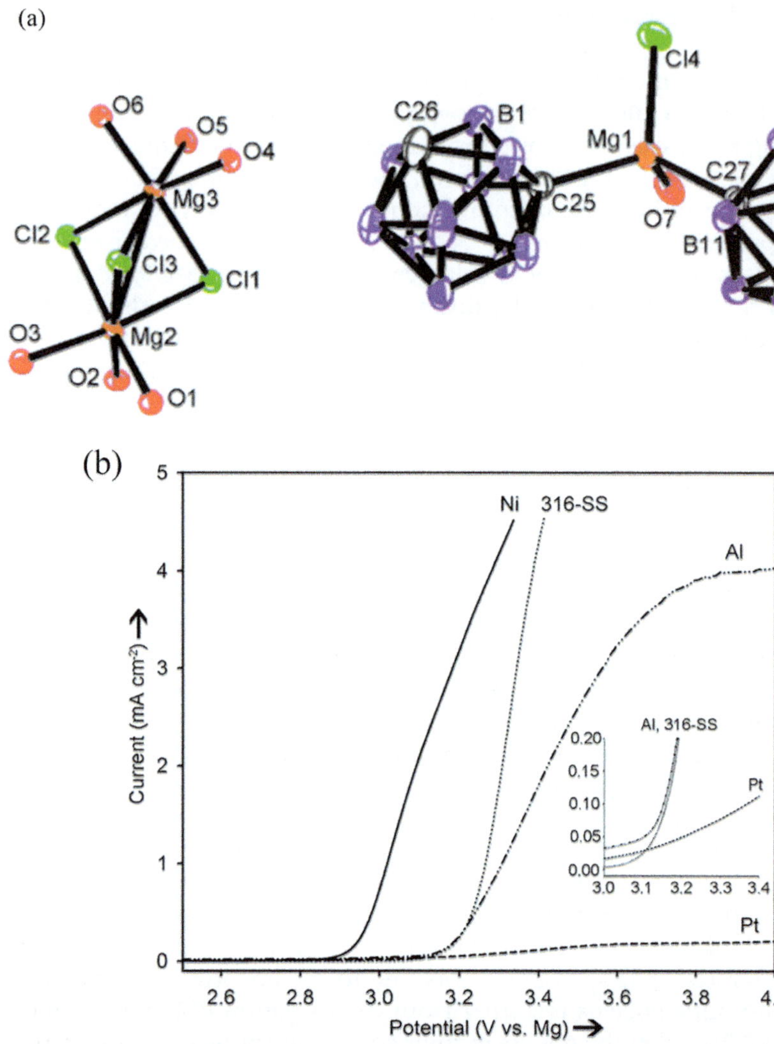

Figure 2.2 (a) Solid-state structure determined X-ray diffraction analysis of a single crystal obtained from a THF solution of 1-(1,7-carboranyl) magnesium chloride. Hydrogen atoms and THF carbon atoms are omitted for clarity (*i.e.* each oxygen atom represents a THF solvent molecule). Thermal ellipsoids are shown at 50% probability. (b) Linear sweep voltammograms on Pt, stainless steel, Ni and Al electrodes of a THF solution of 1-(1,7-carboranyl) magnesium chloride (inset: expanded view of the oxidation onset). Reproduced from ref. 3 with permission from John Wiley and Sons, © 2014 Wiley-VCH Verlag GmbH & Co. KGaA.

2.2.2 Nitrogen-based Anions

A number of amidomagnesium chloride compounds were initially surveyed by Gregory *et al.* showing lower conductivity but higher open-circuit potentials (*i.e.* higher oxidation stability) when compared to Grignard reagents, the latter arising from the higher electronegativity of nitrogen.[6] In a later report, Liebenow *et al.* introduced a hexamethylsilylamide (HMDS) anion and showed that it displays a superior oxidative stability compared to other strongly basic amides.[7] Their report did not receive much initial attention due to the inferior electrochemical properties of the (HMDS)MgCl electrolyte compared to already available organoaluminate electrolytes.[8] More recent work capitalized on the $HMDS^-$ non-nucleophilic character to prepare a series of electrolytes compatible with a sulphur cathode and allowed the furthering of the Mg–S battery system.[9–12] These new HMDS-based systems were obtained by addition of $AlCl_3$ or $MgCl_2$ to either (HMDS)MgCl or $Mg(HMDS)_2$ precursors (Figure 2.3), providing solutions with increased anodic stability (3.3 V *vs.* Mg/Mg^{2+}) and a high coulombic efficiency (98–99%).[9,13,14]

The ratio of performance enhancer to $Mg(HMDS)_2$ plays a critical role towards providing highly performing HMDS-based electrolytes (Figure 2.3). Since the HMDS anion is the species with the lowest oxidation stability in these electrolytes, an excess of performance enhancer is typically used to minimize the amount of free HMDS anions in solution. For instance, increasing the ratio of the $AlCl_3$ to Mg precursor ensures that the reaction between the HMDS anion and $AlCl_3$ is driven to completion to yield more oxidatively stable $[(HMDS)_xAlCl_{(3-x)}]^-$ complex anions.[9,13,15] By the same token, electrolytes with a high $MgCl_2/Mg(HMDS)_2$ ratio provide various HMDS-containing complex anions and display a widened electrochemical window and higher coulombic efficiency.[14] Crystal structure data suggested that the $[Mg_2Cl_3]^+$ dimer was the dominant cationic species in solution,[9,13] but recent mass spectrometry experiments support that the $[MgCl]^+$ cation is also present as one of the predominant species, particularly in solvents with increasing solvation power.[15,16]

Other amidomagnesium chloride salts were recently studied in triglyme at 100 °C in the presence of $MgTFSI_2$ as a solubility enhancer.[17] Their oxidative stability were consistent with those obtained in previous reports.[7]

2.2.3 Oxygen-based Anions

Alkoxide- and phenolate-based magnesium electrolytes have received a great deal of attention recently due to the less nucleophilic nature of oxygen-based anions compared to amides, as well as being poised to provide a higher oxidative stability owing to the higher electronegativity of oxygen. The current body of work on alkoxide-/phenolate-based electrolytes has revealed that sterically hindered anions provide a higher electrochemical performance and established the importance of accurately tuning the electron density in the anion through incorporation of electron-withdrawing groups.

Oxygen-based anions were first introduced as magnesium electrolytes in 2012 by Wang *et al.*, where a number of phenolate-based $(ArO)MgCl/AlCl_3$ systems in THF were shown to support reversible Mg deposition/stripping, an anodic stability of up to 2.6 V *vs.* Mg/Mg^{2+} and a maximum conductivity of 2.56 mS cm^{-1} at 0.5 M.[18] Notably, these electrolytes maintained their anodic stability after exposure to air/moisture for 3 h, in contrast to those derived from C- and N-based anions, albeit with an inevitable slight deterioration of

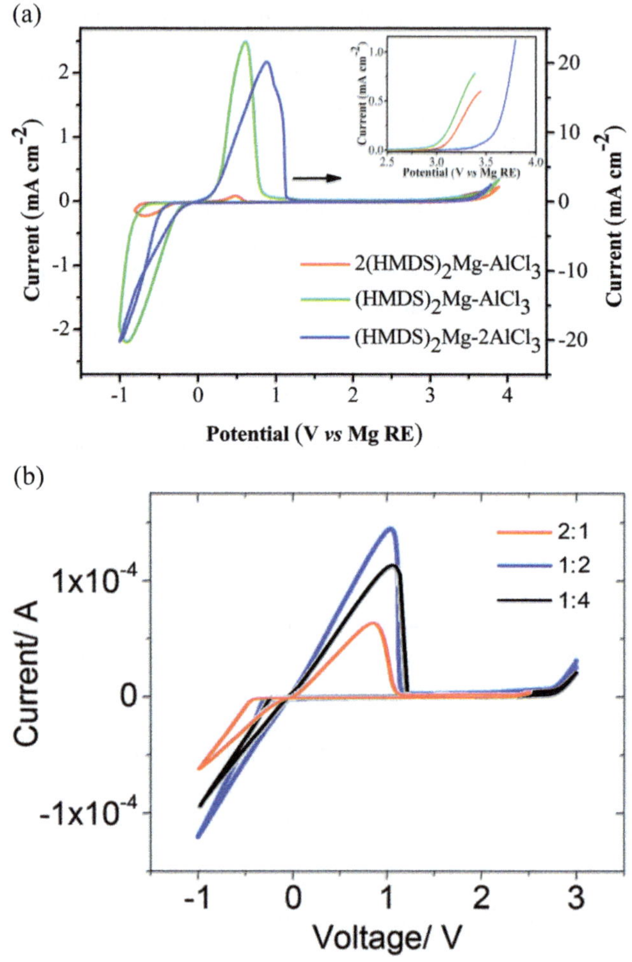

Figure 2.3 Typical cyclic voltammograms of Pt electrodes in THF solutions of selected HMDS-containing electrolytes. (a) $Mg(HMDS)_2/AlCl_3$ electrolyte in 2 : 1 (red), 1 : 1 (green), and 1 : 2 (blue curve using the secondary *y*-axis on the right hand side) ratios at a scan rate of 25 mV s^{-1}; the inset shows linear sweep voltammograms at a scan rate of 5 mV s^{-1} to determine the anodic stability. Reproduced from ref. 13 with permission from the Royal Society of Chemistry, Copyright 2014. (b) $Mg(HMDS)_2/MgCl_2$ in 2 : 1 (red), 1 : 2 (blue) and 1 : 4 (black) ratios at a scan rate of 100 mV s^{-1}. Adapted from ref. 14 with permission from the Royal Society of Chemistry.

their Mg deposition/stripping process. In a follow-up study, phenyl ring substituents were shown to have a significant effect on the electrochemical performance of these types of electrolytes.[19,20] For instance, the highest oxidative stability was achieved by incorporating either electron-withdrawing groups (Figures 2.4a and c; 2.9 V *vs.* Mg/Mg^{2+} for $-CF_3$ substitution) or branching alkyl groups (Figures 2.4b and 2.4d; 2.73 V *vs.* Mg/Mg^{2+} for ^{t}Bu substitution) on the ring *para* position, as a result of charge dissipation from the oxidation

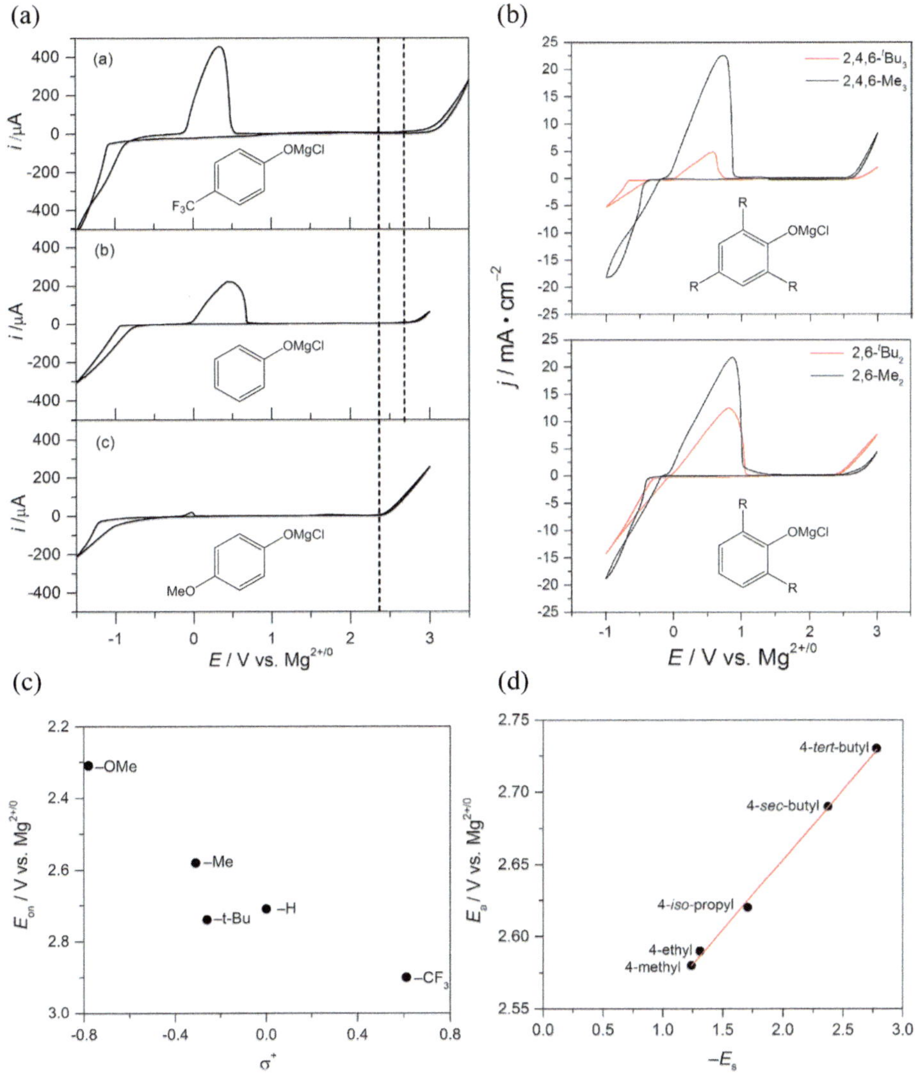

Figure 2.4 (a) and (c) Typical cyclic voltammograms of Pt electrodes at a scan rate of 25 mV s^{-1} in selected phenolate electrolytes at 0.5 M in THF with 2 : 1 ratio (ArO)MgCl/$AlCl_3$. Reproduced from ref. 19 with permission from the Royal Society of Chemistry, Copyright 2014. Dependence of the electrolyte oxidative stability phenol *p*-substituent in terms of the (b) Hammett parameter (σ^+) and (d) Taft's steric bulk parameter ($-E_S$). Reproduced from ref. 20 with permission from the Royal Society of Chemistry.

site. Further, solution conductivity was greatly enhanced by including substituents at the 2 and 6 positions (1.25 mS cm^{-1} *vs.* 2.56 mS cm^{-1} for 4-Me *vs.* 2,4,6-Me_3 phenyl ring substitution at 0.5 M in THF), without affecting anodic stability when alkyl groups were used, presumably owing to the increased anion steric shielding that reduced ion pairing.

Interestingly, when the $AlCl_3$ additive was substituted by $MgCl_2$ in the above phenolate-based magnesium electrolytes, the effect of the phenyl ring substituents on the electrochemical performance was slightly different.[21] Bulky alkyl groups provided the highest oxidative stability, with the electrolyte (DTBP)MgCl/$MgCl_2$ in THF (DTBP = 2,6-di-*tert*-butylphenolate) displaying an anodic stability of 2.3 V *vs.* Mg/Mg^{2+}, although with a low ionic conductivity of 0.66 mS cm^{-1} at 0.5 M. In contrast, the presence of electron-withdrawing groups in the phenyl ring failed to yield high performance systems, with oxidative stabilities of below 2.0 V *vs.* Mg/Mg^{2+}. Crystals obtained from (DTBP) MgCl/$MgCl_2$ solutions in THF yielded a solid-state structure composed of a $[Mg_2Cl_3]^+$ dimer cation and a $[(DTPB)_3Mg]^-$ anion. The authors rationalized that an ionic pair was obtained due to the sterically hindered groups in DTPB, while other less sterically demanding phenolates would fail to stabilize the anionic species, yielding predominantly neutral bridged species.

Compared to phenolates, alkoxides have the advantages of lower molecular weights (*i.e.* higher mobility[22]) and ease of access to a larger selection of alcohol precursors.[23] Alkoxide systems based on several short-chain alcohols were studied in combination with $MgBr_2$.[24] A remarkable maximum ionic conductivity of 4.1 mS cm^{-1} was obtained for 1.0 M $Mg(OEt)_2$/$MgBr_2$ in THF; however, its coulombic efficiency was limited to 90% and the oxidative stability was not determined. Similar-sized alkoxide systems (RO)MgCl were also combined with $Mg(TFSI)_2$ (TFSI = bis(trifluoromethanesulfonyl)imide) to prepare concentrated solutions of up to *ca.* 2.5 M in G3 (G3 = triglyme).[17] At 100 °C, these systems were able to reversibly deposit/strip Mg metal and X-ray diffraction analysis of crystals isolated from a (EtO)MgCl/$Mg(TFSI)_2$/THF (2 : 1 : 30) solution revealed a complex trimagnesium cation containing bridging Cl^- and EtO^- units and an uncoordinated TFSI anion. It is interesting to note that this work also explored alkoxy anions with multiple coordination sites, such as $CH_3O(CH_2)_2O^-$, and those furnished poorly performing electrolytes.

Liao *et al.* studied alkoxides based on slightly larger-chain available alcohols, such as *n*-butanol, *tert*-butanol and trimethylsilanol.[23] Despite the high solubility of those (RO)MgCl salts in THF (up to 2.0 M for ^{t}BuOMgCl), as shown in Figure 2.5a these displayed a relatively low ionic conductivity (1.20 mS cm^{-1} for 1.2 M ^{t}BuOMgCl) and reduced anodic stability (2.0 V *vs.* Mg/Mg^{2+}). As noted for other systems, the addition of $AlCl_3$ greatly improved their performance (Figure 2.5b), with ^{t}BuOMgCl displaying the largest anodic stability (2.5 V *vs.* Mg/Mg^{2+}), albeit an unusually high (RO)MgCl/$AlCl_3$ ratio (6 : 1) was used to attain high solubility in THF (>1.0 M). It is worth noting that when similar alkoxide-based magnesium electrolytes were prepared using $Mg(OR)_2$ and $AlCl_3$ in THF, such as $Mg(O^tBu)_2$/$AlCl_3$ (2 : 3), their oxidative stability was higher (2.8 V *vs.* Mg/Mg^{2+}) but their coulombic efficiency was lower (42%) and the deposits contained 2.6 atom% of Al.[25] When $MgCl_2$ was used as an additive in ^{t}BuOMgCl/THF solutions at a ratio higher than 1 : 1, the electrolyte system displayed a superior performance (Figure 2.5c), with a high oxidative stability (3.3 V *vs.* Mg/Mg^{2+}), coulombic efficiency of

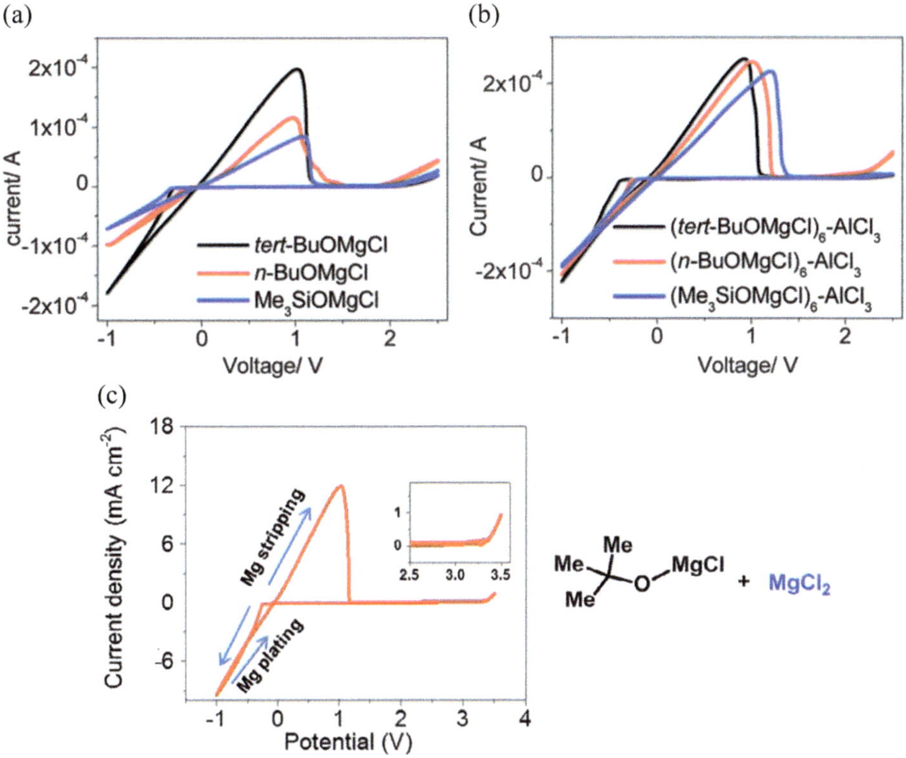

Figure 2.5 Typical cyclic voltammograms of Pt electrodes at a scan rate of 100 mV s^{-1} in THF solutions of selected alkoxide-based electrolytes. (a) 1.0 M ^{t}BuOMgCl (black), 1.0 M *n*-BuOMgCl (red), and 1.0 M $Me_3SiOMgCl$ (blue). Reproduced from ref. 23 with permission from the Royal Society of Chemistry, Copyright 2014. (b) 1.0 M $^{t}BuOMgCl/AlCl_3$ (6 : 1) (black), 1.0 M *n*-$BuOMgCl/AlCl_3$ (6 : 1) (red), and 1.0 M $Me_3SiOMgCl/AlCl_3$ (1 : 6) (blue). Reproduced from ref. 23 with permission from the Royal Society of Chemistry, Copyright 2014. (c) 0.5 M $^{t}BuOMgCl/MgCl_2$ (1 : 1). Reproduced from ref. 26 with permission from John Wiley and Sons, ©2016 Wiley-VCH Verlag GmbH & Co. KGaA.

close to 100% and ionic conductivity of 1.57 mS cm^{-1} at 0.5 M.[26] Other sterically bulky alkoxides/siloxides were also examined within this work and their electrochemical performance was also better with $MgCl_2$ than with $AlCl_3$, in terms of both oxidative stability (up to 3.5 V *vs.* Mg/Mg^{2+}) and coulombic efficiency. A relatively high oxidative stability (~2.8 V *vs.* Mg/Mg^{2+}) was reported on stainless steel for 0.5 M $^{t}BuOMgCl/MgCl_2$ in THF solution, albeit without including any electrochemical data. An active ionic species based on the $[Mg_2Cl_3]^+$ dimer cation and the unsymmetrically bridged homobimetallic anion $[(\mu\text{-DPE})_2Mg_2Cl_3(THF)]^-$ (DPE = diphenylethoxide) was proposed for another bulky alkoxylate system (DPE)$MgCl/MgCl_2$ in THF based on X-ray diffraction analysis of crystallized material and nuclear magnetic resonance (NMR) spectroscopic analyses.[26]

For larger and bulkier alkoxide anions, such as $(Ph_3CO)MgCl$ or $(Ph_3SiO)MgCl$, the improvement afforded by addition of $MgCl_2$ or $AlCl_3$ was reversed compared to smaller anions, with the strong Lewis acid $AlCl_3$ offering a better option.[27] For instance, adding one equivalent of $MgCl_2$ to 0.25 M $(Ph_3SiO)MgCl$ in THF solution improved many aspects of the electrochemical behaviour, but oxidative stability was still limited at around 2 V *vs.* Mg/Mg^{2+} (Figure 2.6a). Based on the non-ionic bridging structure isolated by crystallization of a THF solution of $(Ph_3SiO)MgCl/MgCl_2$, the authors hypothesized that $MgCl_2$ alone is not capable of favouring the formation of an ionic species (Figure 2.6b). Unlike $MgCl_2$, addition of 0.5 equivalents of $AlCl_3$ improves all of the electrochemical properties, including the oxidative stability (to 3.0 V *vs.* Mg/Mg^{2+}); yet, addition of one equivalent of both $AlCl_3$ and $MgCl_2$ brings an even larger improvement than with the use of $AlCl_3$ alone (Figure 2.6b). Because the same ion-separated structure was isolated from THF solutions of 0.25 M $(Ph_3SiO)MgCl/AlCl_3$ and of 0.25 M $(Ph_3SiO)MgCl/AlCl_3/MgCl_2$ (Figure 2.6c), it was suggested that both electrolytes contain the same major

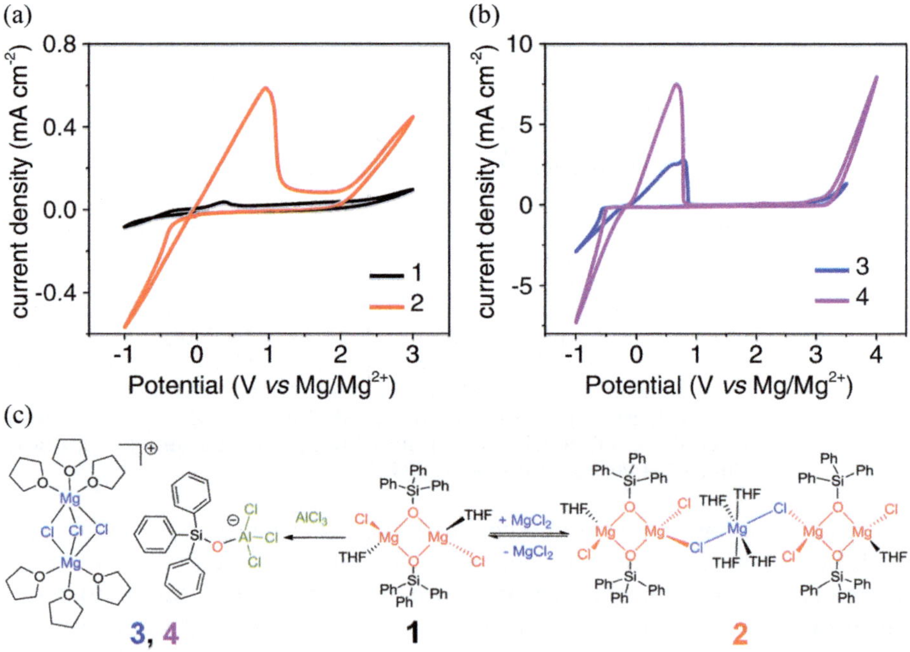

Figure 2.6 Typical cyclic voltammograms of Pt electrodes at a scan rate of 100 mV s^{-1} in THF solutions containing selected sterically-hindered alkoxide-based electrolytes. (a) 0.25 M $(Ph_3SiO)MgCl$ (black) and 0.25 M $(Ph_3SiO)/MgCl_2$ (1 : 1); (b) 0.25 M $(Ph_3SiO)/MgCl_2/AlCl_3$ (1 : 1 : 0.5) (blue) and 0.25 M $(Ph_3SiO)MgCl/MgCl_2/AlCl_3$ (1 : 1 : 1). (c) Proposed solution equilibria of $(Ph_3SiO)MgCl$ with $MgCl_2$ and $AlCl_3$ to furnish compounds obtained by crystallization. Reproduced from ref. 27 with permission from The Electrochemical Society, Copyright 2016.

electroactive component in solution, although in an increased amount in the latter. In a follow-up study, a series of increasingly sterically hindered magnesium dialkoxides $Mg(OCH_xPh_{(3-x)})_2$ were investigated in combination with $AlCl_3$ in THF solution, confirming the higher electrochemical performance of the two most sterically hindered representatives ($x = 0, 1$).[1] Interestingly, the $Mg(OCPh_3)_2/AlCl_3$ electrolyte appears to contain Mg-reactive species in solution as judged by the lower deposition overpotential, lower cell impedance and moderate coulombic efficiency (~80%) compared to the other systems therein studied, as well as the bright red colour appearing upon contact with Mg metal.

Various fluorinated alkoxide anions were also recently explored as a means to develop high performance electrolytes.[28,29] To that end, Crowe *et al.* prepared and studied a series of electrolytes based on *tert*-butoxide anions with different degrees of fluorination.[28] The system 1.2 M $((CF_3)_2(CH_3))COMgCl$ + 0.2 M $AlCl_3$ in THF solution stood out as the best performing electrolyte, exhibiting an anionic stability of 3.2 V *vs.* Mg/Mg^{2+}, a coulombic efficiency of 98% and an ionic conductivity of 3.5 mS cm^{-1},[28] as well as yielding crystalline Mg deposits with a very uniform growth normal to the electrode surface.[30] The authors rationalized the mismatch between the highest degree of fluorination and electrochemical performance based on arguments pertaining to the electrolyte solution composition. As found from ^{27}Al-NMR spectroscopy, a more anodically stable tris(alkoxy)chloroaluminate anion (according to the HOMO values calculated from density functional theory calculations) was the major species in solutions based on the partially fluorinated $[((CF_3)_2(CH_3))CO]^-$ anion, while a less anodically stable tetrakis(alkoxy)aluminate anion was the major species in solutions based on the perfluorinated $[((CF_3)_3)CO]^-$ anion.[28] Another closely related anion, 1,1,1,3,3,3-hexafluoro-2-propanol (HFIP), also yields a highly performing electrolyte; specifically, a 0.25 M mixture $Mg(HFIP)_2/AlCl_3$ (1 : 2) in DME has a conductivity of 4.77 mS cm^{-1}, a high coulombic efficiency (*ca.* 100%), an anodic stability of 3.1 V *vs.* Mg/Mg^{2+} and very low overpotential for Mg plating (−70 mV).[29]

2.2.4 Halides as Anions

Magnesium deposition was shown to be possible from saturated $MgBr_2$ solutions in pyridine[31] and diethyl ether[32] in as early as 1927, although with reduced reversibility in the former and with low current densities in the latter. More recently, $MgCl_2/AlCl_3$ in THF solution was studied in 2005,[33] but the electrolyte showed poor performance, with a high overpotential for Mg deposition (>900 mV) and low stripping efficiency (34%), turning attention away from $MgCl_2$ as a magnesium precursor for almost a decade. The $MgCl_2/AlCl_3$ system was again revisited in 2013, when two groups independently confirmed the viability of this all-inorganic electrolyte exclusively based on chloride anions.[34,35] The magnesium–aluminium chloride complex (MACC), as it is currently known, is typically prepared by mixing $MgCl_2$ and $AlCl_3$ in THF or glymes in $MgCl_2/AlCl_3$ ratios ranging from about

1 : 2 to 3 : 1 and with Mg molarity of between about 0.1 and 0.5 M, depending upon the solvent used.[36] Despite its simplicity, MACC is superior in many aspects to many of the existing electrolytes based on organic Mg^{2+} precursors combined with $AlCl_3$; it exhibits excellent reversibility of Mg deposition and dissolution (coulombic efficiency of 100%), high anodic stability (~3.2 V *vs.* Mg/Mg^{2+}) and low nucleophilicity (compatible with a sulphur cathode). However, one of the most important limitations of MACC (and other halide-ion containing) electrolytes is their incompatibility with typical current collectors due to the very high Cl^- content, with MACC showing relatively low anodic stabilities on stainless steel (2.2 V *vs.* Mg/Mg^{2+}) and aluminium (1.1 V *vs.* Mg/Mg^{2+}).[37]

Another challenging aspect of MACC has been designing a simple procedure yielding an electrochemically competent electrolyte solution, which likely accounts for the 8-year setback following its initial unpromising report. Freshly prepared MACC electrolyte typically displays low Mg deposition and stripping reversibility,[38,39] due to impurities present in reactants (residual water or others) and solvents or introduced by the reaction atmosphere, as well as oligomers formed upon ring-opening polymerization of THF during the initial synthesis of MACC.[37,39] Several approaches have been found to yield high performance MACC. Earlier reports focused on electrochemical conditioning to improve its performance in THF and DME, involving cycling the solutions in a sealed cell, and using an inert working electrode and a Mg counter electrode, between potentials of −1.2 and 2.8 V at a slow scan rate (<1 mV s^{-1}).[38–40] Later, Ha *et al.* developed a conditioning-free method involving dissolution of Mg metal in $AlCl_3$/THF using catalytic amounts of $CrCl_3$. The resulting electrolyte solution was composed of 0.28 M Mg, a Mg/Al ratio of ~13 and <1 ppm Cr,[41] with the high Mg content hypothesized as the origin for the high reversibility of Mg deposition/stripping (coulombic efficiency >99.8%) displayed. A more recent strategy involved preparing MACC in the presence of Mg powder, where the latter appears to act as a reductive scavenger to remove any interfering impurities.[37] This method was successfully applied to prepare MACC electrolyte in THF, DME or diglyme, and a better reversibility was obtained in THF and DME but higher anodic stability in diglyme.

There have been many different proposals for the solution speciation in MACC electrolyte and this is a topic that is still under debate. Crystals isolated from $MgCl_2/AlCl_3$ (2 : 1) in THF were composed of a $[(\mu\text{-Cl})_3Mg_2(THF)_6]^+$ dimer cation and an $[AlCl_4]^-$ anion (Figure 2.7a), which was electrochemically active for reversible Mg deposition/stripping once redissolved, leading Liu *et al.* to propose this salt as the main active species of MACC in THF.[35] The Mg dimer species was proposed to be generated in two steps. In a first step, a $[MgCl(THF)_3]^+$ cation is generated by Cl^- abstraction from $MgCl_2$ by the Lewis acid $AlCl_3$; then $[MgCl(THF)_3]^+$ reacts with $MgCl_2$ to yield $[(\mu\text{-Cl})_3Mg_2(THF)_6]^+$. This mechanism of formation was further supported by detecting $[MgCl(THF)_3]^+$ species in solution by subambient pressure ionization with nanoelectrospray mass spectrometry (SPIN-MS)

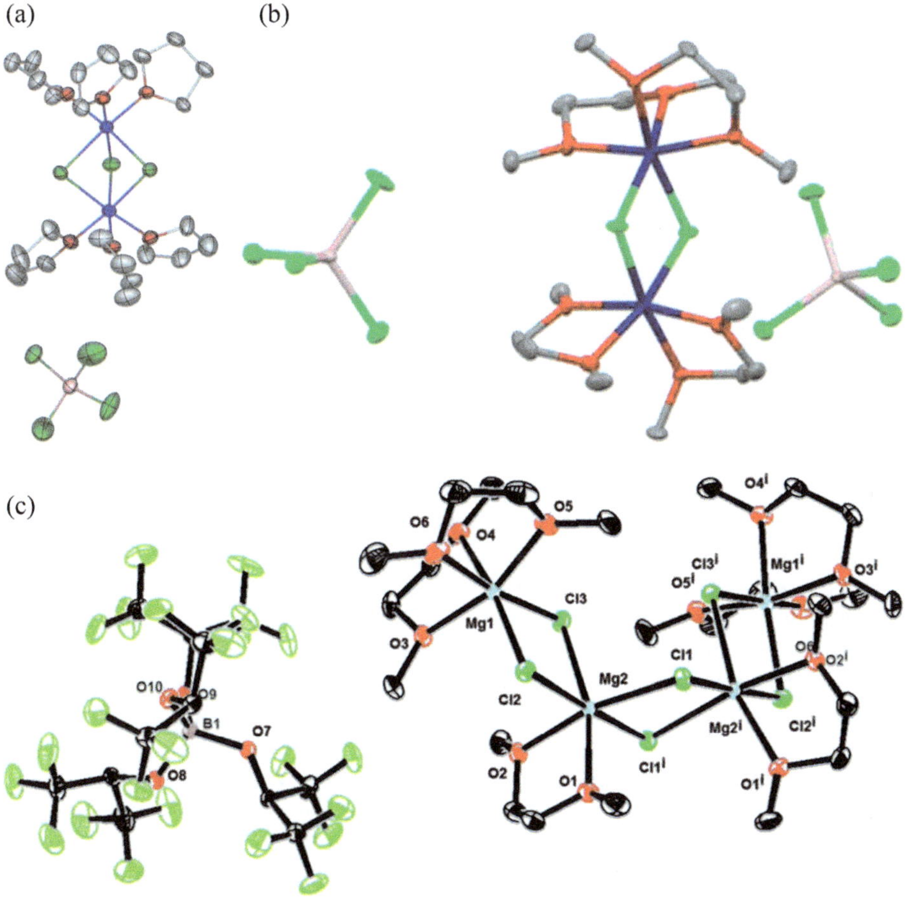

Figure 2.7 Molecular structures of (a) $[(\mu\text{-Cl})_3Mg_2(THF)_6]AlCl_4$. Reproduced from ref. 35 with permission from the Royal Society of Chemistry, Copyright 2014. (b) $[Mg_2(\mu\text{-Cl})_2(DME)_4][AlCl_4]_2$. Reproduced from ref. 44 with permission from the Royal Society of Chemistry. (c) $[Mg_4Cl_6(DME)_6][B(HFP)_4]_2$. Reproduced from ref. 45 with permission from the Royal Society of Chemistry. Hydrogen atoms are omitted for clarity. Thermal ellipsoids at 30% probability for (a) and (b) and 50% for (c). Atom colours for (a) and (b): Mg, blue; Cl, green; Al, pink; B, purple; C, gray; O, red.

analysis of $MgCl_2/AlPh_3$ and $MgCl_2/AlEtCl_2$ solutions in THF,[42] two electrolyte solutions that also yield a Mg dimer cation when crystallized.[35] Computational methods used by Canepa *et al.* depicted a different scenario, pointing to more simple chemical structures present in MACC solutions in THF.[43] Their *ab initio* calculations suggested that the only stable constituents of the electrolyte are $[MgCl(THF)_3]^+$, $[MgCl_2(THF)_2]$, $[AlCl_4]^-$ and $[AlCl_3(THF)]$ in chemical equilibrium, excluding the presence of higher cationic Mg multimers. As such, the authors speculated that solid-state structures containing $[(\mu\text{-Cl})_3Mg_2(THF)_6]^+$ appear only under conditions of crystallization due to agglomeration of $MgCl_2$ and $[MgCl]^+$ and do not represent solution speciation. Beyond THF solvent, the single-crystal XRD analysis of crystals isolated from a MACC in DME solution revealed a structure containing the dicationic Mg dimer $[(\mu\text{-}Cl)_2Mg_2(DME)_4]^{2+}$ (Figure 2.7b).[44,45] This species was proposed to originate from self-dimerization of metastable $[MgCl(DME)_2]^+$ cations, which in turn resulted from chloride abstraction from $MgCl_2$ by $AlCl_3$.[42]

Other Lewis acids beyond $AlCl_3$ have been examined recently in combination with $MgCl_2$ and many yielded suitable electrolyte systems. For instance, during the initial reports on the MACC electrolyte, $AlPh_3$ and $AlEtCl_2$ were also shown to provide *ca.* 100% coulombic efficiencies for Mg deposition/dissolution and similar anodic stability at around 3.0 V *vs.* Mg/Mg^{2+}.[35] A series of 29 inorganic chlorides and $[NBu_4]Cl$ were tested in combination with $MgCl_2$ in DME, where only BCl_3, $InCl_3$, $SnCl_2$, $SbCl_3$, and $BiCl_3$, supported reversible Mg deposition, albeit in all cases with low anodic stability (<2 V *vs.* Mg/Mg^{2+}) and unsatisfactory coulombic efficiency.[40] Importantly, these results suggest that when considering new additives to couple with $MgCl_2$, there are additional considerations beyond their Lewis acidity that determine their ability to yield highly performing electrolytes. The strong Lewis acid tris(hexafluoroisopropyl)borate $[B(HFP)_3]$ was also combined with $MgCl_2$ in different ratios in the presence of excess Mg powder in THF and glymes, aiming at designing an electrolyte compatible with a sulphur cathode.[45] The best electrochemical performance, when the $[B(HFP)_3]$ concentration was fixed at 0.5 M, was obtained at a 2 : 1 ratio of $[B(HFP)_3]/MgCl_2$ in DME solution, as it provided the highest oxidation peak current, the lowest overpotential for Mg deposition (−0.11 V *vs.* Mg/Mg^{2+}), the highest coulombic efficiency (98.67% in the 50th cyclic voltammetry cycle) and a high oxidative stability (3.3 V *vs.* Mg/Mg^{2+}). It is worth noting that the optimized electrolyte displayed a higher anodic stability on aluminium than on stainless steel (3.0 V *vs.* 2.5 V *vs.* Mg/Mg^{2+}, respectively), which was rationalized by a passivation layer forming on the aluminium metal surface due to anodic reactions caused by the electrolyte. Surprisingly, single-crystal XRD analysis of crystals isolated from the electrolyte in DME solution disclosed an unusual tetranuclear salt $[Mg_4Cl_6(DME)_6][B(HFP)_4]_2$ (Figure 2.8c), while those isolated from THF solution contained the common Mg-dimer cation $[(\mu\text{-}Cl)_3Mg_2(THF)_6][B(HFP)_4]$; unfortunately, there is no proposed mechanism of formation for any of the two salts. ^{11}B NMR analysis of crystals and electrolyte in DME solutions confirmed that $[B(HFP)_4]^-$ anions accounted for most of the boron content before and after crystallization and confirmed the lack of any unreacted $[B(HFP)_3]$ precursor. Electrolyte prepared in DME solution at the optimized composition proved to be very suitable for cycling Mg/S cells, providing a high discharge capacity of above 1000 mA h g^{-1} for 100 cycles at a rate of ~C/10.

Relatively competent electrolytes based on $MgCl_2$ can also be obtained in the absence of any Lewis acids or cosalts. For instance, 0.8 M solutions of $MgCl_2$ in DPSO/THF (1/1, v/v) (DPSO = dipropyl sulfone) displayed moderate ionic conductivity (1.1 mS cm^{-1} at 30 °C) and reversible Mg cycling,

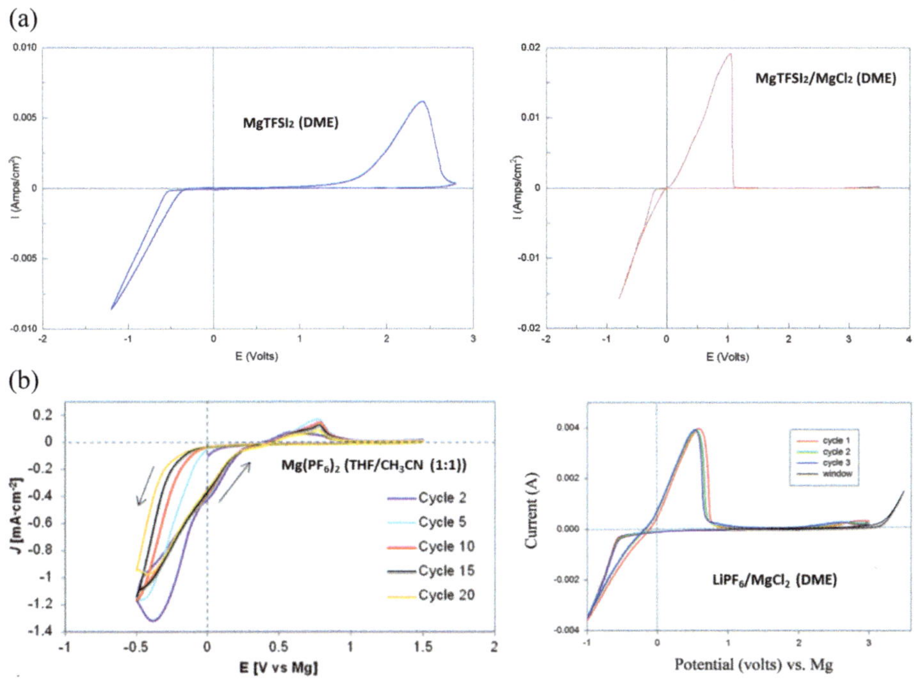

Figure 2.8 Typical cyclic voltammograms of the following electrolytes: (a) 0.5 M $MgTFSI_2$ in DME. Reproduced from ref. 53, http://dx.doi.org/10.1149/2.0161513jes, under the terms of the CC BY 4.0 licence, https://creativecommons.org/licenses/by/4.0/. (b) Conditioned 0.25 M $MgTFSI_2$ + 0.5 M $MgCl_2$ in DME. Reproduced from ref. 53, http://dx.doi.org/10.1149/2.0161513jes, under the terms of the CC BY 4.0 licence, https://creativecommons.org/licenses/by/4.0/. (c) 0.12 M $Mg(PF_6)_2$ in THF/CH_3CN (1:1). Reproduced from ref. 50 with permission from the American Chemical Society, Copyright 2016. (d) Conditioned 0.18 M $LiPF_6$ + 0.045 M $MgCl_2$ in DME. Reproduced from ref. 54 with permission from American Chemical Society, Copyright 2017. The scan rate is 25 mV s^{-1}. Working electrode is Pt in all cases, except for (c) where GC was used.

with a coulombic efficiency of above 90% after electrochemical conditioning (57.3% at the 2nd cyclic voltammetry cycle prior to conditioning).[46] Significantly lower coulombic efficiency (without conditioning, 18.3% at the 2nd cyclic voltammetry cycle) and ionic conductivity (0.3 mS cm^{-1}) were recorded at the same concentration in the absence of THF, pointing at the critical role played by the ether cosolvent. The similar electrolyte system $MgCl_2$/EnPS (1:8 mol ratio) (EnPS = ethyl *n*-propyl sulfone), which was only studied in the absence of ether cosolvent, also displayed low ionic conductivity (0.17 mS cm^{-1} at room temperature) but revealed a high coulombic efficiency (96%).[47] The THF cosolvent effect was rationalized based on Raman spectroscopy and XRD analysis of single crystals isolated from each DPSO-based electrolyte, where a neutral $[Mg_2(\mu\text{-}Cl)_2Cl_2(DPSO)_4]$ complex was found for a DPSO-only solution and an ionic $[Mg(DPSO)_6]^{2+}[MgCl_3(THF)]_2^-$ salt resulted from the DPSO/THF solution.[46] Further confirmation of the electrolyte performance was demonstrated by successfully cycling a Mg/Mo_6S_8 cell for over 300 cycles with high coulombic efficiency (98–99%).

Other magnesium halide salts have also been recently used as Mg precursors, albeit those are far less common. For instance, electrolyte solutions obtained by mixing MgF_2 and the highly-fluorinated Lewis acid tris(2*H*-hexafluoroisopropyl) borate in different ratios in DME displayed high coulombic efficiency (>95%).[48] While spectroscopic analyses did not discount the presence of fluoride ion-containing complex species in solution that could initiate corrosion, this electrolyte displayed a high anodic stability (>3.5 V *vs.* Mg/Mg^{2+}) in stainless steel and aluminium. Bertasi *et al.* studied $[EMIm]^+[AlI_4]^-/(\delta\text{-}MgI_2)_x$ solutions (EMIm = 1-ethyl-3-methylimidazolium) ($0 \le x \le 0.023$) and noted an excellent average coulombic efficiency for Mg deposition/stripping (99.94%), although with a relatively low ionic conductivity (0.3 mS cm^{-1}) and limited anodic stability of 1.2 V *vs.* Mg/Mg^{2+}.[49]

2.2.5 Weakly Coordinating Anions

Magnesium salts of many common weakly coordinating anions typically used in lithium batteries, such as BF_4^-, PF_6^- or ClO_4^-, display limited reversibility of Mg deposition/stripping due to formation of Mg^{2+}-insulating films that inhibit further electrochemical activity.[6,50,51] However, recent work has shown that in some cases, such as for PF_6^- and $TFSI^-$, the coulombic efficiency can be greatly improved by the addition of $MgCl_2$ (Figure 2.8).[52–54]

Pioneering work at Pellion Technologies disclosed that the addition of $MgCl_2$ to $MgTFSI_2$ in ethereal solvents or in their mixtures with certain TFSI-based ionic liquids provide electrolytes capable of displaying high coulombic efficiency of Mg deposition/stripping.[52] In a follow-up study, solutions containing $MgTFSI_2/MgCl_2$ (1 : 2) in DME were shown to exhibit 98% reversibility, 200 mV of deposition overpotential and an anodic stability of 3.1 V *vs.* Mg/Mg^{2+}, following a required electrochemical conditioning similar to that of the MACC electrolyte (Figure 2.8b).[53] Solutions containing $LiTFSI/MgCl_2$ in DME also displayed a similar performance (after a conditioning process), although in this case the optimal molar ratio was 1 : 4. Complex $[Mg_2(\mu\text{-}Cl)_2(DME)_4]^{2+}$ and $[Mg_3(\mu\text{-}Cl)_4(DME)_5]^{2+}$ cations were isolated *via* crystallization from DME solutions of $MgTFSI_2/MgCl_2$ in 1 : 1 and 1 : 2 molar ratios, respectively, suggesting that the bidentate nature of DME coordination may promote the formation of multivalent cations.[55] In addition to DME, $MgTFSI_2/MgCl_2$ solutions in THF and G2 also displayed reversible Mg deposition/stripping, although with a lower coulombic efficiency (max. 93%), which was highly dependent on the degree of chloride concentration.[56]

The enabling effect of chloride to otherwise Mg-passivating solutions was also demonstrated for the PF_6^- anion, in a study mainly motivated to resolving concerns cast over the reported reversibility of Mg deposition/dissolution from $Mg(PF_6)_2$ in CH_3CN/THF solution (Figure 2.8c).[50] Galvanostatic deposition of Li on a Pt electrode from a $LiPF_6$/DME solution was not possible when using a Mg metal counter electrode, strongly implying the incompatibility of pure PF_6^--based solutions with Mg metal.[54] The addition of $MgCl_2$ in a 0.25 molar ratio yielded solutions capable of reversible Mg deposition/dissolution with an average coulombic efficiency of 93.8%, following galvanostatic conditioning (Figure 2.8d).

2.3 Chloride-free Magnesium Electrolytes

Several challenges associated with the presence of the chloride ion as part of the electrolyte make-up were reported. As discussed in Section 2.2, the magnesium ion active species are based on a $[Mg–Cl]^+$ complex cation where the strong binding between Mg^{2+} and Cl^- was recently found to be problematic. For example, the dissociation energy of Mg–Cl in a THF coordination complex is typically greater than 3 eV.[57] This large energy barrier can hinder the formation of Mg^{2+} ions that are available to insert onto host structures thereby prohibiting the insertion cathode function. This problem also extends to conversion type materials that were thought to cycle Mg^{2+} ions. In fact, it was recently revealed that all Mg organic cathodes cycled in electrolytes containing $MgCl^+$ species interact with $MgCl^+$, not Mg^{2+}.[58] The implication for this is that capacities reported for all of these batteries were overestimated as they were calculated based on the assumption that Mg^{2+}, not $MgCl^+$, is the active species. In addition, further lowering of the energy density is caused by the use of excess amounts of electrolyte, which is required to provide sufficient supply of $MgCl^+$. Another obvious challenge is associated with the corrosive properties of Cl^- that induce parasitic side reactions whose effects become pronounced at voltages exceeding 2.5 V *vs.* Mg/Mg^{2+}.[59,60]

Over the past few years, new classes of electrolytes and salts have been reported that are free from Cl^--based salts and reagents. Next, these classes and their properties are discussed.

2.3.1 Halogen-free Simple Salts

Motivated by solving the corrosion problem caused by chloride ions and enabling flexibility in solvent selection, the first inorganic salt that is also halide free was proposed and demonstrated.[61] The salt is based on magnesium borohydride ($Mg(BH_4)_2$) where the premise of this concept is that the BH_4^- ion, being a relatively strong reducing agent, can withstand the reducing environment of the magnesium anode.[61] Highly efficient magnesium deposition/stripping at high coulombic efficiency (94%), high current densities (25 mA cm^{-2} stripping peak current) and low deposition overpotentials (−0.3 V) and stripping (0 V) were demonstrated for the developed $Mg(BH_4)_2:LiBH_4$ (1:3 molar ratio) in DME (Figure 2.9). These findings demonstrated that simple ionic salts could be made compatible with magnesium metal if

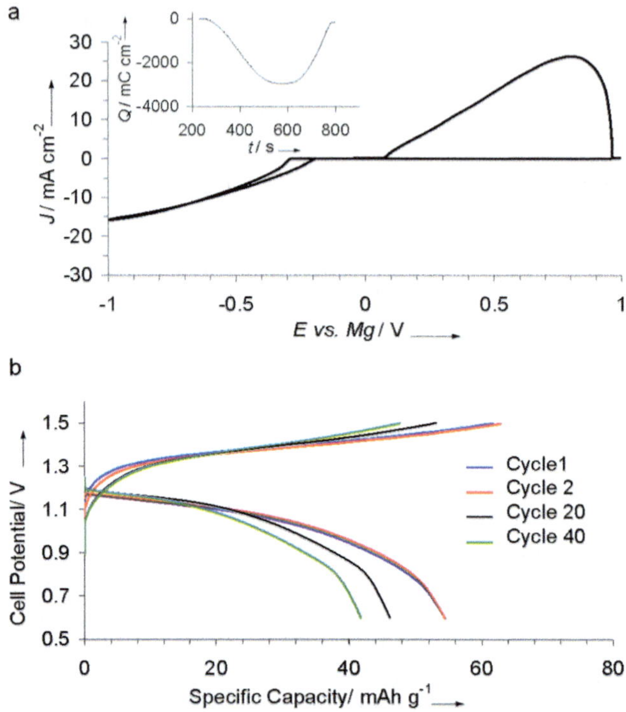

Figure 2.9 Performance in $Mg(BH_4)_2$ electrolytes; (a) cyclic voltammogram (inset shows deposition/stripping charge balance) and (b) charge/discharge profiles with Mg anode/Chevrel phase. Reproduced from ref. 61 with permission from John Wiley and Sons, © 2012 Wiley-VCH Verlag GmbH & Co. KGaA, Weinheim.

the anion in the salt has sufficient reductive stability. The electrochemical performance was suggested to be governed by the extent of salt dissociation per the spectroscopic analyses, which revealed the presence of strong association in these electrolytes where the content of the complex cation $MgBH_4^+$ tracked with improved performance.[61] Key factors that governed the extent of its dissociation were the number of oxygen donors (*i.e.* denticity of the ether) that was demonstrated initially using DME,[61] then later using diglyme[62] and tetraglyme,[63] in addition to the effect of additives such as $LiBH_4$ that compete to associate with the BH_4^- ions.[61] The outcome of combining these factors resulted in close to 100% Mg deposition/stripping efficiency achieved in diglyme and tetraglyme solvents in the presence of $LiBH_4$.[62,63] It is worth mentioning that the highest concentration could be obtained in tetraglyme solvent (0.5 M),[62,63] while it was much lower in shorter chain glymes.[61,62] The presence of $MgBH_4^+$ and its electroreductive stability were later supported by computational results.[64] In addition, computation was used to understand the role of coordination and the effect of the solvent on the solubility and dissociation in these electrolytes.[65,66] Significant and irreversible salt agglomeration in all glymes ranging from THF to tetraglyme was found in all non-dilute borohydride salt solutions.[65] The agglomeration rate and diffusivity of Mg^{2+} in longer chain ethers such as tetraglyme were at their lowest and tracked with the self-diffusivity of the solvent. The low solubility was suggested to be related to the formation of low mobility clusters, which increased in formation rate as the concentration was elevated.[65] Calculations of the free energies to generate singly or doubly charged clusters showed lower values in DME (5.1 and 12.2 kcal mol^{-1}) than in THF (11.5 and 22.2 kcal mol^{-1}), which suggested that cations form more freely in DME[66] and was consistent with experimental observations.[61] It is worth noting that the discovery of this salt and demonstration of the impact of Mg association on the performance of the electrolyte triggered examination and demonstration of association effects on other simple salts such as $Mg(TFSI)_2$, as discussed in Section 2.3.2.

The stability against electrochemical oxidation was however low in all borohydride-based solutions (*i.e.* 1.7, 2.2 and 2.3 V (*vs.* Mg/Mg^{2+}) on platinum, stainless steel and glassy carbon electrodes, respectively).[61] Its worth noting that, as the borohydride electrolytes are not corrosive, these stability trends are opposite to those observed for other magnesium electrolytes. The higher stability of the borohydride on a non-noble metal was suggested to result from catalytic effects of platinum on BH_4^- decomposition.[61] To increase the oxidative stability of the $Mg(BH_4)_2$ electrolytes, strengthening the B–H bond through forming 3-dimensional B–B bonds as in icosahedral boron clusters (boranes and carboranes) was proposed and demonstrated using *closo*-carboranes.[3,67,68] This concept was first demonstrated using the dicarborane anion, where carboranyl magnesium chloride electrolyte (1-(1,7-$C_2B_{10}H_{11}$)MgCl) was reported with high oxidative stability (3.3 V *vs.* Mg). These results also demonstrated the high compatibility of the carborane anion with Mg metal (*i.e.* coulombic efficiency approaching 100%). Based on these findings, a new simple salt based on the more weakly coordinating monocarborane anion $[CB_{11}H_{12}]^-$ was reported.[69] The high solubility of $Mg(CB_{11}H_{12})_2$ (>1 M) in long chain ethers such as tetraglyme and triglyme was reported and the electrolyte efficiently cycled Mg metal (>98%) with very low overpotentials (Figure 2.10). Unlike the case in the borohydride electrolytes, the Mg^{2+} cation in this electrolyte was found be unassociated with the anion, as revealed in the crystal structure of the isolated salt (Figure 2.10). It is worth mentioning that the oxidative stability of $Mg(CB_{11}H_{12})_2$ (measured in acetonitrile solvent), was found to be very high (4.9 V *vs.* Mg/Mg^{2+}), exceeding that of all ether solvents.[70] The salt was found to be non-corrosive and stable to moisture, which was expected from the known low nucleophilicity/electrophilicity of the monocarborane anion. Follow up investigations of different monocarborane-based electrolytes also demonstrated the expected Mg compatibility and high anodic stability.[71,72] Most importantly, the $Mg(CB_{11}H_{12})_2$ salt enabled cycling, for the first time, of a high voltage Mg battery (Mg anode, α-MnO_2 cathode) in a standard stainless steel coin cell (Figure 2.10),[70] which qualified it as the third breakthrough made in Mg electrolytes (the 1st breakthrough was the report on Mg cycling at a voltage of <1.9 V in an organoborate/THF electrolyte and the 2nd was demonstrating a Mg battery prototype in organohaloaluminate/THF electrolyte (2.2 V) and Chevrel phase cathode).[73]

It has been generally understood and accepted that the formation of a film on the magnesium surface would typically passivate the metal and block the diffusion of Mg ions to the surface.[74] However, in the case of the hydrogen boron electrolytes, recent interfacial evidence has revealed the presence of what resembles a solid electrolyte interface (SEI) that forms in the Li-ion battery and is Li ion permeable.[22] The first evidence of a Mg^{2+} permeable SEI in Mg batteries was shown in borohydride solutions.[75] The properties of the interface were demonstrated to play an important role in the performance of the electrolyte. For example, the presence of the SEI was observed in Mg/Mg symmetric cells and was found to be dependent on the type of solvent and borohydride additive used. This was evident in the lower interfacial resistances in $Mg(BH_4)_2$/DME (503 kΩ) compared to those in $Mg(BH_4)_2$/THF (950

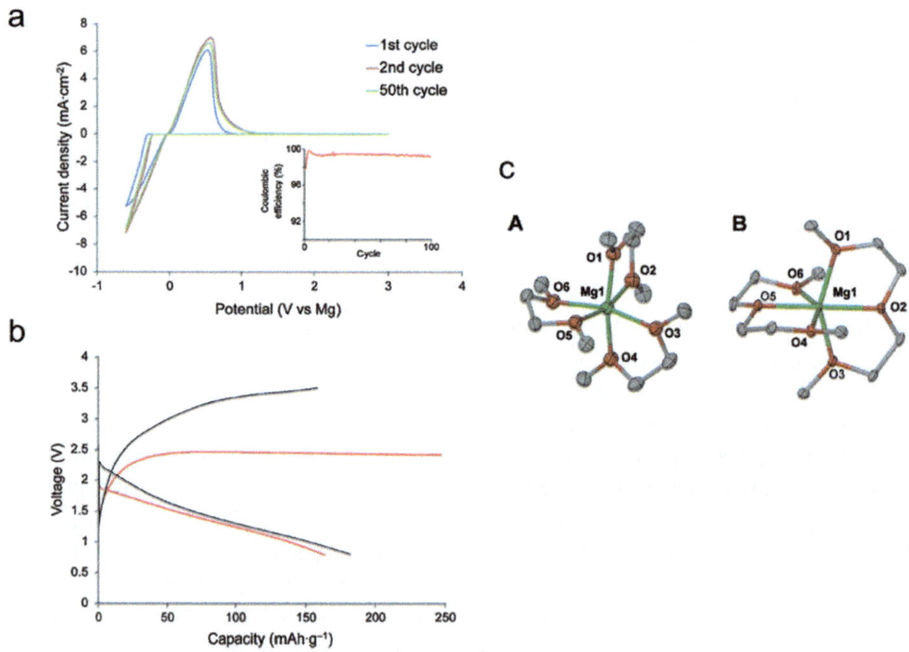

Figure 2.10 Performances of $Mg(CB_{11}H_{12})_2$/tetraglyme electrolyte; (a) cyclic voltammograms of the electrolyte on Pt electrode (inset: cycling efficiencies of Mg deposition and dissolution), (b) initial discharge–charge profiles of a rechargeable Mg battery with $Mg(CB_{11}H_{12})_2$/tetraglyme (black line) and APC (red line) as the electrolyte, a Mg anode, and α-MnO_2 cathode, and (c) Mg cation coordination environment in monoglyme (A) and diglyme (B). Reproduced from ref. 69 with permission from John Wiley and Sons, © 2015 Wiley-VCH Verlag GmbH & Co. KGaA, Weinheim.

kΩ) solutions. In addition, electrolytes containing $LiBH_4$ were found to be constituted of adsorbed species with surface diffusivity on the Mg electrode. These were found to play an important interfacial role in the $Mg(BH_4)_2:LiBH_4$/DME that reduced the bulk (8.75 Ω) and interface resistance of the electrolyte (50 kΩ) and enhanced the surface mobility of adsorbed species at the interface (12.5 kΩ in 1:1 molar Mg:Li solutions, 4.1 kΩ in 1:3 molar Mg:Li solutions).[75] The interfacial and bulk resistances obtained in these solutions tracked well with the overpotentials observed for Mg deposition/stripping, *i.e.* lower resistances resulted in lower overpotentials. Detailed investigation of the nature of this interface and morphology were conducted using operando electrochemical–synchrotron soft X-ray absorption (sXAS) and transmission electron microscopy (TEM).[75] Chemical transformation of the BH_4^-, accompanied by H_2 gas release, into $B_{12}H_{12}^-$, was observed to occur during the Mg deposition process (Figure 2.11). It is worth noting that formation of a Mg rich, Li poor alloy was observed in the 1:6 molar $Mg(BH_4)_2:LiBH_4$/diglyme electrolyte and was suggested to play a role in the Mg deposition process.[76] However, the formation of this alloy seems to be limited thus far to this $LiBH_4$ rich mixture as Li presence was not found in the 1:3 molar $Mg(BH_4)_2:LiBH_4$/DME electrolyte.[75]

To understand the factors that dictate the performance of $Mg(CB_{11}H_{12})_2$ electrolytes, detailed investigations of the interfacial properties and Mg cycling behaviour were conducted.[70] In symmetric Mg/Mg cells, the $Mg(CB_{11}H_{12})_2$/tetraglyme electrolyte was shown to support stable long-term cycling for hundreds of hours (0.034 V overpotential at 0.1 mA cm^{-2}), with observed high overpotential (*ca.* 1 V) in the first cycle if the cells were held under open circuit conditions for an extended period of time (>50 h). The increase in the cell impedance under rest conditions was analogous to that observed in haloaluminate electrolytes and was explained by the adsorption of the electrochemically inactive species onto the Mg metal surface, inducing an increase in interfacial impedance, under polarization conditions, which

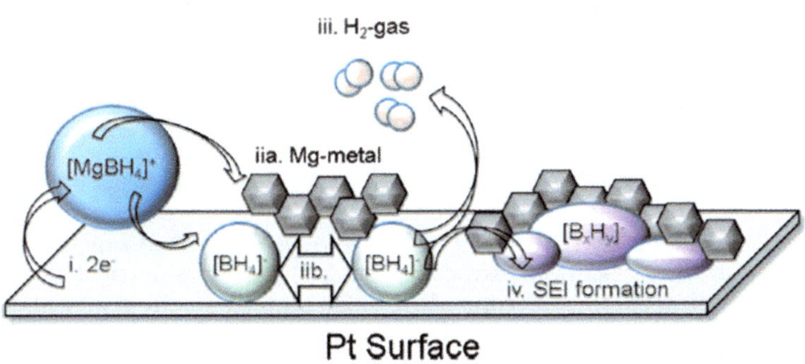

Figure 2.11 Interphase formation from $Mg(BH_4)_2$ electrolytes. Reproduced from ref. 75 with permission from the American Chemical Society, Copyright 2017.

get replaced by the electrochemically active species necessary for Mg deposition/dissolution. The long-term stability of the Mg metal/electrolyte interface was also monitored and compared with that of the Li metal/electrolyte, where Mg metal showed superior tolerance than Li, as demonstrated by the absence of microshort formation with long-term cycling.[70] The cyclic stability of Mg in $Mg(CB_{11}H_{12})_2$/tetraglyme prompted detailed studies of the Mg interface and its morphology following extended cycling. The presence of a SEI was discovered and constituted of Mg, B and C species, in addition to a crystalline phase indexed to be close to a MgB_2O_5 like species. The deposited magnesium was found to be in the form of Mg nanoparticles (*ca.* 10 nm) embedded in an amorphous matrix of SEI material, the growth of which was captured *in situ* in scanning transmission electron microscopy (STEM) measurements (Figure 2.12). The morphology and resultant SEI enabled cycling Mg/Mg symmetric cells under unprecedented high current rate conditions (10 mA cm^{-2}).[77]

Another halogen free electrolyte that was recently reported is based on a bis(η^5-cyclopentadienyl) magnesium (magnesocene, $MgCp_2$) complex.[78] Mg

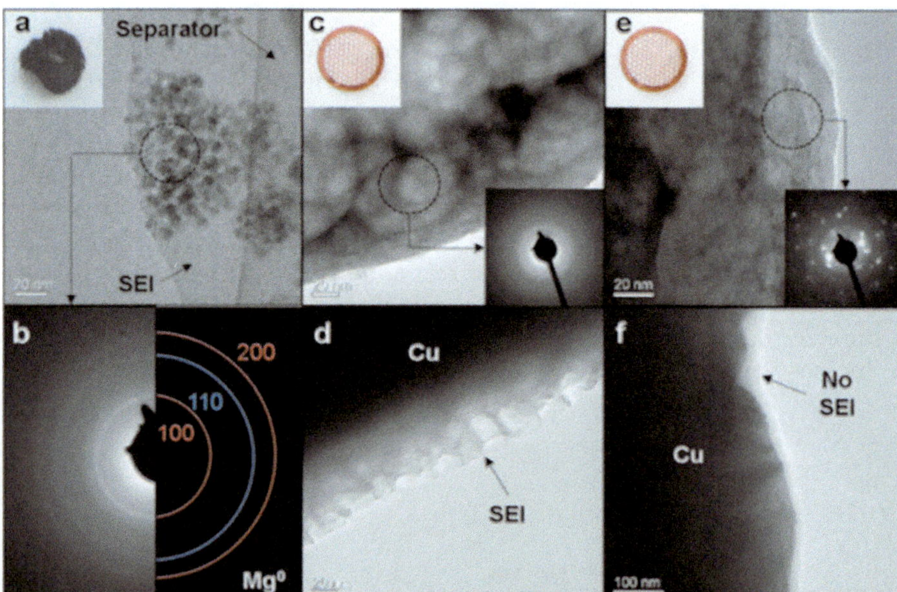

Figure 2.12 In a $Mg(CB_{11}H_{12})_2$/tetraglyme electrolyte: (a) a TEM image of a separator (inset) sample extracted from a Mg/Mg symmetrical cell after 100 cycles; (b) the corresponding SAED pattern for the material circled in (a), (c) a TEM image of a Cu grid sample extracted after 9.5 cycles (after deposition of Mg) at 0.1 mA cm^{-2} (at 0.5 mA h^{-1}), (d) a TEM image of a Cu grid sample extracted after 10 cycles (after dissolution of Mg) at 0.1 mA cm^{-2} (at 0.5 mA h^{-1}). In APC electrolyte: (e) repetition of experiment (c) and (f) repetition of experiment (d). Reproduced from ref. 77 with permission from the American Chemical Society, Copyright 2018.

deposition/stripping was examined in 0.5 M $MgCp_2$/THF solutions, however, it was found to be dependent on the rate applied during the cyclic voltammetry tests. For example, for 0.5 M $MgCp_2$/THF at a 10 mV s^{-1} scan rate, the efficiency was initially 83% and reached 98% after 70 cycles. However, at much lower rates (0.1 mV s^{-1}), the coulombic efficiency was 99% in the first cycle, but decreased with cycling (*i.e.* 80% after 10 cycles, suggested to be due to the weak adhesion of deposited Mg particles). Mg deposits were also observed to have less smooth morphologies compared with the spherical deposits typically observed from Mg compatible electrolytes. As expected from this chemistry, the oxidative stability (1.6 V *vs.* Mg/Mg^{2+}) and the ionic conductivity (12.3 μS cm^{-1} at 0.28 M) were low.

2.3.2 Halogen-based Simple Salts

It is well known and established that salts based on anions analogues to those used typically used in Li or Na ion batteries are incompatible with Mg metal chemistry due to passivation of the surface by the decomposition products of these anions. Until recently, as discussed in Section 2.2, this limited the electrolyte designs exclusively to chloro/organochloro reagents.[51,74,79,80] However, a report on the possibility of enabling high compatibility with Mg metal using borohydride salts, as discussed in Section 2.3.1, contributed towards reviving interests in these inorganic salts. One of the most commonly examined salts for Mg batteries is magnesium bis(trifluoromethanesulfonyl)imide ($Mg[N(SO_2CF_3)_2]_2$), referred to as $Mg(TFSI)_2$. This is owing to the familiarity of the battery community with this anion,[22] as it is successfully applied in Li batteries due to its high anodic stability, its commercial availability, in addition to acceptable solubility in ethers[38] or ethereal solvent mixtures.[81] The possibility of Mg plating/stripping was suggested from $Mg(TFSI)_2$ in 1:1 vol/vol DME/diglyme, however, very large overpotentials in the order of −2 V *vs.* Mg/Mg^{2+} even at very low current densities (*i.e.* 0.01 mA cm^{-2}) were observed in the first cycle and were surprisingly reduced to −10 mV with galvanostatic hold of the cell for about 40 min in a 0.3 M solution (the performance was dependent on the concentration of the salt where lower concentrations led to worsened overpotentials). Indication of possible Mg plating/stripping was also reported in triglyme at 100 °C, however the slow kinetics were suggestive of pseudopassivation, as shown by a very broad Mg stripping peak in the reported cyclic voltammogram (1.5 V).[82] The possibility of Mg plating from $Mg(TFSI)_2$ ethereal solutions was later studied in further details that demonstrated the inefficiency of this salt (Figure 2.13). Namely, examination of the salt in diglyme solvent in three electrode electrochemical cells revealed very high overpotentials in excess of 2 V (Mg deposition and stripping potentials of −0.6 V and above +1.6 V, respectively), low coulombic efficiency below 30% and the presence of trace amounts of F, O, and C, which could be associated with the decomposition of the $TFSI^-$ anion.[83] These studies also revealed that the transference number of Mg^{2+} ions is not very high in the these solutions, where 0.297 was reported for the 0.2 M $Mg(TFSI)_2$/diglyme. This was ascribed

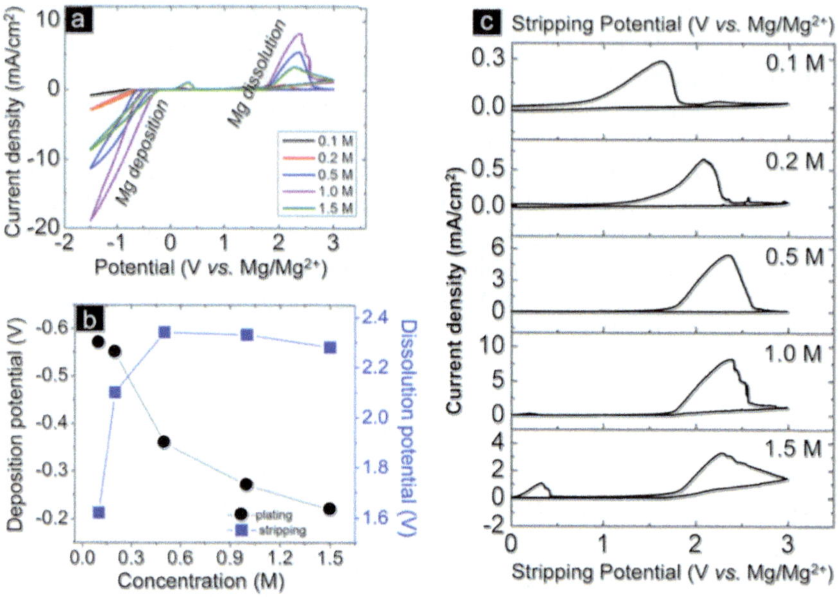

Figure 2.13 (a) Cyclic voltammograms of $Mg(TFSI)_2$/diglyme electrolyte in a concentration range from 0.1 to 1.5 M; (b) concentration dependence of the Mg deposition (black)/dissolution (blue) onset potential; (c) zoomed in graph of the major stripping peak from panel a. Reproduced from ref. 83 with permission from the Royal Society of Chemistry.

Table 2.1 Electrochemical properties of $Mg(TFSI)_2$/G2 electrolyte at concentrations ranging from 0.1 to 1.5 M. Reproduced from ref. 83 with permission from the Royal Society of Chemistry.

Electrolyte	Concentration (M)	Mg (ppm)	ΔC(mM)	t^+	t^-
$Mg(TFSI)_2$/G2	0.2	2.099	2.1	0.297	0.703
$Mg(TFSI)_2$/G2	0.5	2.576	2.58	0.141	0.859
$Mg(TFSI)_2$/G2	1.0	2.571	2.61	0.13	0.87
$Mg(TFSI)_2$/G2	1.5	3.297	2.9	0.036	0.964
$[Mg_2(\text{-}Cl)_3.6(OC_4H_8)]^+$	0.4	35.61	0.73	0.018	0.982
DCC/THF	0.5	2.595	2.62	0.13	0.87
APC/THF	0.5	2.478	2.53	0.159	0.841

to the monodentate association between the Mg^{2+} and the $TFSI^-$ anion, which became more pronounced and more bidentate in nature as the salt concentration increased, thereby further lowering the transference number (Table 2.1).[83] Low coulombic efficiencies (60%) and high overpotentials were also reported in DME solvent.[53]

Recently, a study on depositing/stripping Mg from 0.3 M $Mg(TFSI)_2$/diglyme (following a sudden overpotential drop) revealed the presence of a soft short circuit caused by what is described "globular" dendrites, which comprised interconnected Mg spheres that crossed the separator used in

Mg/Mg symmetric cells.[84] There were various symptoms of this short circuit; such as a current increase of 3 orders in magnitude during cyclic voltammetry studies (*i.e.* cycles 0 to 272 (current density was between −0.19 and 0.37 mA cm^{-2}) and the current rapidly increased at cycle 273 and reached 134 mA cm^{-2} (at 1 V) and −194 mA cm^{-2} (at −1.5 V) during cycles 273 to 320, rate 10 mV s^{-1}). This was concurrent with a linear relationship between the current and voltage, whose slope was independent of the sweep rate, mimicking the behaviour of an ohmic resistor. The reduction in the solution and charge transfer resistance based on impedance measurements also supported the presence of this short circuit. Galvanostatic cycling of symmetric Mg/Mg cells (±0.1 mA cm^{-2} for 1 h) revealed the presence of a short soft circuit that was evident from spikes of very high overpotential (±2 V) that suddenly dropped to ± 0.05 V. This was accompanied by a uniform and rectangular voltage profile. Analysis of the Mg surface showed a preferential growth in the 002 plane, with $TFSI^-$ decomposition products such as fluoride and sulphur, present at the surface of the Mg.[84] Interestingly, the solvent was suggested to alter this behaviour as while short circuits were observed in DME and diglyme solvents, they were absent in electrolytes based on the long chain ethers triglyme or tetraglyme.[84]

The poor performance of $Mg(TFSI)_2$ electrolytes was attributed to the instability of the $TFSI^-$ anion,[64] passivation of the Mg metal by trace water impurities in the electrolytes (<3 ppm)[56] and instability of the diglyme itself in the presence of trace water impurities.[85] Note that in the latter case, the poor performance of the electrolyte persisted even in the presence of very small amounts of trace water, which suggests that water cannot be the main contributor to the Mg passivating behaviour. This was further supported from studies on the monocarborane electrolytes discussed above, where the presence of trace water impurities did not inhibit the high compatibility of these salts. In the case of $TFSI^-$ instability, computations using first-principles quantum mechanics attributed the source of this instability to the presence of Mg^{2+}–$TFSI^-$ association, which weakened the C–S bond and made the anion vulnerable to breakage during the charge transfer process.[64] That is, the computational results showed drastic reduction of the bond dissociation energy (BDE) of C–S in the $[MgTFSI]^+$ complex cation (−0.85 eV) compared to the non-associated $TFSI^-$ anion (2.41 eV), which also suggested a high inherent stability of the non-associated $TFSI^-$ anion.[64] Experimental efforts that minimized/eliminated the Mg^{2+}–TFSI association were reported using crown ether chelates for solutions studied in a tetrahydrofuran solvent.[86] The properties of an electrolyte consisting of equimolar amounts of 18-crown-6 ether and $Mg(TFSI)_2$ (0.5 M) in THF were examined where the FTIR results supported the presence of interaction between the Mg^{2+} and the crown ether and a good conductivity was reported (10^{-3} mS cm^{-1}). However, Mg deposition/stripping efficiency was very low (27%), the overpotentials remained high, and the Mg deposits were not uniform and described as being dendritic.[86] Examination of the deposited Mg surface using energy dispersive spectroscopy (EDS) also showed the presence of F, C, N, and O, as observed previously for $Mg(TFSI)_2$ in glymes.[86] Interesting, experimental studies on $Mg(TFSI)_2$ in DME using Raman and NMR spectroscopies showed the presence of mostly free $TFSI^-$ anions and Mg^{2+} fully coordinated to three DME molecules, where the proposed coordination structure in solution resembled the material crystallized out of the electrolyte.[87] The reports of mainly free anions in some $Mg(TFSI)_2$ electrolytes suggested that the lower observed stability of the $TFSI^-$ anion is not necessarily caused by the presence of a $[MgTFSI]^+$ complex cation. Recently, ambient pressure X-ray photoelectron spectroscopy (APXPS) was used for *in situ* observation of the interfacial processes that occur upon contact of the Mg metal with diglyme solvents in order to assess the possible role of the solvent and impurities on the electrolyte behaviour.[85] These results revealed a more complex picture regarding the stability of the $Mg(TFSI)_2$ electrolytes. Namely, the presence of (O=C)–O moieties was observed when contacting diglyme solvent with Mg metal, which were attributed to bond cleavage along the backbone of the diglyme. *Ab initio* molecular dynamics (AIMD) simulations demonstrated stability of a single diglyme molecule on the Mg(0001), however, in the evitable presence of MgO and $Mg(OH)_2$ on the Mg metal surface, the destabilization of diglyme was explained through OH^- nucleophilic attack. Although diglyme is generally stable against OH^-, it was reported to become compromised by the glyme's backbone strain caused with the chelation of undercoordinated acidic Mg^{2+} cations present at the Mg surface (Figure 2.14). This instability was proposed to occur at the interface and not in the bulk. This was explained by the low bulk OH^- concentrations in addition to the presence of a fully chelated Mg^{2+} cation. Another finding was that the instability of the $[MgTFSI]^+$ complex cation occurred under open circuit conditions where decomposition products of the $TFSI^-$ anion were found. However, the complex was stable upon physical contact with the Mg metal. AIMD simulations supported the stability of the cation towards Mg metal, however, suggested that destabilization can occur in the proximity of the OH^- present at the Mg surface.[85] It is worth mentioning though that soaking of Mg metal in $Mg(TFSI)_2$/DME electrolyte for about 12 hours resulted in decomposition of the electrolyte, as revealed by the strong presence of F species on the Mg surface detected using X-ray photoelectron spectroscopy (XPS).[70] In any case, these results demonstrated that defining the true stability limits of $Mg(TFSI)_2$ is very difficult as it is dependent on various complex factors that include the Mg^{2+}–$TFSI^-$ association nature in ethers and $TFSI^-$ interaction with native interfacial species at the Mg interface, such as $Mg(OH)_2$.

There have been several proposals to enhance the stability of the $TFSI^-$ by eliminating the C–S bond; proven to be a weak link in the molecule, and replacing it with more robust bonds such as C–P or C–Si. Based on this concept, several structures with variable functionalities were computationally screened as a function of their calculated electrochemical window, chemical stability, resistance to hydrolysis, liquid structure and transport properties. After eliminating molecules that were deemed highly toxic or unstable, the two proposed alternate anions (based on the C–P bond) were bis-[(ditrifluoromethyl)phosphoryl]azanide and bis[methyl(ditrifluoromethyl)phosphoryl]azanide.[88] However, it remains unclear how the stability

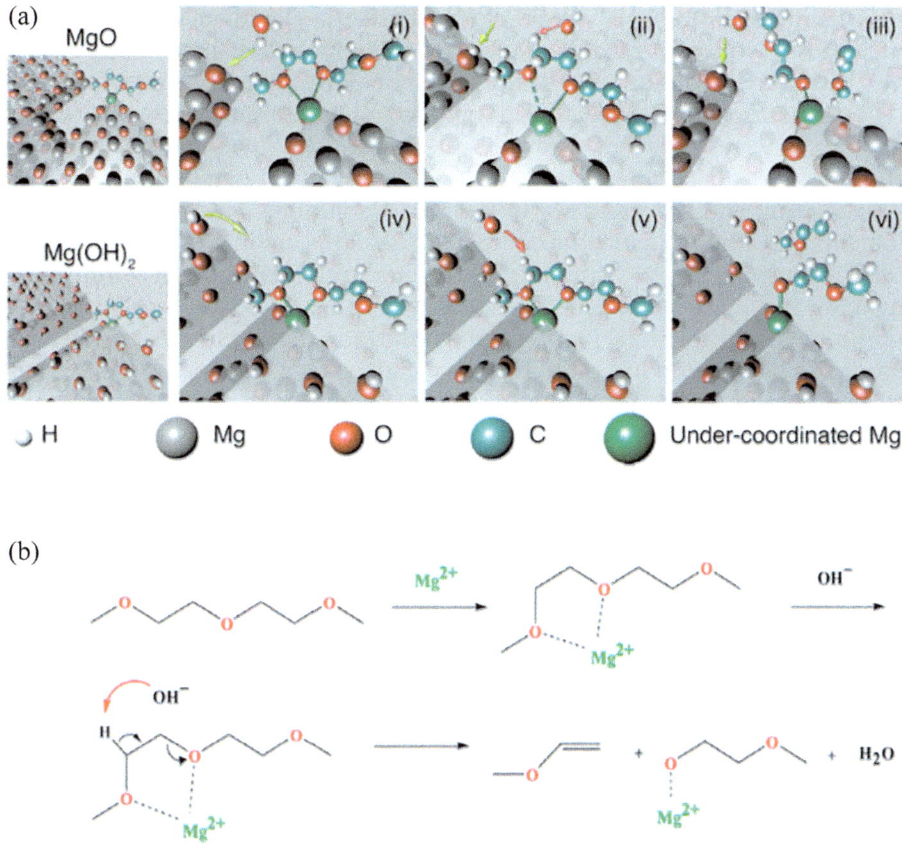

Figure 2.14 (a) Schematics of diglyme decomposition on MgO and $Mg(OH)_2$ surfaces with defects and water impurities. (i) Diglyme chelating the under-coordinated Mg at the MgO kink and H_2O dissociating near the active O site. (ii) Free OH^- deprotonating the deformed diglyme. (iii) Decomposition and the regeneration of water. (iv) Diglyme chelating the open Mg site at the $Mg(OH)_2$ kink and the hopping of native surface OH^-. (v) Deprotonation of the strained diglyme by OH^- and the induced bond cleavage. (vi) Decomposition products and the regeneration of water. (b) A plausible mechanism of diglyme decomposition and the initial decomposition product. Reproduced from ref. 85 with permission from the American Chemical Society, Copyright 2017.

of these anions will be impacted in the presence of unavoidable bulk or surface impurities, as discussed above. To date, despite all these efforts, no good performance could be achieved using $Mg(TFSI)_2$ electrolytes.

The ability to demonstrate some Mg deposition/stripping in $Mg(TFSI)_2$ electrolytes, albeit being highly inefficient, motivated investigating $Mg(PF_6)_2$,[50] which is another salt that was previously deemed to be incompatible with Mg metal. This interest was also partly driven by the success of the PF_6^- anion in Li-ion batteries, primarily due to its protection of the Al current collectors used on the positive electrode side.[22] $Mg(PF_6)_2$ was prepared in acetonitrile solvent, resulting in an acetonitrile solvate that was used for electrolyte studies (acetonitrile is known to be unstable under a low potential environment below −2.5 V *vs.* SHE). Cyclic voltammetry examination of $Mg(PF_6)_2$ (0.12 M) in a 1:1 THF/CH_3CN mixture showed pseudo reversibility, where extremely low coulombic efficiencies were observed. Given the presence of Mg incompatible acetonitrile solvent in these solutions, a follow up study investigated the possibility of depositing and stripping Mg in DME/PF_6^- solutions. Since the Mg salt is not commercially available, the compatibility of this anion with Mg metal was examined using the $LiPF_6$ salt in a three electrode electrochemical cell employing a Mg metal counter electrode of which the electrochemical response was monitored during Li metal plating and stripping on the working electrode. The strong Mg passivation indicated by a very high voltage response (50 V) settled the unsuitability of $Mg(PF_6)_2$ solutions for Mg batteries.[54]

Different types of Mg salts were prepared and isolated utilizing bulky and weakly coordinating anions based on fluorinated alkoxyaluminates and alkoxyborates, of which some were used to study Mg–S batteries.[89] Note that electrolytes based on fluorinated alkoxyaluminates[90] and alkoxyborates anions[48] that were examined without the necessity of isolating the Mg salt are discussed below in Section 2.3.3. Magnesium tetrakis(hexafluoroisopropyloxy)borate $Mg[B(hfip)_4]_2$ was synthesized from a one pot acid (HOC(H)$(CF_3)_2$)–base ($Mg(BH_4)_2$) process and isolated in the DME solvate form.[89] It was found to undergo melting at 53 °C, is stable in moist air, has a high thermal stability up to 150 °C and good solubility in a variety of ethers (molarity was calculated based on the amount of $Mg(BH_4)_2$ used in the synthesis reaction as $Mg[B(hfip)_4]_2$ was not initially isolated for the solubility electrochemistry studies). Studies in three electrode cyclic voltammetry cells showed that $Mg[B(hfip)_4]_2$ (0.6 M) in DME supported Mg plating with relatively high overpotentials (−0.72 V in the 1st cycle) and Mg stripping was observed from 0.43 V. The electrolyte/electrode setup seemed to undergo activation evident from reduction in the plating overpotential in subsequent cycles, which reduced to −0.4 V in 2nd cycle, and an increase in the peak stripping current density up to 45 mA cm^{-2} during the Mg stripping (Figure 2.15) (a high conductivity of 10 mS cm^{-1} was determined for 0.3 M at 23 °C which is comparable to that of electrolytes in Li ion batteries). The coulombic efficiency was not reported in the three electrode cell study but it was measured using a galvanostatic method at 98% and XRD measurements confirmed that Mg was plated from this electrolyte. As expected from this highly fluorinated anion, the oxidative stability of this electrolyte was high (4.3 V on stainless steel) (Figure 2.15).[89] A follow up study improved the synthesis process of $Mg[B(hfip)_4]_2$ and electrochemical examination of the isolated salt, rather than the reaction mixture, enhanced the Mg reversibility apparent from the absence of voltage spikes during the galvanostatic cycling of symmetric Mg/Mg cells; however, electrolyte activation still occurred as the overpotentials were relatively high in the first few cycles (±0.3 V), decreasing to 0.2 V upon cycling.[91] It is noteworthy that the cyclic stability reported in symmetric cells seemed to be dependent on the salt concentration apparent from a continuous increase in the cell overpotentials when the electrolyte concentration was 0.4 M.[91] The report

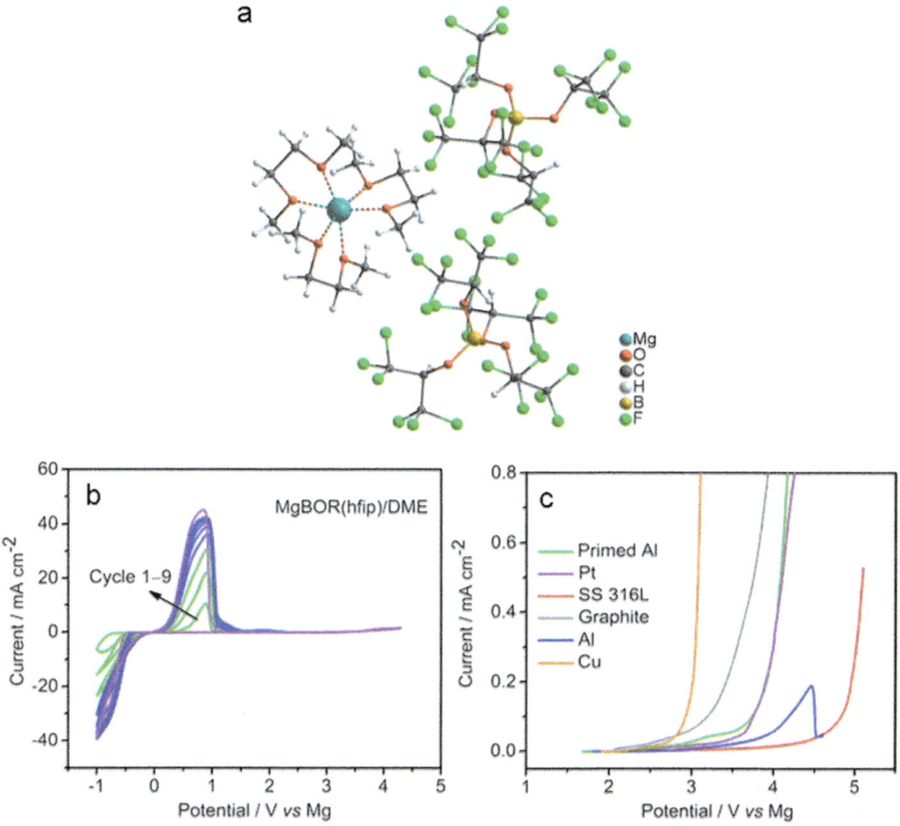

Figure 2.15 Magnesium tetrakis(hexafluoroisopropyloxy)borate electrolyte: (a) structure of the DME solvate, (b) cyclic voltammograms using Pt as the working electrode, and (c) linear sweep voltammograms of various electrodes. Reproduced from ref. 89 with permission from the Royal Society of Chemistry.

of Mg discoloration after symmetric cells cycling suggested decomposition of electrolyte at the interface of what was described as possibly a SEI, which could help to explain the increased overpotential at higher concentrations.[91]

Given the known robustness of the C–F bond towards oxidation, fluorination of the C–H bond in the $[CB_{11}H_{12}]^-$ anion, discussed in Section 2.3.1, was recently demonstrated to further increase its oxidative stability (by 0.3 V). In this case, fluorination of the C–H bond did not impact its compatibility with the Mg metal, apparent from high coulombic efficiencies (96%) and low overpotentials (measured in 5 mM salt concentrations in triglyme).[92]

2.3.3 Halogen-based Reagents

Several studies were conducted in alkoxyaluminate and alkoxyborate electrolyte systems, where the identity of Mg salts present in these electrolytes cannot be solidly determined, as they were neither isolated nor analysed.

The alkoxyaluminate system based on combining aluminium hexafluoroisopropoxide $Al(HFIP)_3$ with Mg hexafluoroisopropoxide $Mg(HFIP)_2$ in a 1:2 ratio in DME was reported.[90] The formation of $Mg(Al(HFIP)_4)_2$ was suggested based on ^{27}Al NMR results; however, the salt was neither isolated nor identified (a full spectrum of other constituents was not provided, and the presence of trace unreacted HFIP was proposed).[90] The electrolyte exhibited low coulombic efficiency (58%) and high overpotential (−0.8 V) in the first cyclic voltammogram cycle, however the coulombic efficiency increased to 99% with further cycling. It is important to note that the electrolyte performance drastically changed as a function of the solvent, *i.e.* it was not successful in diglyme solvent, where the coulombic efficiency remained very low (43%). In addition, the oxidative stability of this electrolyte was strongly dependent on the substrate, where it was about 2 V on Pt and up to 3.5 V *vs.* Mg/Mg^{2+} when measured on glassy carbon. The source of the large discrepancy is currently unknown.

Electrolytes based on alkoxyborates were reported by mixing the acidic tris(2*H*-hexafluoroisopropyl) borate (THFPB) with MgF_2[48] or MgO[45] in DME. Large excess amounts of THFPB (20 times that of MgF_2 (0.05 M)) were used to form the electrolyte mixture that exhibited low coulombic efficiency (*ca.* 79%) and high overpotentials (>0.5 V) that improved with cycling (95% at 20 cycles). Diffusion limitations were apparent in the Mg deposition where atypical large hysteresis was observed. A substantial increase in the Mg plating/stripping current densities with cycling by more than an order of magnitude was reported, which strongly suggests electrolyte transformation during the cyclic voltammogram.[45] The identity of all present charge carriers and speciation in these solution was not reported or explored. The use of MgO (with 40 times excess THFPB) led to more sluggish kinetics evident from high overpotentials in symmetric Mg/Mg cells (*ca.* +0.45 V for Mg stripping at 10 μA cm^{-2}), the source of the Mg cycling performance and activation procedure remain unknown.[45]

As follow up to the aforementioned studies, THFPB was recently combined with $Mg(BH_4)_2$ to study Mg–S batteries, see Chapter 10.[93] The nucleophilic nature of the BH_4^- was suggested to be minimized in its solutions with THFBP. Note that cycling of the sulphur cathode in the Li–S battery using $LiBH_4$ solid state electrolyte (initial discharge capacity of 1140 mA h g^{-1} that remained as high as 730 mA h g^{-1} with nearly 100% coulombic efficiency over the next 45 cycles) was reported before which suggested the presence of important interfacial factors that enable this performance.[94] In the case of a Mg–S battery, the $Mg(BH_4)_2$:THFPB (1:2 molar) mixture was examined (in diglyme solvent) and was transformed through a Lewis acid:Lewis base reaction; however, the resultant salts were not isolated and the electrochemical studies were conducted using the reaction mixture which was indicated to include $[B(OC(H)(CF_3)_2)_4]^-$, $[MgBH_4(Diglyme)_2]^+$, free BH_4^- and other molecular fragments from reacted THFPB. As expected from $Mg(BH_4)_2$, the solutions displayed high compatibility with Mg metal with low overpotentials and >99% Mg stripping/plating efficiency after about 20 cycles. Surprisingly, despite the presence of BH_4^-, 2.8 V oxidative stability on a stainless steel substrate was indicated.[93]

To improve the compatibility of $Mg(TFSI)_2$ with Mg metal, one of the approaches followed was to combine it with $Mg(BH_4)_2$, known for its high reductive stability ($TFSI^-$-based ionic liquids were also combined with $Mg(BH_4)_2$ and this is explained in the next section discussing the development of non-flammable Mg electrolytes). In one report, $Mg(BH_4)_2$ was used to alter the solvation structure of $Mg(TFSI)_2$ so it becomes more resistant to decomposition through incorporating the reductively stable BH_4^- in the Mg^{2+} solvation sphere.[95] This mixed solvation structure was corroborated through ^{25}Mg NMR and computations (quantum chemistry calculations of ^{25}Mg NMR chemical shifts, and classical molecular dynamics, MD). For example, for the mixture consisting of 0.01 M $Mg(BH_4)_2$ and 0.001 M $Mg(TFSI)_2$, the ^{25}Mg NMR peak centre was located at 12.2 ppm, which differed from the 13.5 ppm value measured for pure 0.01 $Mg(BH_4)_2$. Based on varying the salts ratios and examining the ^{25}Mg NMR shifts; coupled with computational support, the presence of a mixed BH_4^-–$TFSI^-$ solvation shell of Mg^{2+} was proposed, in addition to the increased presence of free $TFSI^-$ anions, reported to be more stable than the $[MgTFSI]^+$ complex. These findings were used to explain improvements in the performance of $Mg(TFSI)_2$-/diglyme-based electrolytes (*i.e.* in a 0.01 M $Mg(BH_4)_2$, 0.001 M $Mg(TFSI)_2$ diglyme solution, the overpotential was 0.16 V with 89% coulombic efficiency, measured using cyclic voltammogram in a three electrode cell). Another study showed that 0.5 M $Mg(TFSI)_2$/tetraglyme electrolyte with small concentrations of $Mg(BH_4)_2$ (6 mM) allowed for reversible Mg deposition/stripping (coulombic efficiency initially *ca.* 84% that dropped to *ca.* 75% after 500 cycles) and low initial deposition (−0.36 V) and stripping (0 V) overpotentials)).[96] This study proposed that the borohydride acted as a moisture scavenger in these solutions; however, information and analysis of the borohydride reaction products present in the electrolyte solution were not addressed. A different approach that was applied to modify $Mg(TFSI)_2$ solutions with BH_4^- was using a BH_4^- anion with hydride groups that were partially substituted with phenol.[97] 0.5 M $Mg(TFSI)_2$ combined with 0.15 M $Mg[B(OPh)_3H]_2$ in diglyme exhibited low deposition potentials (*ca.* −0.64 V); however, in this case the coulombic efficiency remained as low as 64%, even after extended cycling. The occurrence of a short circuit was reported during galvanostatic cycling in two electrode cells; however, the observation of dendrites was not reported.[97]

To enable better performance of $Mg(TFSI)_2$/diglyme, the incorporation of an anthracene–Mg coordinating agent was proposed as one possible solution.[98] The idea was to introduce what was described as π stabilization during the Mg deposition process. The addition of anthracene in the range of 0.0075–0.07 M (Figures 2.16) introduced a remarkable effect on the Mg deposition/stripping process where reduced overpotentials, increased coulombic efficiency, and increased current densities were observed. A small reduction peak that preceded Mg deposition in anthracene-containing electrolytes was suggested to result from an anthracene magnesium complex involving the electron carrier properties of anthracene; however, no direct evidence was provided. Interestingly, in stainless steel/Mg cells, galvanostatic cycling in 0.5 M $Mg(TFSI)_2$ with 0.07 M anthracene underwent high polarization of about

2 V during the first 15 cycles, which was lowered with further cycling indicating the occurrence of a conditioning process. In addition, the presence of deposited Mg that was inhomogeneous and in the form of aggregates was found at the surface of the stainless steel working electrode (Figure 2.16). These deposits were suggested to have highly resistive blocking layers that may contribute to short circuiting of the cells during cycling. This challenge was overcome using a $MgCl_2$ additive, as discussed in detail in the previous section.[98]

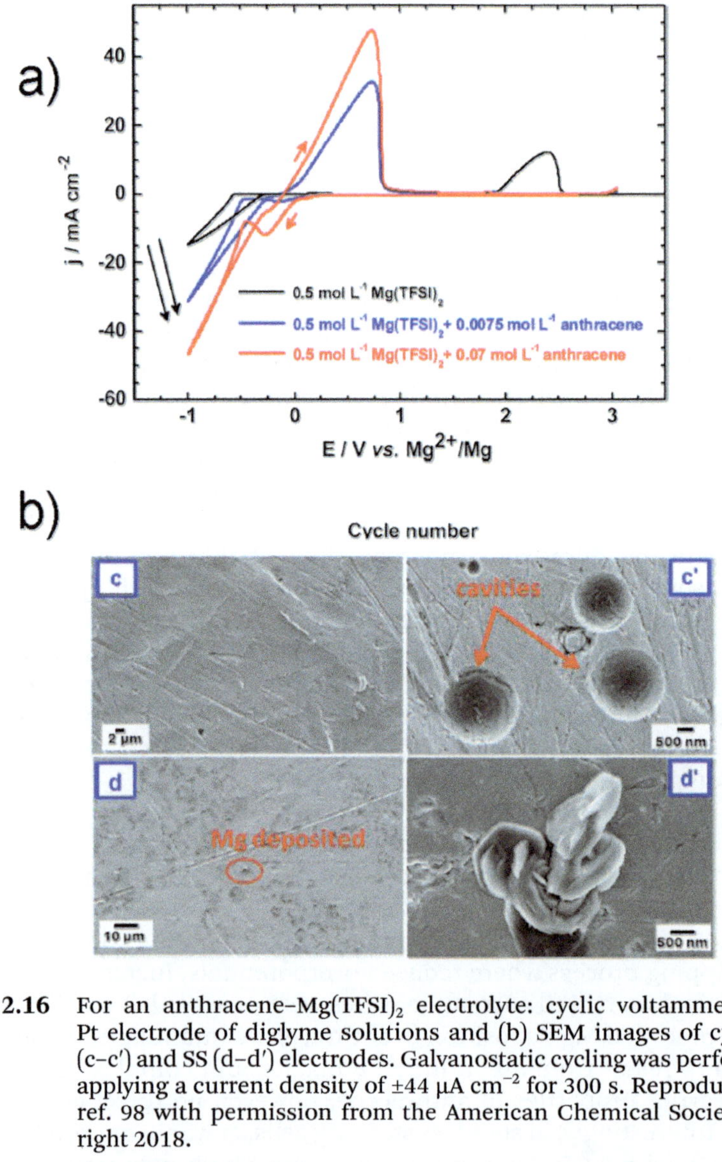

Figure 2.16 For an anthracene–$Mg(TFSI)_2$ electrolyte: cyclic voltammetry on a Pt electrode of diglyme solutions and (b) SEM images of cycled Mg (c–c′) and SS (d–d′) electrodes. Galvanostatic cycling was performed by applying a current density of ±44 μA cm^{-2} for 300 s. Reproduced from ref. 98 with permission from the American Chemical Society, Copyright 2018.

2.3.4 Electrolytes Based on Non-ethereal Solvents

Electrolytes based on typical aprotic solvents such carbonates, sulfonates and nitriles were proven to be incompatible with Mg metal chemistry due to their instability at the Mg deposition potential.[80] This is the case for many simple Mg salts; with few exceptions where non-blocking SEI forms,[75,77] instability and decomposition caused passivation that blocks transport of Mg^{2+}, as discussed above in Section 2.3.1. There have been a few recent reports on utilizing sulfone solvents as the main solvent[47] or mixed with THF;[16] however, this was only possible in the presence of reducing agents or chloride-based systems such as $MgCl_2$, and exhibited high Mg deposition/stripping overpotentials (*i.e.* >2 V Mg stripping peak width in cyclic voltammetry).[47] A recent report suggested the introduction of a polymer-based artificial SEI on Mg as a means to reversibly strip/deposit Mg in a carbonate/water electrolyte; however, it was only shown in symmetric Mg/Mg cells cycled under an extremely low capacity (although it was not explicitly reported, the capacity could be discerned from the low current density and short hold time, *i.e.* 0.01 mA cm^{-2}, *ca.* 30 min per charge/discharge cycle). This, coupled with the absence of Mg cycling voltammetry (in the carbonate/water electrolyte blend) and coulombic efficiency measurements makes it hard to establish that Mg was truly cycling in these solutions.[99]

Given the undesired flammability of most common aprotic solvents, an attractive choice for eliminating the safety hazards would be using ionic liquids due to their very low volatility, non-flammability, and high thermal stability in addition to a wide electrochemical window.[100] Promising ionic liquids studied for Li and Na batteries are typically composed of an organic cation such as phosphonium, pyrrolidinium or piperidinium coupled with a fluorinated inorganic anion, such as bis(fluorosulfonyl)imide ($(FSO_2)_2N^-$, FSI) or bis(fluorosulfonyl)imide ($(CF_3SO_2)_2N^-$, TFSI).[100] The incompatibility of these common anions with Mg metal, as discussed in the previous section, has hindered the practical use of ionic liquids. It is worth noting that ionic liquids (mixed with aprotic solvents) were used to examine Mg^{2+} insertion in cathodes in electrochemical cells that use alternative anode materials to Mg metal.[101]

It was earlier claimed that reversible magnesium plating was possible with the conventional salt magnesium triflate ($Mg(CF_3SO_3)_2$) in an imidazolium based ionic liquid.[102] However, magnesium metal passivation was later on confirmed to take place in these ionic liquids.[103–105] Reversible magnesium deposition/dissolution was however shown in ionic liquid/ether solvent blends from solutions containing either strong nucleophilic reagents such as phenyl magnesium bromide[104] and alkyl (or allyl)magnesium bromide[106,107] or when used as additives in an electrolyte mixture based on $MgCl_2$ (combined with $Mg(HMDS)_2$) and THF in *N*,*N*-diethyl-*N*-methyl-*N*-(2-methoxyethyl)ammonium TFSI.[108] In ether-free ionic liquid solutions, Mg disposition/stripping was reported in the Cl^- based system

1-ethyl-3-methylimidazolium chloride (EMImCl) doped with $AlCl_3$ and highly amorphous δ-$MgCl_2$.[109] Recently, Mg deposition/stripping was shown to be possible in chloride- and ether-free electrolytes based on $Mg(BH_4)_2$ salt and $TFSI^-$-based ionic liquids.[110,111] Reported systems consisted of a pyrrolidinium cation with dangling PEGylated chains[111] or with ether-functionalized ammonium cation $N_{2(20201)3}^+$.[110] The incorporation of chelating ethereal moieties was inspired by previous studies on enhanced $Mg(BH_4)_2$ dissociation and performance in ethers with increased number of oxygen electron donors.[61,62] For example, ionic liquids based on methyl polyethylene glycol $MPEG_3PyrTFSI$ (three ether oxygens) and $MPEG_7PyrTFSI$ were superior to *N*-butyl-*N*-methylpyrrolidinium BMPyrTFSI.[111] Further studies showed that ionic liquid solutions containing small concentrations of $Mg(BH_4)_2$ (*i.e.* 18 mM) in 0.3 M $Mg(TFSI)_2$/tetraglyme, *N*-methyl-*N*-butyl pyrrolidinium TFSI (1 : 2 molar) allowed some reversible Mg deposition/plating (initial coulombic efficiency *ca.* 87% that degraded with cycling). Despite the presence of $Mg(BH_4)_2$, the anodic stability was reported to be up to 3 V, however it dropped to *ca.* 2.1 V in cycled solutions (no linear scanning voltammetry that measured the anodic stability was shown).[112]

In all of the aforementioned studies, the oxidative stabilities of these solutions were low (<2.5 V) and those based on strong reducing agents are expected to be incompatible with cathodes prone to nucleophilic attack, such as sulfur. Therefore, to eliminate the challenges associated with the use of chlorides and strong reducing agents, a new recent alternate approach was used to design, for the first time, ionic liquids that are Mg compatible. This was achieved by forming ionic liquids that incorporate the highly stable *closo*-monocarborane anion (this proof of concept was demonstrated by coupling the anion with the ammonium cations $N_{2(20201)3}^+$ and $N_{4(20201)3}^+$).[113] Reversible Mg deposition/stripping was shown where a coulombic efficiency of 70% was sustained over 50 cycles for $Mg(CB_{11}H_{12})_2$ in an ionic liquid/tetraglyme mixture (3 : 1 molar ratio) (Figure 2.17) and smooth morphologies of deposited Mg were confirmed. The oxidative stabilities of these ionic liquids were also higher than ethers by about 0.5 V.[113] They were also reported to remain in the liquid phase down to −52 °C, had good conductivity (in the order of 10^{-4} mS cm^{-1}) and were highly dissociated.[113]

2.3.5 Solid State Electrolytes

To help reduce the risks associated with the use of volatile solvents, one strategy would be trapping the electrolyte within polymeric matrices or porous hosts.[114–121] However, the applicability of this approach remains questionable as in most of these electrolytes, the magnesium salts applied were known to be incompatible with the magnesium metal.[114,115,117,118,120]

Until very recently, the observation of solid state magnesium ion conduction in inorganic salts occurred only at temperatures exceeding 500 °C.[122–124] However, the presence of good Mg^{2+} mobility has been known and demonstrated in sulphide-based Mg cathodes such as Chevrel phase Mo_3S_4,[125,126] spinel Ti_2S_4,[127] and a recent report showed high Mg mobility in spinel $MgSc_2Se_4$.[128]

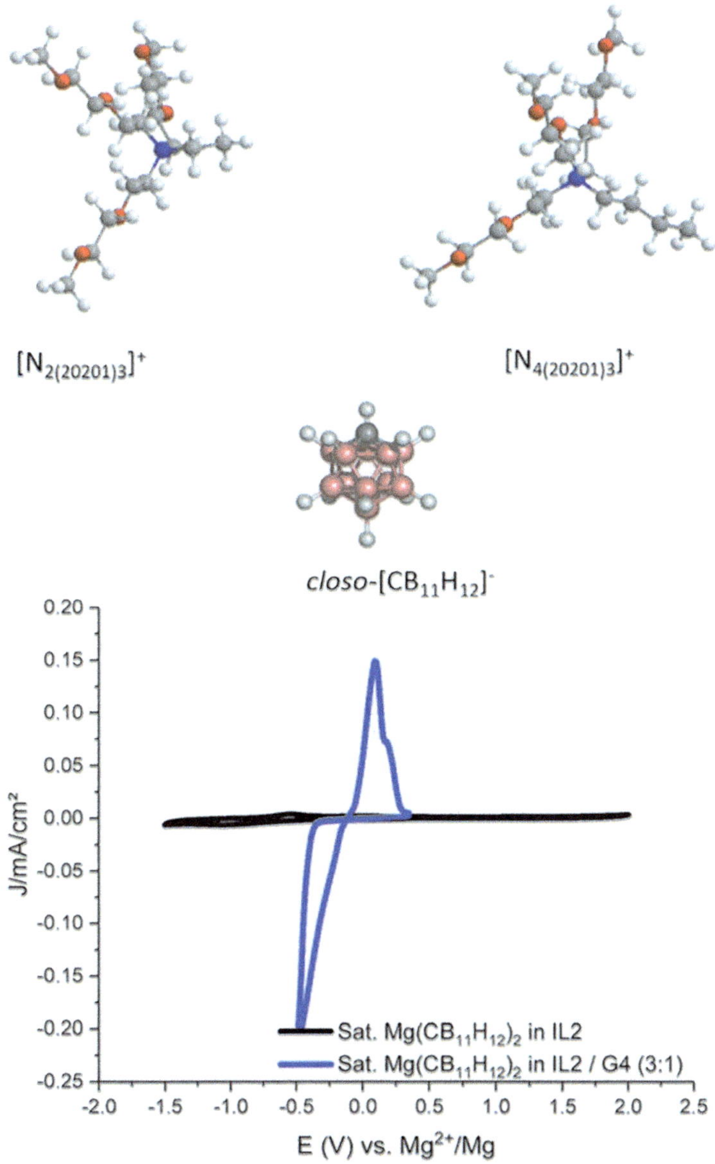

Figure 2.17 Structures of the closo-carborane ionic liquids (N = blue, O = red, C = grey, H = white, and B = pink.); Cyclic voltammogram on a Pt electrode of saturated $Mg[CB_{11}H_{12}]_2$ in the ionic liquid. Reproduced from ref. 113 with permission from the Royal Society of Chemistry.

Note that the latter is also an electronic conductor so it does not qualify as an electrolyte (it was reported to be extremely reactive even towards inert current collectors). The first study that demonstrated Mg diffusion in solid state electrolytes at lowered temperatures was in a borohydride-based salt.[129] Studies on magnesium ion mobility in $Mg(BH_4)(NH_2)$ were motivated by the reported compatibility of borohydride-type liquid electrolytes with Mg metal in addition to the presence of cavities large enough to enable magnesium ion conduction through a hopping mechanism.[130] A conductivity of about 10^{-3} mS cm^{-1} at 150 °C for $Mg(BH_4)(NH_2)$ was obtained (Figure 2.18), which is three

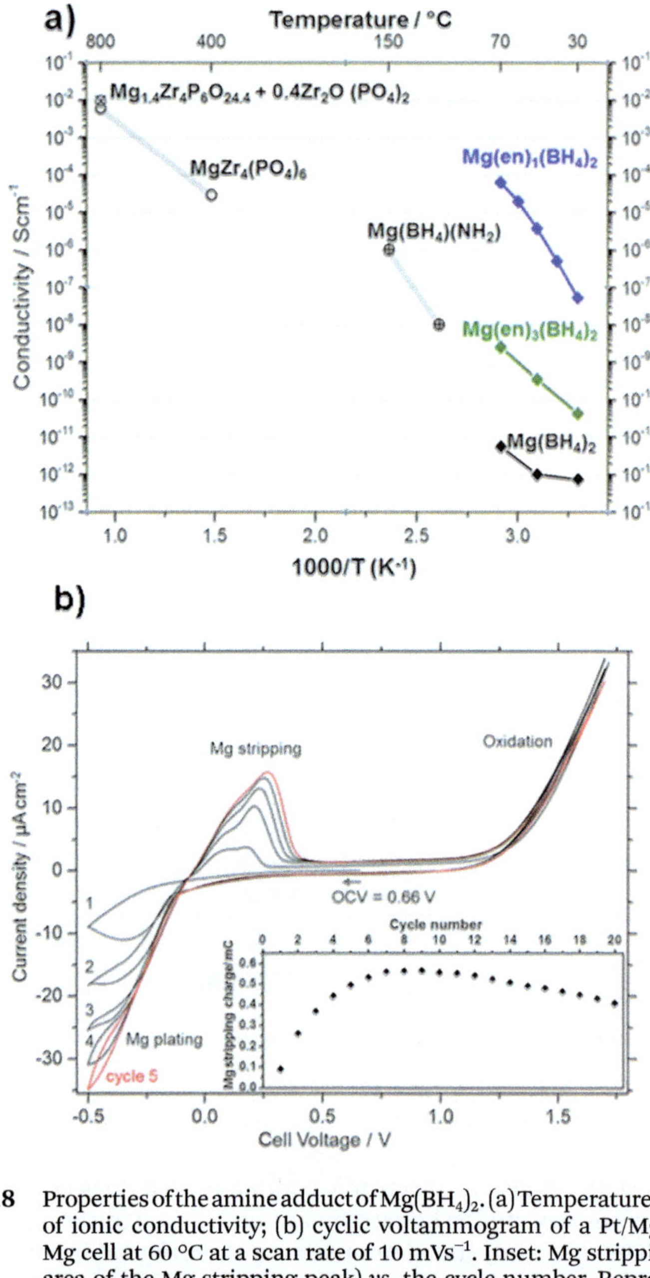

Figure 2.18 Properties of the amine adduct of $Mg(BH_4)_2$. (a) Temperature dependence of ionic conductivity; (b) cyclic voltammogram of a Pt/$Mg(en)_1(BH_4)_2$/Mg cell at 60 °C at a scan rate of 10 mVs^{-1}. Inset: Mg stripping charge (= area of the Mg stripping peak) *vs.* the cycle number. Reproduced from ref. 132, https://doi.org/10.1038/srep46189, under the terms of the CC BY 4.0 licence, https://creativecommons.org/licenses/by/4.0/.

orders of magnitude higher than that of $Mg(BH_4)_2$, suggested to be partly as a result of the shorter distance between the two nearest Mg atoms (3.59 Å in $Mg(BH_4)(NH_2)$ *vs.* 4.32 Å in $Mg(BH_4)_2$). Interestingly, the oxidative stability of the $Mg(BH_4)(NH_2)$ salt was found to be in excess of 3 V *vs.* Mg at 423 K.[129] Mg conductivity was recently reported in $Mg(BH_4)_2$-diglyme$_{0.5}$ (10^{-5} S cm^{-1} at 350 K), where the results suggested the presence of mixed phases that undergo partial melting at 330 K.[131] The use of a borohydride amine adduct was shown to dramatically improve the ionic conductivity of the borohydride, where the highest reported conductivity of Mg^{2+} (6×10^{-5} S cm^{-1} at 343 K, Figure 2.18) in a solid state electrolyte to date was achieved in an ethylene diamine adduct of $Mg(BH_4)_2$ *cis*-$Mg(en)(BH_4)_2$, en = $NH_2(CH_2)_2NH_2$ (it was however found that the amine adduct limited the potential window to 1.2 V).[132]

Acknowledgement

RM and OT acknowledge the support from Dr T. Matsunaga at Toyota Research Institute of North America during the preparation of this chapter. Funding was provided by Toyota Motor Corporation.

References

1. C. A. Nist-Lund, J. T. Herb and C. B. Arnold, *J. Power Sources*, 2017, **362**, 308.
2. P. Saha, M. K. Datta, O. I. Velikokhatnyi, A. Manivannan, D. Alman and P. N. Kumta, *Prog. Mater. Sci.*, 2014, **66**, 1.
3. T. J. Carter, R. Mohtadi, T. S. Arthur, F. Mizuno, R. Zhang, S. Shirai and J. W. Kampf, *Angew. Chem., Int. Ed.*, 2014, **53**, 3173.
4. E. G. Nelson, S. I. Brody, J. W. Kampf and B. M. Bartlett, *J. Mater. Chem. A*, 2014, **2**, 18194.
5. A. J. Crowe, K. K. Stringham, J. L. DiMeglio and B. M. Bartlett, *J. Phys. Chem. C*, 2017, **121**, 7711.
6. T. D. Gregory, R. J. Hoffman and R. C. Winterton, *J. Electrochem. Soc.*, 1990, **137**, 775.
7. C. Liebenow, Z. Yang and P. Lobitz, *Electrochem. Commun.*, 2000, **2**, 641.
8. D. Aurbach, Z. Lu, A. Schechter, Y. Gofer, H. Gizbar, R. Turgeman, Y. Cohen, M. Moshkovich and E. Levi, *Nature*, 2000, **407**, 724.
9. H. S. Kim, T. S. Arthur, G. D. Allred, J. Zajicek, J. G. Newman, A. E. Rodnyansky, A. G. Oliver, W. C. Boggess and J. Muldoon, *Nat. Commun.*, 2011, **2**, 427.
10. Z. Zhao-Karger, X. Zhao, D. Wang, T. Diemant, R. J. Behm and M. Fichtner, *Adv. Energy Mater.*, 2014, **5**, 1401155.
11. X. Yu and A. Manthiram, *ACS Energy Lett.*, 2016, **1**, 431.
12. Z. Zhao-Karger, X.-M. Lin, C. Bonatto Minella, D. Wang, T. Diemant, R. J. Behm and M. Fichtner, *J. Power Sources*, 2016, **323**, 213.
13. Z. Zhao-Karger, X. Zhao, O. Fuhr and M. Fichtner, *RSC Adv.*, 2013, **3**, 16330.

14. C. Liao, N. Sa, B. Key, A. K. Burrell, L. Cheng, L. A. Curtiss, J. T. Vaughey, J.-J. Woo, L. Hu, B. Pan and Z. Zhang, *J. Mater. Chem. A*, 2015, **3**, 6082.
15. L. C. Merrill and J. L. Schaefer, *Langmuir*, 2017, **33**, 9426.
16. L. C. Merrill and J. L. Schaefer, *Chem. Mater.*, 2018, **30**, 3971.
17. T. Mandai, Y. Akita, S. Yagi, M. Egashira, H. Munakata and K. Kanamura, *J. Mater. Chem. A*, 2017, **5**, 3152.
18. F.-f. Wang, Y.-s. Guo, J. Yang, Y. Nuli and S.-i. Hirano, *Chem. Commun.*, 2012, **48**, 10763.
19. E. G. Nelson, J. W. Kampf and B. M. Bartlett, *Chem. Commun.*, 2014, **50**, 5193.
20. A. J. Crowe and B. M. Bartlett, *J. Mater. Chem. A*, 2016, **4**, 368.
21. B. Pan, J. Zhang, J. Huang, J. T. Vaughey, L. Zhang, S.-D. Han, A. K. Burrell, Z. Zhang and C. Liao, *Chem. Commun.*, 2015, **51**, 6214.
22. K. Xu, *Chem. Rev.*, 2004, **104**, 4303.
23. C. Liao, B. Guo, D.-e. Jiang, R. Custelcean, S. M. Mahurin, X.-G. Sun and S. Dai, *J. Mater. Chem. A*, 2014, **2**, 581.
24. I.-T. Kim, K. Yamabuki, M. Sumimoto, H. Tsutsumi, M. Morita and N. Yoshimoto, *J. Power Sources*, 2016, **323**, 51.
25. J. T. Herb, C. Nist-Lund, J. Schwartz and C. B. Arnold, *ECS Electrochem. Lett.*, 2015, **4**, A49.
26. B. Pan, J. Huang, M. He, S. M. Brombosz, J. T. Vaughey, L. Zhang, A. K. Burrell, Z. Zhang and C. Liao, *ChemSusChem*, 2016, **9**, 595.
27. B. Pan, J. Huang, N. Sa, S. M. Brombosz, J. T. Vaughey, L. Zhang, A. K. Burrell, Z. Zhang and C. Liao, *J. Electrochem. Soc.*, 2016, **163**, A1672.
28. A. J. Crowe, K. K. Stringham and B. M. Bartlett, *ACS Appl. Mater. Interfaces*, 2016, **8**, 23060.
29. J. T. Herb, C. A. Nist-Lund and C. B. Arnold, *J. Mater. Chem. A*, 2017, **5**, 7801.
30. A. J. Crowe, J. L. DiMeglio, K. K. Stringham and B. M. Bartlett, *J. Phys. Chem. C*, 2017, **121**, 20613.
31. R. Müller, F. Hölzl, W. Knaus, F. Planiszig and K. Prett, *Monatsh. Chem.*, 1923, **44**, 219.
32. J. H. Connor, W. E. Reid and G. B. Wood, *J. Electrochem. Soc.*, 1957, **104**, 38.
33. Y. Viestfrid, M. D. Levi, Y. Gofer and D. Aurbach, *J. Electroanal. Chem.*, 2005, **576**, 183.
34. R. E. Doe, R. Han, J. Hwang, A. J. Gmitter, I. Shterenberg, H. D. Yoo, N. Pour and D. Aurbach, *Chem. Commun.*, 2014, **50**, 243.
35. T. Liu, Y. Shao, G. Li, M. Gu, J. Hu, S. Xu, Z. Nie, X. Chen, C. Wang and J. Liu, *J. Mater. Chem. A*, 2014, **2**, 3430.
36. R. E. Doe, R. Han, Y. Gofer, D. Aurbach, N. Pour and I. Sterenberg, *International Pat. App.*, PCT/US2012/071350, 2013.
37. J. Luo, S. He and T. L. Liu, *ACS Energy Lett.*, 2017, **2**, 1197.
38. I. Shterenberg, M. Salama, Y. Gofer, E. Levi and D. Aurbach, *MRS Bull.*, 2014, **39**, 453.
39. C. J. Barile, E. C. Barile, K. R. Zavadil, R. G. Nuzzo and A. A. Gewirth, *J. Phys. Chem. C*, 2014, **118**, 27623.

40. C. J. Barile, R. G. Nuzzo and A. A. Gewirth, *J. Phys. Chem. C*, 2015, **119**, 13524.
41. J. H. Ha, B. Adams, J.-H. Cho, V. Duffort, J. H. Kim, K. Y. Chung, B. W. Cho, L. F. Nazar and S. H. Oh, *J. Mater. Chem. A*, 2016, **4**, 7160.
42. T. Liu, J. T. Cox, D. Hu, X. Deng, J. Hu, M. Y. Hu, J. Xiao, Y. Shao, K. Tang and J. Liu, *Chem. Commun.*, 2015, **51**, 2312.
43. P. Canepa, S. Jayaraman, L. Cheng, N. N. Rajput, W. D. Richards, G. S. Gautam, L. A. Curtiss, K. A. Persson and G. Ceder, *Energy Environ. Sci.*, 2015, **8**, 3718.
44. Y. Cheng, R. M. Stolley, K. S. Han, Y. Shao, B. W. Arey, N. M. Washton, K. T. Mueller, M. L. Helm, V. L. Sprenkle, J. Liu and G. Li, *Phys. Chem. Chem. Phys.*, 2015, **17**, 13307.
45. H. Xu, Z. Zhang, Z. Cui, A. Du, C. Lu, S. Dong, J. Ma, X. Zhou and G. Cui, *Electrochem. Commun.*, 2017, **83**, 72.
46. S.-J. Kang, S.-C. Lim, H. Kim, J. W. Heo, S. Hwang, M. Jang, D. Yang, S.-T. Hong and H. Lee, *Chem. Mater.*, 2017, **29**, 3174.
47. Y. Nakayama, R. Matsumoto, K. Kumagae, D. Mori, Y. Mizuno, S. Hosoi, K. Kamiguchi, N. Koshitani, Y. Inaba, Y. Kudo, H. Kawasaki, E. C. Miller, J. N. Weker and M. F. Toney, *Chem. Mater.*, 2018, **30**, 6318.
48. Z. Zhang, Z. Cui, L. Qiao, J. Guan, H. Xu, X. Wang, P. Hu, H. Du, S. Li, X. Zhou, S. Dong, Z. Liu, G. Cui and L. Chen, *Adv. Energy Mater.*, 2017, 7, 1602055.
49. F. Bertasi, F. Sepehr, G. Pagot, S. J. Paddison and V. Di Noto, *Adv. Funct. Mater.*, 2016, **26**, 4860.
50. E. N. Keyzer, H. F. J. Glass, Z. Liu, P. M. Bayley, S. E. Dutton, C. P. Grey and D. S. Wright, *J. Am. Chem. Soc.*, 2016, **138**, 8682.
51. Z. Lu, A. Schechter, M. Moshkovich and D. Aurbach, *J. Electroanal. Chem.*, 1999, **466**, 203.
52. R. E. Doe, G. H. Lane, R. E. Jilek and J. Hwang, *US Pat.*, US9293790, 2013.
53. I. Shterenberg, M. Salama, H. D. Yoo, Y. Gofer, J.-B. Park, Y.-K. Sun and D. Aurbach, *J. Electrochem. Soc.*, 2015, **162**, A7118.
54. I. Shterenberg, M. Salama, Y. Gofer and D. Aurbach, *Langmuir*, 2017, **33**, 9472.
55. M. Salama, I. Shterenberg, L. J. W. Shimon, K. Keinan-Adamsky, M. Afri, Y. Gofer and D. Aurbach, *J. Phys. Chem. C*, 2017, **121**, 24909.
56. J. G. Connell, B. Genorio, P. P. Lopes, D. Strmcnik, V. R. Stamenkovic and N. M. Markovic, *Chem. Mater.*, 2016, **28**, 8268.
57. L. F. Wan, B. R. Perdue, C. A. Apblett and D. Prendergast, *Chem. Mater.*, 2015, **27**, 5932.
58. H. Dong, Y. Liang, O. Tutusaus, R. Mohtadi, Y. Zhang, F. Hao and Y. Yao, *Joule*, 2019, **3**, 782–793.
59. J. Muldoon, C. B. Bucur, A. G. Oliver, T. Sugimoto, M. Matsui, H. S. Kim, G. D. Allred, J. Zajicek and Y. Kotani, *Energy Environ. Sci.*, 2012, **5**, 5941.

60. D. Lv, T. Xu, P. Saha, M. K. Datta, M. L. Gordin, A. Manivannan, P. N. Kumta and D. Wang, *J. Electrochem. Soc.*, 2013, **160**, A351.
61. R. Mohtadi, M. Matsui, T. S. Arthur and S.-J. Hwang, *Angew. Chem., Int. Ed.*, 2012, **51**, 9780.
62. Y. Shao, T. Liu, G. Li, M. Gu, Z. Nie, M. Engelhard, J. Xiao, D. Lv, C. Wang, J.-G. Zhang and J. Liu, *Sci. Rep.*, 2013, **3**, 3130.
63. F. Tuerxun, Y. Abulizi, Y. NuLi, S. Su, J. Yang and J. Wang, *J. Power Sources*, 2015, **276**, 255.
64. N. N. Rajput, X. Qu, N. Sa, A. K. Burrell and K. A. Persson, *J. Am. Chem. Soc.*, 2015, **137**, 3411.
65. D. Samuel, C. Steinhauser, J. G. Smith, A. Kaufman, M. D. Radin, J. Naruse, H. Hiramatsu and D. J. Siegel, *ACS Appl. Mater. Interfaces*, 2017, **9**, 43755.
66. J. D. Deetz, F. Cao, Q. Wang and H. Sun, *J. Electrochem. Soc.*, 2018, **165**, A61.
67. O. Tutusaus and R. Mohtadi, *ChemElectroChem*, 2015, **2**, 51.
68. R. Mohtadi, T. J. Carter and T. S. Arthur, *US Pat.*, US9240613, 2013.
69. O. Tutusaus, R. Mohtadi, T. S. Arthur, F. Mizuno, E. G. Nelson and Y. V. Sevryugina, *Angew. Chem., Int. Ed.*, 2015, **54**, 7900.
70. O. Tutusaus, R. Mohtadi, N. Singh, T. S. Arthur and F. Mizuno, *ACS Energy Lett.*, 2017, **2**, 224.
71. S. G. McArthur, L. Geng, J. Guo and V. Lavallo, *Inorg. Chem. Front.*, 2015, **2**, 1101.
72. S. G. McArthur, R. Jay, L. Geng, J. Guo and V. Lavallo, *Chem. Commun.*, 2017, **53**, 4453.
73. J. W. Choi and D. Aurbach, *Nat. Rev. Mater.*, 2016, **1**, 16013.
74. D. Aurbach, I. Weissman, Y. Gofer and E. Levi, *Chem. Rec.*, 2003, **3**, 61.
75. T. S. Arthur, P.-A. Glans, N. Singh, O. Tutusaus, K. Nie, Y.-S. Liu, F. Mizuno, J. Guo, D. H. Alsem, N. J. Salmon and R. Mohtadi, *Chem. Mater.*, 2017, **29**, 7183.
76. J. Chang, R. T. Haasch, J. Kim, T. Spila, P. V. Braun, A. A. Gewirth and R. G. Nuzzo, *ACS Appl. Mater. Interfaces*, 2015, **7**, 2494.
77. N. Singh, T. S. Arthur, O. Tutusaus, J. Li, K. Kisslinger, H. L. Xin, E. A. Stach, X. Fan and R. Mohtadi, *ACS Appl. Energy Mater.*, 2018, **1**, 4651.
78. R. Schwarz, M. Pejic, P. Fischer, M. Marinaro, L. Jörissen and M. Wachtler, *Angew. Chem., Int. Ed.*, 2016, **55**, 14958.
79. R. Mohtadi and F. Mizuno, *Beilstein J. Nanotechnol.*, 2014, **5**, 1291.
80. H. D. Yoo, I. Shterenberg, Y. Gofer, G. Gershinsky, N. Pour and D. Aurbach, *Energy Environ. Sci.*, 2013, **6**, 2265.
81. S.-Y. Ha, Y.-W. Lee, S. W. Woo, B. Koo, J.-S. Kim, J. Cho, K. T. Lee and N.-S. Choi, *ACS Appl. Mater. Interfaces*, 2014, **6**, 4063.
82. Y. Orikasa, T. Masese, Y. Koyama, T. Mori, M. Hattori, K. Yamamoto, T. Okado, Z.-D. Huang, T. Minato, C. Tassel, J. Kim, Y. Kobayashi, T. Abe, H. Kageyama and Y. Uchimoto, *Sci. Rep.*, 2014, **4**, 5622.
83. N. Sa, N. N. Rajput, H. Wang, B. Key, M. Ferrandon, V. Srinivasan, K. A. Persson, A. K. Burrell and J. T. Vaughey, *RSC Adv.*, 2016, **6**, 113663.

84. M. S. Ding, T. Diemant, R. J. Behm, S. Passerini and G. A. Giffin, *J. Electrochem. Soc.*, 2018, **165**, A1983.
85. Y. Yu, A. Baskin, C. Valero-Vidal, N. T. Hahn, Q. Liu, K. R. Zavadil, B. W. Eichhorn, D. Prendergast and E. J. Crumlin, *Chem. Mater.*, 2017, **29**, 8504.
86. F. Sagane, K. Ogi, A. Konno, M. Egashira and K. Kanamura, *Electrochemistry*, 2016, **84**, 76.
87. M. Salama, I. Shterenberg, H. Gizbar, N. N. Eliaz, M. Kosa, K. Keinan-Adamsky, M. Afri, L. J. W. Shimon, H. E. Gottlieb, D. T. Major, Y. Gofer and D. Aurbach, *J. Phys. Chem. C*, 2016, **120**, 19586.
88. X. Qu, Y. Zhang, N. N. Rajput, A. Jain, E. Maginn and K. A. Persson, *J. Phys. Chem. C*, 2017, **121**, 16126.
89. Z. Zhao-Karger, M. E. Gil Bardaji, O. Fuhr and M. Fichtner, *J. Mater. Chem. A*, 2017, **5**, 10815.
90. J. T. Herb, C. A. Nist-Lund and C. B. Arnold, *ACS Energy Lett.*, 2016, **1**, 1227.
91. Z. Zhao-Karger, R. Liu, W. Dai, Z. Li, T. Diemant, B. P. Vinayan, C. Bonatto Minella, X. Yu, A. Manthiram, R. J. Behm, M. Ruben and M. Fichtner, *ACS Energy Lett.*, 2018, **3**, 2005.
92. N. T. Hahn, T. J. Seguin, K.-C. Lau, C. Liao, B. J. Ingram, K. A. Persson and K. R. Zavadil, *J. Am. Chem. Soc.*, 2018, **140**, 11076.
93. H. Xu, Z. Zhang, J. Li, L. Qiao, C. Lu, K. Tang, S. Dong, J. Ma, Y. Liu, X. Zhou and G. Cui, *ACS Appl. Mater. Interfaces*, 2018, **10**, 23757.
94. A. Unemoto, S. Yasaku, G. Nogami, M. Tazawa, M. Taniguchi, M. Matsuo, T. Ikeshoji and S.-i. Orimo, *Appl. Phys. Lett.*, 2014, **105**, 083901.
95. J. Z. Hu, N. N. Rajput, C. Wan, Y. Shao, X. Deng, N. R. Jaegers, M. Hu, Y. Chen, Y. Shin, J. Monk, Z. Chen, Z. Qin, K. T. Mueller, J. Liu and K. A. Persson, *Nano Energy*, 2018, **46**, 436.
96. Z. Ma, M. Kar, C. Xiao, M. Forsyth and D. R. MacFarlane, *Electrochem. Commun.*, 2017, **78**, 29.
97. S. Hebié, H. P. K. Ngo, J.-C. Leprêtre, C. Iojoiu, L. Cointeaux, R. Berthelot and F. Alloin, *ACS Appl. Mater. Interfaces*, 2017, **9**, 28377.
98. S. Hebié, F. Alloin, C. Iojoiu, R. Berthelot and J.-C. Leprêtre, *ACS Appl. Mater. Interfaces*, 2018, **10**, 5527.
99. S.-B. Son, T. Gao, S. P. Harvey, K. X. Steirer, A. Stokes, A. Norman, C. Wang, A. Cresce, K. Xu and C. Ban, *Nat. Chem.*, 2018, **10**, 532.
100. D. R. MacFarlane, M. Forsyth, P. C. Howlett, M. Kar, S. Passerini, J. M. Pringle, H. Ohno, M. Watanabe, F. Yan, W. Zheng, S. Zhang and J. Zhang, *Nat. Rev. Mater.*, 2016, **1**, 15005.
101. M. M. Huie, C. A. Cama, P. F. Smith, J. Yin, A. C. Marschilok, K. J. Takeuchi and E. S. Takeuchi, *Electrochim. Acta*, 2016, **219**, 267.
102. Y. NuLi, J. Yang and R. Wu, *Electrochem. Commun.*, 2005, **7**, 1105.
103. N. Amir, Y. Vestfrid, O. Chusid, Y. Gofer and D. Aurbach, *J. Power Sources*, 2007, **174**, 1234.
104. G. T. Cheek, W. E. O'Grady, S. Z. El Abedin, E. M. Moustafa and F. Endres, *J. Electrochem. Soc.*, 2008, **155**, D91.

105. G. Vardar, A. E. S. Sleightholme, J. Naruse, H. Hiramatsu, D. J. Siegel and C. W. Monroe, *ACS Appl. Mater. Interfaces*, 2014, **6**, 18033.
106. N. Yoshimoto, M. Matsumoto, M. Egashia and M. Morita, *J. Power Sources*, 2010, **195**, 2096.
107. J.-H. Cho, J. H. Ha, S.-H. Lee, B. W. Cho, K. Y. Chung, B.-K. Na, K.-B. Kim and S. H. Oh, *Electrochim. Acta*, 2017, **231**, 379.
108. B. Pan, K.-C. Lau, J. T. Vaughey, L. Zhang, Z. Zhang and C. Liao, *J. Electrochem. Soc.*, 2017, **164**, A902.
109. F. Bertasi, C. Hettige, F. Sepehr, X. Bogle, G. Pagot, K. Vezzù, E. Negro, S. J. Paddison, S. G. Greenbaum, M. Vittadello and V. Di Noto, *ChemSusChem*, 2015, **8**, 3069.
110. M. Kar, Z. Ma, L. M. Azofra, K. Chen, M. Forsyth and D. R. MacFarlane, *Chem. Commun.*, 2016, **52**, 4033.
111. T. Watkins, A. Kumar and D. A. Buttry, *J. Am. Chem. Soc.*, 2016, **138**, 641.
112. Z. Ma, M. Forsyth, D. R. MacFarlane and M. Kar, *Green Energy Environ.*, 2019, **4**, 146–153.
113. M. Kar, O. Tutusaus, D. R. MacFarlane and R. Mohtadi, *Energy Environ. Sci.*, 2019, **12**, 566–571.
114. G. G. Kumar and N. Munichandraiah, *Electrochim. Acta*, 1999, **44**, 2663.
115. M. Morita, N. Yoshimoto, S. Yakushiji and M. Ishikawa, *Electrochem. Solid-State Lett.*, 2001, **4**, A177.
116. O. Chusid, Y. Gofer, H. Gizbar, Y. Vestfrid, E. Levi, D. Aurbach and I. Riech, *Adv. Mater.*, 2003, **15**, 627.
117. G. P. Pandey, R. C. Agrawal and S. A. Hashmi, *J. Solid State Electrochem.*, 2011, **15**, 2253.
118. M. L. Aubrey, R. Ameloot, B. M. Wiers and J. R. Long, *Energy Environ. Sci.*, 2014, **7**, 667.
119. S. S. Park, Y. Tulchinsky and M. Dincă, *J. Am. Chem. Soc.*, 2017, **139**, 13260.
120. S. Song, M. Kotobuki, F. Zheng, Q. Li, C. Xu, Y. Wang, W. D. Z. Li, N. Hu and L. Lu, *J. Electrochem. Soc.*, 2017, **164**, A741.
121. Y. Shao, N. N. Rajput, J. Hu, M. Hu, T. Liu, Z. Wei, M. Gu, X. Deng, S. Xu, K. S. Han, J. Wang, Z. Nie, G. Li, K. R. Zavadil, J. Xiao, C. Wang, W. A. Henderson, J.-G. Zhang, Y. Wang, K. T. Mueller, K. Persson and J. Liu, *Nano Energy*, 2015, **12**, 750.
122. N. Imanaka, Y. Okazaki and G.-y. Adachi, *J. Mater. Chem.*, 2000, **10**, 1431.
123. S. Ikeda, M. Takahashi, J. Ishikawa and K. Ito, *Solid State Ionics*, 1987, **23**, 125.
124. J. Kawamura, K. Morota, N. Kuwata, Y. Nakamura, H. Maekawa, T. Hattori, N. Imanaka, Y. Okazaki and G.-y. Adachi, *Solid State Commun.*, 2001, **120**, 295.
125. E. Levi, G. Gershinsky, D. Aurbach, O. Isnard and G. Ceder, *Chem. Mater.*, 2009, **21**, 1390.
126. F. Thöle, L. F. Wan and D. Prendergast, *Phys. Chem. Chem. Phys.*, 2015, **17**, 22548.

127. X. Sun, P. Bonnick, V. Duffort, M. Liu, Z. Rong, K. A. Persson, G. Ceder and L. F. Nazar, *Energy Environ. Sci.*, 2016, **9**, 2273.
128. P. Canepa, S.-H. Bo, G. Sai Gautam, B. Key, W. D. Richards, T. Shi, Y. Tian, Y. Wang, J. Li and G. Ceder, *Nat. Commun.*, 2017, **8**, 1759.
129. S. Higashi, K. Miwa, M. Aoki and K. Takechi, *Chem. Commun.*, 2014, **50**, 1320.
130. R. Mohtadi and S.-i. Orimo, *Nat. Rev. Mater.*, 2016, **2**, 16091.
131. T. Burankova, E. Roedern, A. E. Maniadaki, H. Hagemann, D. Rentsch, Z. Łodziana, C. Battaglia, A. Remhof and J. P. Embs, *J. Phys. Chem. Lett.*, 2018, **9**, 6450.
132. E. Roedern, R.-S. Kühnel, A. Remhof and C. Battaglia, *Sci. Rep.*, 2017, **7**, 46189.

CHAPTER 3

Solid-state Magnesium-ion Conductors

S. PAYANDEH, A. REMHOF* AND C. BATTAGLIA

Empa, Swiss Federal Laboratories for Materials Science and Technology, Materials for Energy Conversion, Überlandstrasse 129, Dübendorf, CH-8600, Switzerland
*E-mail: arndt.remhof@empa.ch

3.1 Introduction

The last decade has witnessed growing activities in the field of solid-state magnesium-ion conductors, spurred on by the prospect of a future competitive all-solid-state magnesium-ion battery. All-solid-state batteries employ solid-state electrolytes rather than the liquid electrolytes commonly used in commercial lithium-ion batteries and magnesium-ion batteries that are in development as described in previous chapters. They promise to overcome the challenges and limitations resulting from the use of liquid electrolytes with respect to energy and power density, thermal and (electro-) chemical stability, and safety. Recent advances in solid-state lithium-ion conductors has triggered intense research efforts focused on developing a competitive all-solid-state battery technology based on lithium. The high expectations are also reflected in the technology roadmaps

Energy and Environment Series No. 23
Magnesium Batteries: Research and Applications
Edited by Maximilian Fichtner

Published by the Royal Society of Chemistry, www.rsc.org

of leading car manufacturers, which forecast all-solid-state batteries to replace current lithium-ion batteries in the next five to ten years.[1]

Research activities focusing on all-solid-state magnesium batteries are comparatively scarce and at a lower level of technological readiness at present. To date, there have been only a few reports demonstrating the stable long-term cycling of an all-solid-state magnesium battery,[2,3] while several proof-of-concept all-solid-state lithium-ion and sodium-ion batteries have been reported in recent years.[4–7] Nevertheless, today, ion conductivities in solid-state magnesium-ion conductors reach values of 10^{-4} S cm^{-1} at room temperature in several material classes, approaching the values of liquid electrolytes.[8,9] However, besides providing a sufficiently high ionic conductivity, candidate electrolyte materials need to fulfill a series of additional criteria to be valid candidates, including low electronic conductivity, (electro-) chemical compatibility with anode and cathode materials, suitable mechanical properties, low cost, *etc.*[6]

All-solid-state magnesium batteries may benefit from the lower tendency of magnesium to form dendrites, which is a major challenge for lithium metal anodes.[10] A magnesium metal anode would be ideal for an all-solid-state battery to reach a high volumetric energy density. While the gravimetric capacity of magnesium (2205 A h kg^{-1}) is lower than that of lithium (3862 A h kg^{-1}), divalent magnesium possesses a higher volumetric capacity than lithium (3833 *vs.* 2046 A h l^{-1}), which (partially) compensates for its lower redox potential (−2.4 V for magnesium *vs.* −3.0 V for lithium *vs.* a standard hydrogen electrode, SHE).[11] Many of the reported liquid electrolytes for magnesium-ion batteries contain halides and are highly corrosive. While all-solid-state batteries are intrinsically leak proof, replacing the liquid electrolytes with a non-corrosive solid-state electrolyte additionally contributes towards improved operational safety and reduces the environmental impact. Future efforts also need to concentrate on developing functional composites by combining a suitable magnesium cathode material with a compatible solid-state electrolyte material. Challenges associated with (electro-) chemical interface stability and mechanical stability arising from the volume expansion and contraction of the cathode material during electrochemical cycling need to be addressed in the future, similar to the current ongoing efforts in the all-solid-state lithium-ion battery community.

In this chapter, we review the existing literature on solid-state magnesium-ion conductors. These can be classified into phosphates, chalcogenides, complex metal hydrides, and metal–organic frameworks (MOFs). A considerable body of results has been reported for $Mg_{0.5}Zr_2(PO_4)_3$ (MZP) and related compounds. These phosphates were studied in analogy to the sodium super ionic conductors (NASICON) with $Na_{1+x}Zr_2Si_xP_{3-x}O_{12}$ ($0 < x < 3$). Studies on other material classes including sulfides, complex metal hydrides, and metal–organic frameworks have appeared more recently and are discussed in subsequent sections.

3.2 Phosphate-based Solid-state Magnesium-ion Conductors

In 1987, magnesium-ion conductivity in the solid state has already been observed in the Mg–Zr–PO_4 system. In this system, the highest conductivity was found for the composition $Mg_{0.5}Zr_2(PO_4)_3$ (*i.e.* MZP) with conductivity values of 2.9×10^{-5} and 6.1×10^{-3} S cm^{-1} at 400 and 800 °C, respectively (Figure 3.1).[12] The crystal structure of MZP is of the scandium tungstate type, in which isolated ZrO_6 octahedra and PO_4 tetrahedra are connected through common vertices forming a mixed three-dimensional framework (Figure 3.2a).[13] Within this structure, the magnesium ions randomly occupy 50% of the tetrahedral interstitial sites between the ZrO_6 and PO_4 units. This partial occupation leads to vacancies that enable ion conduction. Magnesium diffusion proceeds through windows formed by the edges of three PO_4 tetrahedra and three ZrO_6 octahedra. The sizes of these openings vary between 4.5–4.7 Å, which is larger than the edges of MgO_4 tetrahedra (3.3–3.7 Å).

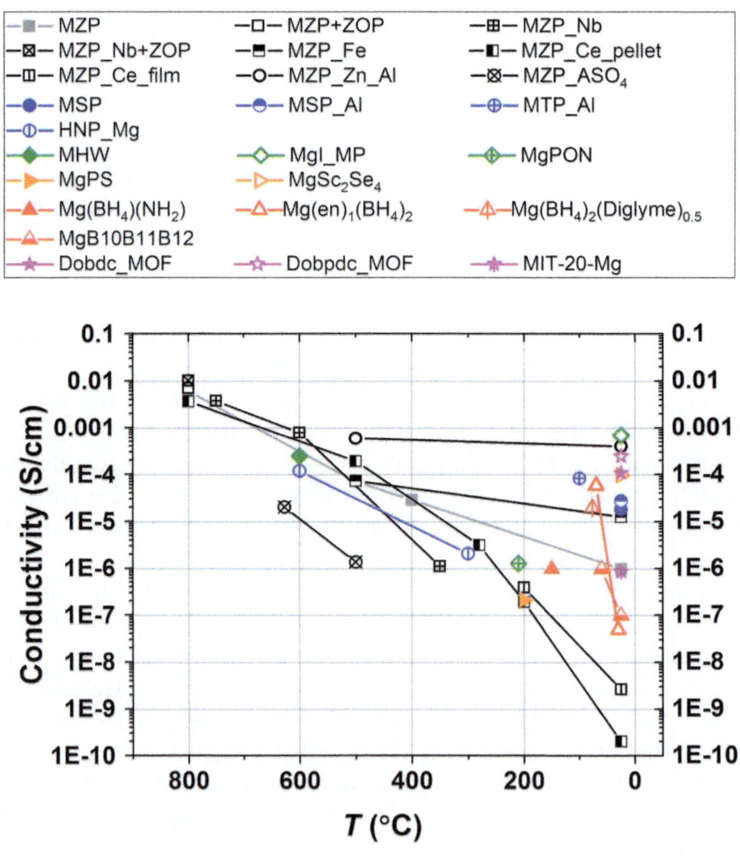

Figure 3.1 Ionic conductivities of inorganic solid-state magnesium-ion conductors.

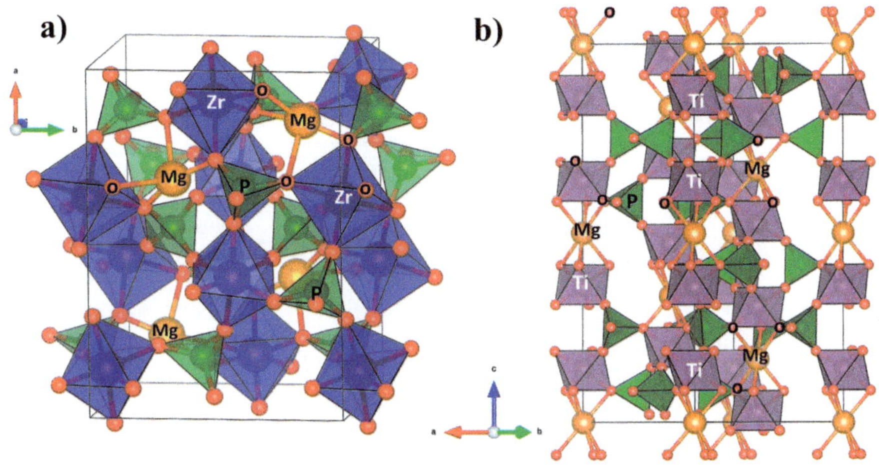

Figure 3.2 Crystal structures of (a) $Mg_{0.5}Zr_2(PO_4)_3$ (MZP), space group $P2_1/n$ and (b) $Mg_{0.5}Ti_2(PO_4)_3$ (MTP), space group R-$3c$.

Upon heating, the unit cell expands predominantly along the *b* axis and to a smaller extent along the *a* axis, leaving the *c* axis almost unchanged. It was suggested that this anisotropic expansion is accompanied by a phase transition with a change in the unit cell symmetry at temperatures above 750 °C.[14]

However, the formation of a new phase at elevated temperatures is still under debate. Although the presence of an exothermic event at 670 °C was reported earlier as an indication of the phase transition, this has not been confirmed in more recent investigations.[14,15]

An activation energy for ionic conduction in MZP of between 0.82 and 0.87 eV was reported by different research groups.[12,16] A significantly lower activation energy was reported in ref. 3, but has not yet been confirmed by independent reports. Activation energies below 0.3 eV are desirable as they lead to high conductivity, reduce the dependency of the conductivity on temperature, and therefore enable application over a wider temperature range.[17]

The electrochemical stability window of MZP was evaluated by linear sweep voltammetry, revealing a stability of 2.5 V *vs.* Mg/Mg^{2+} at ambient temperature (Figure 3.3). MZP is thermally stable up to ~1000 °C, it melts at 1165 °C, and decomposes into $Zr_2(PO_4)_2O$ at a higher temperature.[15,16]

In the presence of insulating $Zr_2O(PO_4)_2$ (*i.e.* ZOP), the conductivity of MZP increases slightly from 6.1×10^{-3} to 6.9×10^{-3} S cm^{-1} at 800 °C (Table 3.1). This is attributed to the higher density of the pellets and the microscopic dispersion of the ZOP secondary phase in the composite that enhances conductivity along the grain boundaries.[18] However, this composite showed a much higher activation energy of 1.41 eV compared to ~0.84 eV observed for MZP.

Several strategies to improve the ionic conductivity of MZP have been followed by increasing the number of charge carriers and/or tuning of the crystal structure.

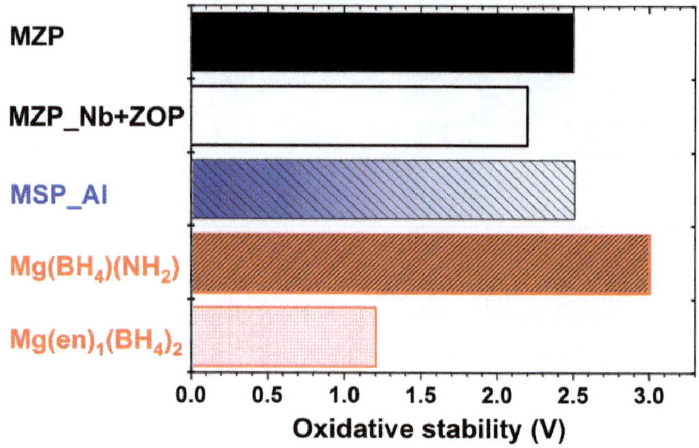

Figure 3.3 Oxidative stability of inorganic solid-state magnesium-ion conductors.

3.2.1 Cation and Anion Substitution in MZP

The relatively large size of the conduction channels in the MZP structure has been identified as an obstacle for magnesium-ion migration. Therefore, a promising strategy is to substitute Zr^{4+} with smaller cations to reduce the unit cell size. These may be ions with a higher charge, such as Nb^{5+}, or ions with a lower charge, such as Fe^{2+}, Zn^{2+} or Al^{3+}. Substituting 15% of the Zr^{4+} with aliovalent Nb^{5+}, which has a smaller ionic radius of 0.78 Å (6 coordinate) compared to 0.86 Å for Zr^{4+} (6 coordinate) results in an ionic conductivity of 3.7×10^{-3} S cm^{-1} at 750 °C.[19–21] To maintain charge neutrality, the occupancy of the interstitial sites with magnesium ions is reduced from 0.5 in MZP to 0.35 in $Mg_{0.35}(Zr_{0.85}Nb_{0.15})_2(PO_4)_3$ (MZP_Nb). Further Nb^{5+} substitution decreases the ionic conductivity, which is ascribed to the reduced number of charge carriers. Activation energies of 0.95 and 1.18 eV were reported for this electrolyte and more recently the electrochemical stability of MZP_Nb was confirmed against Mg metal.[19,20]

The conductivity of MZP_Nb could be further improved to 10^{-2} S cm^{-1} at 800 °C in the presence of insulating ZOP (MZP_Nb + ZOP). As in the case of pure MPZ, the enhancement in the ionic conductivity was ascribed to the improvement in the magnesium-ion conduction at the grain boundaries. An activation energy of 1.26 eV was observed, which is higher than the 0.95 eV for MZP_Nb, but is still lower compared to the 1.41 eV for MZP + ZOP the composite. Linear sweep voltammetry measurements indicated an electrochemical stability of up to ~2.2 V.[22]

When Zr^{4+} is replaced with lower charge cations, the concentration of magnesium ions is increased accordingly to maintain the charge balance. Fe^{2+} has a smaller ionic radius of 0.75 Å, leading to a decrease in the unit cell volume for $Mg_{0.5+y}(Zr_{1-y}Fe_y)_2(PO_4)_3$ (MZP_Fe). Zn^{2+} on the other hand has a slightly larger size of 0.88 Å and is therefore used together with Al^{3+} with a size of 0.68 Å (6 coordinate), causing contraction in the unit cell volume

Table 3.1 List of solid-state electrolytes for magnesium-ion batteries.

Compound	Conductivity S cm^{-1}	T (°C)	Activation energy eV (kJ mol^{-1})	Electrochemical stability (V) (transport no.)	Thermal stability (°C)	Reference
Phosphate/Oxygen containing electrolytes						
$Mg_{0.5}Zr_2(PO_4)_3$ (MZP)	2.9×10^{-5} 6.1×10^{-3}	400 800	0.82 (79)	—	—	12
	$2.0–30 \times 10^{-7}$ 7.1×10^{-5}	RT 500	0.098 (9.4)	2.50 (0.69)	1165, Melt	15
	7.23×10^{-3}	725	$0.84 \le E_a \le 0.87$ (81–84)	—	~1000	16
$Mg_{1.4}Zr_4P_6O_{24.4} + 0.4Zr_2O(PO_4)_2$ (MZP + ZOP)	6.92×10^{-3}	800	1.41 (136)	(~1)	—	18
$Mg_{0.5-x}(Zr_{1-x}Nb_x)_2(PO_4)_3$ (MZP_Nb)	7.7×10^{-4} 3.7×10^{-3} ($x = 0.15$)	600 750	0.95 (92)	(~1)	—	19
	~1.1×10^{-6} ($x = 0.15$) Mg electrodes	350	1.18 (114)	(1)	—	20
$Mg_{1.1}(Zr_{0.85}Nb_{0.15})_4P_6O_{24.4} + 0.4Zr_2O(PO_4)_2$ (MZP_Nb + ZOP)	~10^{-2}	800	1.26 (122)	2.2	—	22
$Mg_{0.5+y}(Zr_{1-y}Fe_y)_2(PO_4)_3$ (MZP_Fe)	1.25×10^{-5} 7.18×10^{-5} ($y = 0.4$)	RT 500	0.14 (14)	—	—	23
$Mg_{0.5}Ce_xZr_{2-x}(PO_4)_3$ Thin film (MZP_Ce_film)	2.0×10^{-10} 1.9×10^{-7} 3.1×10^{-6} 1.9×10^{-4} 3.65×10^{-3}	25 200 280 500 800	0.31 (30) $T \ge 60$ °C, films	—	—	58
$Mg_{0.5}Ce_xZr_{2-x}(PO_4)_3$Pellet (MZP_Ce_pellet)	2.6×10^{-9} 3.8×10^{-7} ($x = 0.2$)	25 200				

(*continued*)

Table 3.1 (*continued*)

Compound	Conductivity S cm^{-1}	T (°C)	Activation energy eV (kJ mol^{-1})	Electrochemical stability (V) (transport no.)	Thermal stability (°C)	Reference
$Mg_{0.9+0.5y}Zn_{0.4}Al_yZr_{1.62-y}(PO_4)_3$ (MZP_Zn_Al)	3.97×10^{-4} 5.82×10^{-4} ($y = 0.3$)	RT 500	0.038 (3.7)	—	—	8
$Mg_{0.5}Zr_2(AsO_4)_x(PO_4)_{3-x}$ (MZP_AsO$_4$)	$1–1.7 \times 10^{-6}$ $1–3 \times 10^{-5}$ ($0 \le x \le 2$)	500 627	—	—	$T < 800$ $x \ge 1$	32
$Mg_{0.5}Si_2(PO_4)_3$ (MSP)	1.83×10^{-5}	RT[a]	—	3.21	$T > 1000$	24
$Mg_{0.625}Si_{1.75}Al_{0.25}(PO_4)_3$ (MSP_Al)	2.78×10^{-5}	RT	—	2.51(0.82)	—	25
$Mg_{0.6}Al_{0.2}Ti_{1.8}(PO_4)_3$ (MTP_Al)	8.4×10^{-5}	600	1.33 (128)	—	—	27
$(Mg_xHf_{1-x})_{4/(4-2x)}Nb(PO_4)_3$ ($0.05 \le x \le 0.3$) (HNP_Mg)	2.1×10^{-6} 1.2×10^{-4} ($x = 0.1$)	300 600	0.66 (64)	(0.99)	—	26
$MgHf(WO_4)_3$ (MHW)	2.5×10^{-4}	600	0.84 (81)	—	—	33
MgI_2: $Mg_3(PO_4)_2$ 0.3:0.7 wt% (MgI–MP)	7.0×10^{-4}	RT		—	—	34
Magnesium phosphorous oxynitride (MgPON)	1.3×10^{-6}	210	1.3 (125)	—	—	35
Chalcogenides						
$MgSc_2Se_4$	~10^{-4}	RT	~0.37 (35.7)	—	—	38
$(100-x)(0.6MgS\cdot 0.4P_2S_5)\cdot xMgI_2$ (MgPS)	2.1×10^{-7} $x = 20$	200	—	—	—	41
Metal complex hydrides						
$Mg(BH_4)_2$	10^{-9}	150	—	—	—	51

$Mg(BH_4)(NH_2)$	10^{-6}	150	1.31(127)	3		51
$Mg(en)_1(BH_4)_2$	5×10^{-8}	30	1.6 (154)	1.2	Phase transition $T \sim 75$	52
	6×10^{-5}	70			Decomposition $T > 100$	
$Mg(en)_3(BH_4)_2$	$\sim 5 \times 10^{-11}$	30	0.9 (87)		$T \sim 100$	52
	$\sim 5 \times 10^{-9}$	70				53
$Mg(BH_4)_2(Diglyme)_{0.5}$	2×10^{-5}	77	—	—	Phase transition $T \sim 57$ Decomposition $T \sim 247$	54
$MgB_{12}H_{12}(Diglyme)$	10^{-7}	25	—	—	—	55
$MgB_{11}H_{11}(Diglyme)$	10^{-6}	60				
$Mg(B_{11}H_{14})_2(diglyme)$						
Mixture – MB10B11B12						
MOF						
$Mg(OPhCF_3)_2$-$Mg(TFSI)_2$ salts in $Mg_2(dobdc)$ MOF (Dobdc_MOF)	10^{-4}	RT	0.15 (14)	—	$T < 100$	9
$Mg(OPhCF_3)_2$–$Mg(TFSI)_2$ Salts in Mg_2(dob-pdc) MOF (Dobpdc_MOF)	2.5×10^{-4}	RT	0.13 (13)		$T < 100$	
MIT–20–Mg	8.8×10^{-7}	RT	0.37 (36)	—	—	56

[a]Not explicitly mentioned, inferred from the context.

for $Mg_{0.9+0.5y}Zn_{0.4}Al_yZr_{1.62-y}(PO_4)_3$ (MZP_Zn_Al). Increasing the number of mobile magnesium ions and decreasing the unit cell volume in these two compounds results in a high ionic conductivity of 1.25×10^{-5} S cm^{-1} at room temperature for MZP_Fe and 3.97×10^{-5} S cm^{-1} for MZP_Zn_Al.[8,23]

Completely replacing Zr^{4+} (0.86 Å) with smaller isovalent cations such as Si^{4+} and Ti^{4+} with crystal ionic radii of 0.54 and 0.75 Å (6 coordinate), respectively, is another strategy to facilitate ionic transport. In both cases, the crystal structure changes. $Mg_{0.5}Si_2(PO_4)_3$ (MSP) crystalizes in a monoclinic unit cell with $P2_1/c$ space group, showing an ionic conductivity of 1.83×10^{-5} S cm^{-1} at room temperature. From linear sweep voltammetry measurements, an oxidative stability of 3.2 V was deduced, but it is not clear against which reference this value was measured and how the onset current was defined.[24] Partial substitution of Si^{4+} with Al^{3+} reduces the unit cell volume further and increases the number of charge carriers ($Si^{4+} \leftrightarrow Al^{3+} + 1/2\ Mg^{2+}$). Consequently, an ionic conductivity of 2.78×10^{-5} S cm^{-1} was observed for $Mg_{0.625}Si_{1.75}Al_{0.25}(PO_4)_3$ (MSP_Al) at room temperature, but the oxidative stability was reported to be reduced to 2.5 V (Table 3.1, Figure 3.3).[25]

$Mg_{0.5}Ti_2(PO_4)_3$ (MTP) crystalizes in a rhombohedral NASICON type crystal structure with an R-$3c$ space group (Table 3.2). This structure is more ordered compared to that of MZP suggesting the facilitated migration of magnesium ions in the structure and a higher ionic conductivity (Figure 3.2a and b).[26] However, an ionic conductivity of 8.4×10^{-5} S cm^{-1} (Figure 3.1) at 600 °C and an activation energy of 1.33 eV were measured for MTP doped with 10 mol% of Al^{3+} (MTP_Al), which is similar to the conductivity values of MZP.[27,28] Further doping of this structure with Fe, Mn, Co, Nb, and Cr was also carried out, and ionic conductivities of ~10^{-4} S cm^{-1} at 600 °C were determined.[29,30]

MTP can also be employed as an anode for lithium- and sodium-ion batteries, showing a specific capacity of 268.6 mA h g^{-1} when cycled between 0.01–3.0 V *vs.* Na^+/Na^0 and a capacity of 629.2 mAh g^{-1} in the voltage window of 0.01–3.0 V *vs.* Li^+/Li^0. This capacity results from the intercalation of lithium and sodium ions into the MTP structure and suggests the application of this type of material as an electrode for both lithium- and sodium-ion batteries.[31]

Alternatively, the stabilization of the NASICON-type structure with the space group R-$3c$ was achieved by partial substitution of Hf^{4+} in $HfNb(PO_4)_3$ by Mg^{2+}. $(Mg_xHf_{1-x})_{4/(4-2x)}Nb(PO_4)_3$, $x = 0.1$ (HNP_Mg) shows an ionic conductivity of 1.2×10^{-4} S cm^{-1} at 600 °C and an activation energy of 0.66 eV (Table 3.1).[26]

Anion substitution in MZP was studied using AsO_4^{3-} to replace PO_4^{3-} with AsO_4^{3-} and an ionic conductivity of 1–3×10^{-5} S cm^{-1} was reported for the solid solution $Mg_{0.5}Zr_2(AsO_4)_x(PO_4)_{3-x}$ ($0 \le x \le 2$) at 627 °C.[32]

To summarize, in the MZP family, cation substitution was found to be the most effective strategy to improve the ionic conductivity. Substitution of Zr^{4+} with smaller cations decreases the unit cell volume and when Zr^{4+} is replaced by lower charge cations such as Fe^{2+}, the concentration of magnesium ions

Table 3.2 Crystal structure information of inorganic magnesium-ion conductors.

Compound	Crystal system	Space group	Volume (Å^3)	Unit cell parameters (Å)	T (°C)	Reference
$Mg_{0.5}Zr_2(PO_4)_3$ (MZP)	Monoclinic	$P2_1/n$	975.53	a = 12.4218(2) b = 8.9025(2) c = 8.8218(2) β = 90.466(1)	25	13
$Mg_{1.05}Zn_{0.4}Al_{0.3}Zr_{1.3}(PO_4)_3$ (MZP_Zn_Al)	Monoclinic	$P2_1/n$	738.76	a = 7.1844(–) b = 8.9173(–) c = 11.5313(–) β =^{a}	25	8
$Mg_{0.35}(Zr_{0.85}Nb_{0.15})_2(PO_4)_3$ (MZP_Nb)	Monoclinic	$P2_1/c$	968.68	a = 8.8139(2) b = 8.8912(1) c = 15.1215(4) β = 125.170(1)°	^{a}	20
$Mg_{0.5}Ti_2(PO_4)_3$ (MTP)	Trigonal	R-3c	1309.53	a = 8.4898(2) b = 8.4898(2) c = 0.9793(7) γ = 120.00	25	27
$(Mg_{0.1}Hf_{0.9})_{1.05}Nb(PO_4)_3$ (MNP_Hf)	Trigonal	R-3c	1478.7	a = 8.734 (–) b = 8.734 (–) c = 2.238(–) γ = 120.00	^{a}	26
$MgSc_2Se_4$	Cubic	Fd-3m	1374.38	a = 11.11823(3)	^{a}	38
$Mg_{0.94}(BH_4)(NH_2)_{0.88}$	Tetragonal	$I\,4_1$	692.15	a = 5.792 (1) b = 5.792 (1) c = 20.632 (4)	27	59
$Mg(en)_3(BH_4)_2$	Orthorhombic	$Pbca$	6553.0	a = 6.4571(17) b = 5.8484(15) c = 5.1248(92)	RT	53

^{a}Not explicitly mentioned.

increases to maintain charge balance. Applying these two strategies in the MZP_Fe and MZP_Zn_Al compounds results in high ionic conductivities of approaching 10^{-4} S cm^{-1} at room temperature (Table 3.1).

3.2.2 Other Oxygen Containing Solid-state Magnesium-ion Conductors

Magnesium hafnium tungstate, $MgHf(WO_4)_3$ (MHW) shows a conductivity of 2.5×10^{-4} S cm^{-1} at 600 °C and an activation energy of 0.84 eV. The crystal structure of this compound is related to $Sc_2W_3O_{12}$ but with ordering of Mg^{2+} and Hf^{4+}.[33]

The ionic conductivity in the $MgI_2:Mg(PO_4)_3$ (MgI–MP) mixture was investigated and a high conductivity of 7.0×10^{-4} S cm^{-1} was reported at room temperature for the sample containing 30 wt% of MgI_2.[34]

Magnesium phosphorous oxynitride (MgPON) was synthesized in analogy to lithium phosphorous oxynitride, known as LiPON, and an ionic conductivity of 1.3×10^{-6} S cm^{-1} and an activation energy of 1.3 eV were observed at 210 °C (Table 3.1).[35]

The magnesium-ion conductivity in the $(1-x)MgSO_4-xAl_2O_3$ composite system was determined to be 1.9×10^{-6} S cm^{-1} at elevated temperatures. The addition of Al_2O_3 was suggested to enhance the charge carrier concentration at the interface and enhance the conductivity.[36] These classes of materials have rarely been studied as magnesium-ion conductors and the conduction mechanisms as well as basic electrochemical properties are still poorly understood.

3.3 Chalcogenide-based Solid-state Magnesium-ion Conductors

Phosphates and oxygen-containing compounds are hard materials that often require high temperature processing for compaction. In contrast, chalcogenides (sulfides in particular) are typically much softer. They can be processed at a lower temperature, and consequently have gained considerable attention as solid-state lithium- and sodium-ion conductors.[37]

Magnesium-ion conductivity was recently investigated in sulfides and selenides spinels with the stoichiometry MgX_2Z_4 (X = In, Y, Sc, Z = S, Se, Te). In these spinel structures, the ion migrates between two tetrahedral sites (*tet*) connected by a vacant octahedral site (*oct*) through shared triangular faces (Figure 3.4). The magnitude of the potential barrier depends on the size of the triangular face, and on the anion species forming this triangle (Figure 3.4b).[38]

Density functional theory calculations indicate that low migration barriers are achieved by choosing large anions, which guarantees large triangular faces for Mg^{2+} cations to diffuse through (S < Se < Te). At the same time, the metal at the X position determines the energy difference between the octahedral and tetrahedral sites. The lowest activation energies were calculated for MgY_2S_4 (~0.360 eV), MgY_2Se_4 (~0.361 eV) and $MgSc_2Se_4$ (~0.375 eV).[38]

$MgSc_2Se_4$ possesses an ionic conductivity of ~10^{-4} S cm^{-1} at room temperature and an activation energy of 0.37 eV, experimentally determined by ^{25}Mg spin–lattice relaxation magic angle spinning nuclear magnetic resonance (MAS-NMR) (Table 3.1).[38] However 0.04% of the total conductivity was ascribed to the electronic conductivity. This percentage is higher than in state-of-the-art solid-state electrolytes (10^{-4}–10^{-6}%).[39,40] The origin of the relatively high electronic conductivity is still under investigation, but is related to either the presence of intrinsic defects or the existence of electron conducting impurity phases. So far, proposed strategies to lower the electronic conductivity by implementation of Se-rich phases and aliovalent doping have led to no sufficient improvement.[60]

Glass and glass-ceramic electrolytes with the composition $MgS-P_2S_5-MgI_2$ (MgPS) have also been explored. The highest conductivity of 2.1×10^{-7} S cm^{-1} at 200 °C within this material class was observed for $0.8(0.6MgS \cdot 0.4P_2S_5)-0.2MgI_2$.[41]

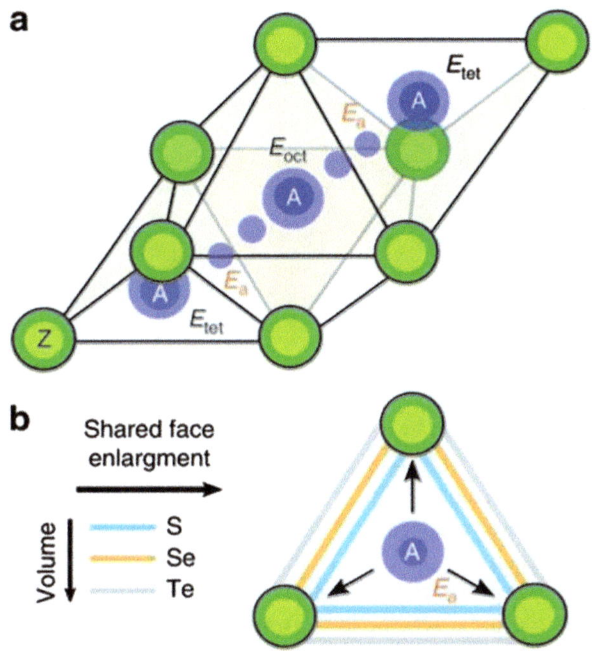

Figure 3.4 (a) *Tet–oct–tet* migration path of Mg^{2+} in a MgX_2Z_4 spinel framework. (b) Effect of the anion size on the conduction channel size shared (triangular) face between *tet* and *oct* sites. Reproduced from ref. 38, https://doi.org/10.1038/s41467-017-01772-1, under the terms of the CC BY 4.0 licence, http://creativecommons.org/licenses/by/4.0/.

3.4 Solid-state Magnesium-ion Conductors Based on Complex Metal Hydrides

The discovery of superionic conductivity for Li ions of ~10^{-3} S cm^{-1} in the high temperature phase of $LiBH_4$ at ~120 °C initiated research on complex metal hydrides as solid-state ion conductors.[42] Today, several borohydride-based materials are known that combine high room temperature lithium- and sodium-ion conductivities with high thermal stability, negligible electronic conductivity, and favorable mechanical properties. In $LiBH_4$–LiI solid solutions as well as in $LiBH_4$–$LiNH_2$ compounds, phase transitions are suppressed and high ionic conductivities of ~10^{-4} and 6.4 × 10^{-3} S cm^{-1} at room temperature have been observed for $LiBH_{4-x}I_x$ and $Li(BH_4)_{1-x}(NH_2)_x$ with $x = 2/3$, respectively.[43,44] Furthermore, high sodium-ion conductivities of 0.9 × 10^{-3} and 0.5 × 10^{-3} S cm^{-1} at 20 °C and room temperature have been observed for $Na_2(B_{12}H_{12})_{0.5}(B_{10}H_{10})_{0.5}$ and $Na_3BH_4B_{12}H_{12}$, respectively.[45,46] Stable cycling of a 3 V all-solid-state sodium–ion battery was demonstrated based on $Na_2(B_{12}H_{12})_{0.5}(B_{10}H_{10})_{0.5}$ as a solid-state electrolyte.[4]

In this class of materials, magnesium borohydride, $Mg(BH_4)_2$ has received significant attention as the first inorganic halide-free compound, enabling

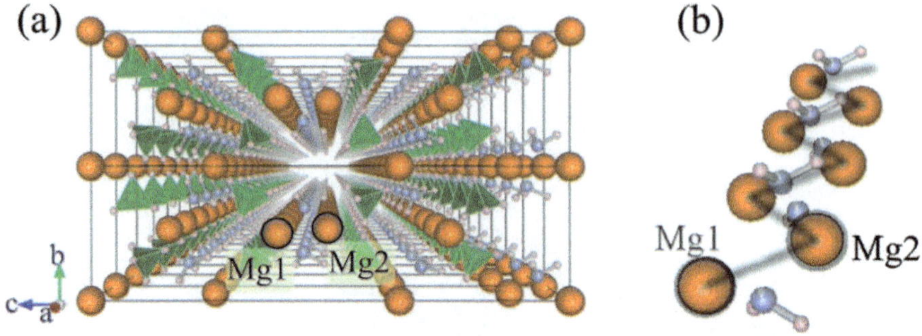

Figure 3.5 (a) Crystal structure of $Mg(BH_4)(NH_2)$ and (b) Mg zigzag structure. Reproduced from ref. 51 with permission from the Royal Society of Chemistry.

reversible magnesium plating and stripping when dissolved in organic solvents such as dimethoxyethane.[47,48] Furthermore, one of the few prototype all-solid-state magnesium-ion batteries is based on a nanocomposite polymer electrolyte containing poly(ethyleneoxide) (PEO), $Mg(BH_4)_2$ and MgO nanoparticles. This electrolyte shows an electrochemical stability of up to ~2.2 V *vs.* Mg/Mg^{2+}, enabling reversible magnesium plating and stripping. Moreover, a battery operated at 100 °C combining a magnesium metal anode, Mo_6S_8 cathode and this electrolyte showed a stable capacity of ~100 mAh g^{-1} for 150 cycles.[2]

In the solid state, pure $Mg(BH_4)_2$ exhibits a low ionic conductivity of 1×10^{-9} S cm^{-1} at 150 °C that can be attributed to the crystal structure of $Mg(BH_4)_2$ in which magnesium ions are localized in firm tetrahedral cages formed by four BH_4^- units.[49]

In $Mg(BH_4)(NH_2)$ on the other hand, which was studied in analogy to the $Li(BH_4)(NH_2)$ system,[41,50] this framework is broken and the structure contains Mg zigzag chains and conduction pathways along the *a* and *b* planes, as shown in Figure 3.5. Consequently, this compound exhibits a higher ionic conductivity of 1×10^{-6} S cm^{-1} at 150 °C and an electrochemical stability of 3 V *vs.* Mg/Mg^{2+} (Figure 3.3).[51]

To further reduce the Mg coordination and to enable ionic conduction at lower temperatures, neutral chelating ligands such as ethylenediamine ($NH_2CH_2CH_2NH_2$, abbreviated as "en") have been added to $Mg(BH_4)_2$.[52] $Mg(en)_3(BH_4)_2$ crystalizes in an orthorhombic unit cell in the space group *Pbca*. In this structure, each magnesium ion is surrounded by three ethylenediamine groups, which coordinate as bidentate ligands, leading to octahedral coordination of Mg^{2+} (Table 3.2).[53] Conductivities of ~5×10^{-11} and ~5×10^{-9} S cm^{-1} at 30 and 70 °C were observed for this compound, with an activation energy of 0.9 eV.[52] Reducing the number of chelating ethylenediamine ligands to one for $Mg(en)_1(BH_4)_2$ leads to a partially chelated magnesium ion and enhanced ionic conductivities of 5×10^{-8} and 6×10^{-5} S cm^{-1} at 30 and 70 °C, respectively. The activation energy was determined to be 1.6 eV

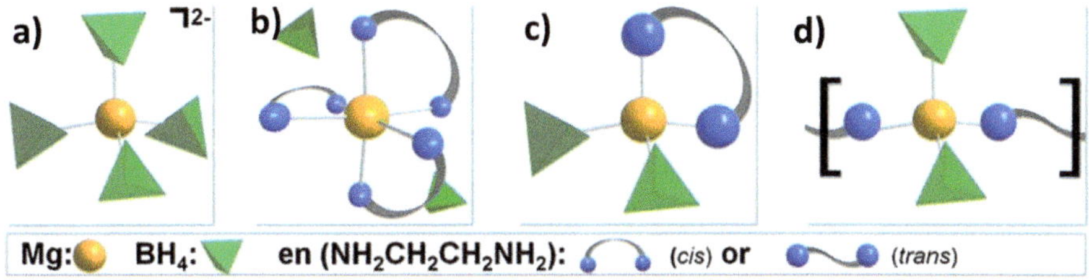

Figure 3.6 Conceptual sketch of the local coordination of Mg^{2+} in (a) $Mg(BH_4)_2$, (b) $Mg(en)_3(BH_4)_2$, and two likely possibilities in $Mg(en)_1(BH_4)_2$ (c and d). Adapted from ref. 52, https://doi.org/10.1038/srep46189, under the terms of the CC BY 4.0 licence, http://creativecommons.org/licenses/by/4.0/.

and an electrochemical stability of 1.2 V *vs.* Mg/Mg^{2+} was observed (Figure 3.3).[52] Two possible configurations exist for the coordination of the ethylenediamine ligand to the Mg^{2+} cation. (i) a chelating coordination of ethylenediamine, similar to $Mg(en)_3(BH_4)_2$, in which the ethylenediamine ligand is non-bridging with *cis* configuration and can be formulated as $[Mg(en)(BH_4)_2]$. (ii) Ethylenediamine acts as a bridging ligand between two magnesium ions in the complex, forming chains and can be formulated as [~$Mg(BH_4)_2$–en–$Mg(BH_4)_2$–en–]. The *cis* configuration has been suggested based on infrared and Raman spectroscopic data (Figure 3.6).[52]

The $Mg(en)(BH_4)_2$ complexes show high magnesium-ion conductivity, proving the validity of using chelating ligands to enhance ionic mobility. However, they suffer from limited electrochemical and thermal stability. This can be explained by the close proximity of hydridic hydrogen atoms in the $[BH_4]^-$ groups to protic hydrogen atoms in the $[NH_2]^+$ groups, which promotes the decomposition of the compounds *via* the evolution of H_2. Therefore, other coordinating ligands such as bis(2-methoxyethyl)ether (diglyme) were investigated.

$Mg(BH_4)_2(diglyme)_{0.5}$ is thermally stable up to ~250 °C and possesses an ionic conductivity of 2×10^{-5} S cm^{-1} at 77 °C. The activation energy for this compound is in the same range as that of $Mg(en)(BH_4)_2$ (~1.2 eV) (Figure 3.3).[54] Recently, mixtures of $MgB_{12}H_{12}$(diglyme), $MgB_{11}H_{11}$(diglyme) and $Mg(B_{11}H_{14})_2$(diglyme) were studied, where magnesium-ion conductivities of 10^{-7} and 10^{-6} S cm^{-1} were observed at 25 and 60 °C, respectively.[55]

3.5 Solid-state Magnesium-ion Conductors Based on Metal–Organic Frameworks

Metal–organic frameworks (MOFs) are composed of metal ions coordinated to multidentate organic ligands to form three-dimensional structures. The structure of a MOF is characterized by an open porous framework that can host magnesium ions. The open metal sites promote cation mobility but hinder the migration of nucleophilic anions.

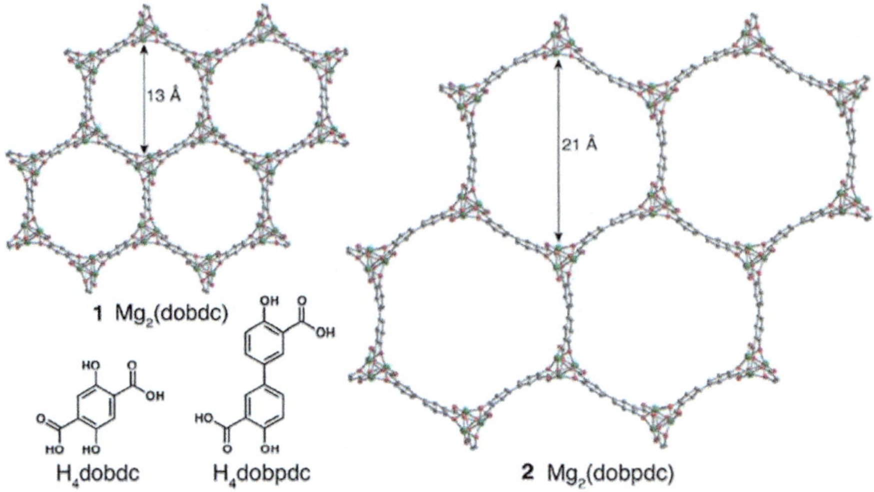

Figure 3.7 Structures of the metal–organic frameworks $Mg_2(dobdc)$ (1) and $Mg_2(dobpdc)$ (2), as viewed along the *c*-axis. The vertices of the one-dimensional (1-D) channels are formed by 1-D chains of magnesium ions linked together by the respective ligands (bottom left) that form the pore walls. Reproduced from ref. 9 with permission from the Royal Society of Chemistry.

A series of MOF-based magnesium electrolytes were prepared and the magnesium-ion conductivity was optimized by varying the pore sizes of the MOFs and tuning the anion basicity of the guest electrolyte salts (Figure 3.7). The highest magnesium-ion conductivity of 2.5×10^{-4} S cm^{-1} at room temperature with an activation energy of 0.13 eV was obtained for mixtures of $Mg(OPhCF_3)_2$–$Mg(TFSI)_2$ salts in a $Mg_2(dobpdc)$ MOF.[9]

Magnesium-ion conductivity in a mesoporous anionic Cu-azolate metal–organic framework was recently investigated and an ionic conductivity of 8.8×10^{-7} S cm^{-1} at room temperature was reported for $Mg_{0.5}[Cu_2Cl_2BrBTD\text{-}D].8(PC)$ (MIT-20-Mg), with an activation energy of 0.37 eV.[56]

3.6 Conclusion

Within the last few years, magnesium-ion conductivities in the range of 10^{-4} S cm^{-1} at room temperature have been reached in various material classes such as phosphates,[8] chalcogenides,[38] and MOFs,[9] reducing the gap to conventional solution-based electrolytes used in commercial lithium-ion batteries (Figure 3.1). Reports on solid-state magnesium-ion conductors are still comparatively scarce and solid-state magnesium-ion conductors are often not fully characterized with respect to their electrochemical properties. Also, very few operational electrochemical cells have been realized so far.

The ionic conductivity strongly depends on the measurement conditions and sample preparation, where impurities and the degree of compaction play a crucial role. For example, reported values for classic solid-state electrolytes such as NASICON or sodium β" alumina vary by more than one order of magnitude.[57] Therefore, unified measurement protocols and careful characterization of the structure and morphology as well as of the phase purity are desirable to facilitate the comparison of ionic conductivities and electrochemical stabilities.

Strategies followed so far to discover and tune the magnesium-ion conductivity have been to transfer the knowledge from the lithium and sodium counterparts and to investigate the magnesium analogues of lithium- and sodium-ion conductors. Examples are MTP, being the structural analogue to NASICON, spinel-type sulfides and selenides, complex metal hydride-based compounds, and MOFs. Significant advancements have been achieved in reducing the temperature necessary to reach superionic conductivity. It turns out that the high charge density of the bivalent magnesium ion, leading to strong electrostatic interactions, is the main challenge to achieving high ionic conductivities. However, high ionic conductivity is not the only criterion for a solid-state ion conductor to qualify as an electrolyte for all-solid-state batteries. The conductor needs to be chemically and electrochemically compatible with the anode and cathode materials, should sustain high current densities to enable reasonable charge and discharge rates, and should have favorable mechanical properties.

Moreover, an emerging magnesium-ion battery technology will compete with more mature lithium-ion and sodium-ion technologies that are aiming for higher energy and power densities, while avoiding the use of rare or critical raw materials. Besides all-solid-state lithium- and sodium-ion batteries, metal–sulfur and metal–air/metal–oxygen batteries are also in development. So far, lithium and sodium-based technologies lead the race for the next battery generation. However, magnesium batteries are a young field and despite the challenges, surprises are to be expected. The discovery of a reversible high voltage, high capacity cathode for example could be a game changer. The race is open for a competitive magnesium-ion battery technology and the promise of a high energy density battery, exclusively made out of earth abundant materials might be fulfilled by an all-solid-state magnesium battery.

References

1. R. Schmuch, R. Wagner, G. Hörpel, T. Placke and M. Winter, *Nat. Energy*, 2018, **3**, 267.
2. Y. Shao, N. N. Rajput, J. Hu, M. Hu, T. Liu, Z. Wei, M. Gu, X. Deng, S. Xu, K. S. Han, J. Wang, Z. Nie, G. Li, K. R. Zavadil, J. Xiao, C. Wang, W. A. Henderson, J.-G. Zhang, Y. Wang, K. T. Mueller, K. Persson and J. Liu, *Nano Energy*, 2015, **12**, 750.
3. E. Sheha, F. Ahmad, P. Zhang, H. Wang and Z. Guo, *Energy Environ. Focus*, 2016, **5**, 125.

4. L. Duchêne, R. S. Kühnel, E. Stilp, E. Cuervo Reyes, A. Remhof, H. Hagemann and C. Battaglia, *Energy Environ. Sci.*, 2017, **10**, 2609.
5. C. Sun, J. Liu, Y. Gong, D. P. Wilkinson and J. Zhang, *Nano Energy*, 2017, **33**, 363.
6. A. Manthiram, X. Yu and S. Wang, *Nat. Rev. Mater.*, 2017, **2**, 1.
7. K. Suzuki, N. Mashimo, Y. Ikeda, T. Yokoi, M. Hirayama and R. Kanno, *ACS Appl. Energy Mater.*, 2018, **1**, 2373.
8. N. K. Anuar and N. S. Mohamed, *J. Sol-Gel Sci. Technol.*, 2016, **80**, 249.
9. M. L. Aubrey, R. Ameloot, B. M. Wiers and J. R. Long, *Energy Environ. Sci.*, 2014, **7**, 667.
10. J. Muldoon, C. B. Bucur and T. Gregory, *Chem. Rev.*, 2014, **114**, 11683.
11. R. Mohtadi and F. Mizuno, *Beilstein J. Nanotechnol.*, 2014, **5**, 1291.
12. S. Ikeda, M. Takahashi, J. Ishikawa and K. Ito, *Solid State Ionics*, 1987, **23**, 125.
13. E. R. Gobechiya, M. V. Sukhanov, V. I. Pet'kov and Y. K. Kabalov, *Crystallogr. Rep.*, 2008, **53**, 53.
14. K. Nomura, S. Ikeda, K. Ito and H. Einaga, *J. Electroanal. Chem.*, 1992, **326**, 351.
15. N. K. Anuar, S. B. R. S. Adnan and N. S. Mohamed, *Ceram. Int.*, 2014, **40**, 13719.
16. M. Adamu and G. M. Kale, *J. Phys. Chem. C*, 2016, **120**, 17909.
17. G. Sahu, E. Rangasamy, J. Li, Y. Chen, K. An, N. Dudney and C. Liang, *J. Mater. Chem. A*, 2014, **2**, 10396.
18. N. Imanaka, Y. Okazaki and G. Y. Adachi, *J. Mater. Chem.*, 2000, **10**, 1431.
19. N. Imanaka, Y. Okazaki and G. Adachi, *Electrochem. Solid-State Lett.*, 2000, **3**, 327.
20. K. Kajihara, H. Nagano, T. Tsujita, H. Munakata and K. Kanamura, *J. Electrochem. Soc.*, 2017, **164**, A2183.
21. R. D. Shannon, *Acta Crystallogr.*, 1976, **A32**, 751.
22. N. Imanaka, Y. Okazaki and G. Adachi, *Ionics*, 2001, **7**, 440.
23. N. K. Anuar, S. B. R. S. Adnan, M. H. Jaafar and N. S. Mohamed, *Ionics*, 2016, **22**, 1125.
24. Z. A. Halim, S. B. R. S. Adnan and N. S. Mohamed, *Ceram. Int.*, 2016, **42**, 4452.
25. Z. A. Halim, S. B. R. S. Adnan, F. M. Salleh and N. S. Mohamed, *J. Magnesium Alloys*, 2017, **5**, 439.
26. S. Tamura, M. Yamane, Y. Hoshino and N. Imanaka, *J. Solid State Chem.*, 2016, **235**, 7.
27. H. Takahashi and H. Takamura, *Mater. Trans.*, 2012, **53**, 932.
28. S. Barth, R. Olazcuaga, P. Gravereau, G. Le Flem and P. Hagenmuller, *Mater. Lett.*, 1993, **16**, 96.
29. H. Takahashia and H. Takamura, *Key Eng. Mater.*, 2012, **508**, 291.
30. K. Makino, Y. Katayama, T. Miura and T. Kishi, *J. Power Sources*, 2002, **112**, 85.
31. Y. Zhao, Z. Wei, Q. Pang, Y. Wei, Y. Cai, Q. Fu, F. Du, A. Sarapulova, H. Ehrenberg, B. Liu and G. Chen, *ACS Appl. Mater. Interfaces*, 2017, **9**, 4709.

32. V. I. Pet'kov, A. S. Shipilov, E. Y. Borovikova, A. M. Kovalskii, I. A. Steninad and A. B. Yaroslavtsev, *Inorg. Mater.*, 2018, **54**, 1021.
33. A. Omote, S. Yotsuhashi, Y. Zenitani and Y. Yamada, *J. Am. Ceram. Soc.*, 2011, **94**, 2285.
34. A. Hanom Ahmad and F. S. Abdul Ghani, *AIP Conf. Proc.*, 2009, **1136**, 31.
35. Y. Nishitania, S. Shibataa, T. Tsujitaa, O. Tetsuyukia and A. Omotea, in *ECS Meeting Abstract MA2016-02*, 2016, p. 676.
36. N. Che Su, M. Sulaiman and N. S. Mohamed, *J. New Mater. Electrochem. Syst.*, 2016, **19**, 27.
37. C. K. Moon, H. J. Lee, K. H. Park, H. Kwak, J. W. Heo, K. Choi, H. Yang, M. S. Kim, S. T. Hong, J. H. Lee and Y. S. Jung, *ACS Energy Lett.*, 2018, **3**, 2504.
38. P. Canepa, S. H. Bo, G. Sai Gautam, B. Key, W. D. Richards, T. Shi, Y. Tian, Y. Wang, J. Li and G. Ceder, *Nat. Commun.*, 2017, **8**, 1759.
39. T. Holzmann, L. M. Schoop, M. N. Ali, I. Moudrakovski, G. Gregori, J. Maier, R. J. Cava and B. V. Lotsch, *Energy Environ. Sci.*, 2016, **9**, 2578.
40. L. Zhang, K. Yang, J. Mi, L. Lu, L. Zhao, L. Wang, Y. Li and H. Zeng, *Adv. Energy Mater.*, 2015, **5**, 1501294.
41. T. Yamanaka, A. Hayashi, A. Yamauchi and M. Tatsumisago, *Solid State Ionics*, 2014, **262**, 601.
42. M. Matsuo, Y. Nakamori and S. Orimo, *Appl. Phys. Lett.*, 2007, **91**, 224103.
43. H. Maekawa, M. Matsuo, H. Takamura, M. Ando, Y. Noda, T. Karahashi and S. I. Orimo, *J. Am. Chem. Soc.*, 2009, **131**, 894.
44. Y. Yan, R. S. Kühnel, A. Remhof, L. Duchêne, E. C. Reyes, D. Rentsch, Z. Łodziana and C. Battaglia, *Adv. Energy Mater.*, 2017, **7**, 1700294.
45. L. Duchêne, R.-S. Kühnel, D. Rentsch, A. Remhof, H. Hagemann and C. Battaglia, *Chem. Commun.*, 2017, **53**, 4195.
46. Y. Sadikin, M. Brighi, P. Schouwink and R. Černý, *Adv. Energy Mater.*, 2015, **5**, 1501016.
47. R. Mohtadi, M. Matsui, T. S. Arthur and S. J. Hwang, *Angew. Chem., Int. Ed.*, 2012, **51**, 9780.
48. J. Muldoon, C. B. Bucur, A. G. Oliver, T. Sugimoto, M. Matsui, H. S. Kim, G. D. Allred, J. Zajicek and Y. Kotani, *Energy Environ. Sci.*, 2012, **5**, 5941.
49. T. Ikeshoji, E. Tsuchida, S. Takagi, M. Matsuo and S. I. Orimo, *RSC Adv.*, 2014, **4**, 1366.
50. M. Matsuo, A. Remhof, P. Martelli, R. Caputo, M. Ernst, Y. Miura, T. Sato, H. Oguchi, H. Maekawa, H. Takamura, A. Borgschulte, A. Züttel and S. I. Orimo, *J. Am. Chem. Soc.*, 2009, **131**, 16389.
51. S. Higashi, K. Miwa, M. Aoki and K. Takechi, *Chem. Commun.*, 2014, **50**, 1320.
52. E. Roedern, R.-S. Kühnel, A. Remhof and C. Battaglia, *Sci. Rep.*, 2017, **7**, 46189.
53. J. Chen, Y. S. Chua, H. Wu, Z. Xiong, T. He, W. Zhou, X. Ju, M. Yang, G. Wu and P. Chen, *Int. J. Hydrogen Energy*, 2015, **40**, 412.
54. T. Burankova, E. Roedern, A. E. Maniadaki, H. Hagemann, D. Rentsch, Z. Łodziana, C. Battaglia, A. Remhof and J. P. Embs, *J. Phys. Chem. Lett.*, 2018, **9**, 6450.

55. F. Mizuno, R. Mohtadi, O. Tutusaus, M. Fichtner and Z. Zhao-Karger, *US Pat.*, US009716289B1, 2017, DOI: 10.1016/S0141.
56. S. S. Park, Y. Tulchinsky and M. Dinca, *J. Am. Chem. Soc.*, 2017, **139**, 13260.
57. J. W. Fergus, *Solid State Ionics*, 2012, **227**, 102.
58. B. Liang, V. Keshishian, S. Liu, E. Yi, D. Jia, Y. Zhou, J. Kieffer, B. Ye and R. M. Laine, *Electrochim. Acta*, 2018, **272**, 144.
59. T. Noritake, K. Miwa, M. Aoki, M. Matsumoto, S. I. Towata, H. W. Li and S.-I. Orimo, *J. Alloys Compd.*, 2013, **580**, S85.
60. L. Wang, Z. Zhao-Karger, F. Klein, J. Chable, T. Braun, A. R. Schür, C. Wang, Y. Guo and M. Fichtner, *ChemSusChem*, 2019, 2286–2293.

CHAPTER 4

Theoretical Modelling of Multivalent Ions in Inorganic Hosts

GOPALAKRISHNAN SAI GAUTAM*[a] AND PIEREMANUELE CANEPA*[b]

[a]Department of Mechanical and Aerospace Engineering, Princeton University, Princeton, New Jersey 08544, USA; [b]Department of Materials Science and Engineering, National University of Singapore, Singapore 117575, Singapore
*E-mail: gautam91@princeton.edu, pcanepa@nus.edu.sg

4.1 Introduction

The design of multivalent batteries, specifically Mg batteries, has proven to be challenging for a number of reasons: (i) a lack of high voltage cathode materials with good mobility of Mg ions;[1–5] (ii) a lack of electrolytes that are compatible with the Mg metal anode and high-voltage cathode materials;[6–8] (iii) poor understanding of metal plating/stripping at the anode;[6,9,10] and (iv) a lack of standard protocols to perform experiments.[5]

Theoretical calculations, which can be thought of as controlled experiments, are ideal to tackle some of the aforementioned issues, as they limit the number of variables affecting the observables. This chapter will focus on the application of theoretical tools, especially using density functional theory (DFT), to identify Mg cathodes with good intrinsic ionic mobility and

Energy and Environment Series No. 23
Magnesium Batteries: Research and Applications
Edited by Maximilian Fichtner

Published by the Royal Society of Chemistry, www.rsc.org

high voltage. We will begin by discussing the thermodynamic and kinetic principles that are used in conjunction with DFT to accurately predict: (i) the intercalation (and conversion) voltages of multivalent electrodes, (ii) the migration barrier of Mg^{2+} ions in host materials, and (iii) the identification of good electrode coatings. We conclude by providing a number of examples of successful applications to topical materials in the Mg chemical space.

4.1.1 Thermodynamics of Multivalent Electrodes

Here, we explore the thermodynamic quantities accessible using DFT in a typical intercalating battery system. We define the Nernst equation for calculating voltages and the subsequent extensions of it to solvent co-intercalating systems, conversion reactions, and electrolyte stabilities.

A reversible reaction of Mg intercalation, where the source of intercalant atoms is from a Mg-metal anode, within a MO_z cathode host (M = redox-active transition metal) is written as:

$$MO_z + x\text{Mg} \rightarrow \text{Mg}_x\text{MO}_z \quad (4.1)$$

The intercalation reaction of eqn (4.1) is thermodynamically favoured, if its Gibbs energy, ΔG, is negative (see below).

$$\Delta G \equiv G_{\text{Mg}_x\text{MO}_z} - G_{\text{MO}_z} - xG_{\text{Mg}} \leq 0 \quad (4.2)$$

Subsequently, the Nernst equation of eqn (4.3) can be used to define a change in the electrochemical potential (ΔV) of any reaction, given its Gibbs energy change (ΔG), the number of electrons transferred (n) and the Faraday constant (F).

$$\Delta V = -\frac{\Delta G}{nF}, \quad (4.3)$$

Combining eqn (4.2) and (4.3), we define the average equilibrium voltage for intercalating x moles of Mg into the MO_z host as:

$$\Delta V = -\frac{G_{\text{Mg}_x\text{MO}_z} - G_{\text{MO}_z} - xG_{\text{Mg}}}{2xF} \quad (4.4)$$

Note that each Mg atom corresponds to a transfer of 2 electrons, which explains the factor of 2 in the denominator of eqn (4.4). Additionally, eqn (4.4) is referenced to as the Gibbs energy of Mg metal (G_{Mg}), which sets a natural reference to ΔV as the standard reduction potential (SRP) of Mg metal, which is −2.37 V *vs.* the standard hydrogen electrode (SHE). Thus, ΔV in eqn (4.4) is always defined with respect to the SRP of Mg^{2+}/Mg. Cathode frameworks that thermodynamically favour Mg intercalation will exhibit a positive ΔV in eqn (4.4). In other words, the anode–(electrolyte) –cathode combination will pump electrical energy into the external circuit, or undergo "discharge", during Mg intercalation, if ΔV is positive. Subsequently, the battery will undergo "charge" or will absorb energy from an external power source (equivalent to ΔV), while Mg ions are de-intercalated. In battery literature, Mg_xMO_z and MO_z are often to as discharged and charged compositions, respectively. Further, the difference between the Gibbs energies of the intercalated and empty cathode in eqn (4.4), *i.e.* $G_{Mg_xMO_z} - G_{MO_z}$, is the change in the chemical potential of Mg (μ^{Mg}) within the cathode host, across the Mg_x-MO_z and MO_z concentrations.[11] Hence, eqn (4.4) can be generalized to define the equilibrium voltage ($V(x)$ *vs.* Mg^{2+}/Mg) of a cathode host at a given Mg concentration (x), using eqn (4.5):

$$V(x) = -\frac{\mu^{Mg}_{Mg_xMO_z}(x) - \mu^{Mg}_{Mg\text{-anode}}}{2F} \quad (4.5)$$

First-principles calculations, such as those based on DFT,[12,13] are used to calculate the average intercalation voltage (ΔV from eqn (4.4)), for intercalating compositions, x and $x + \Delta x$, while approximating the Gibbs energies of the cathode and the anode to the internal energies at 0 K ($G \approx E$), which are directly obtained from total energy calculations using DFT, resulting in eqn (4.6).

$$\Delta V \approx -\frac{E_{Mg_{x+\Delta x}MO_z} - E_{Mg_xMO_z} - \Delta x E_{Mg}}{2\Delta x F} \quad (4.6)$$

Eqn (4.6) includes important approximations, which are often justified, but do play a role while directly comparing DFT calculated with an experimental ΔV, where the experiment is typically done at 298 K and 1 atm. For example, the contributions arising from the volume change as intercalation occurs within the cathode (or an intercalation anode) host, *i.e.* $p\Delta V$, are typically neglected since such contributions are in the order of 10^{-5} eV for most materials, whereas the change in internal energies are in the order of a few (1–10) eV.[11] Similarly, the changes in entropy within the cathode host, *i.e.* $T\Delta S \approx 0.01 - 0.1$ eV, are also often neglected. However, entropy changes can influence the phase behaviour of a given cathode host during an intercalation reaction, as discussed in Section 4.1.2. In particular, entropy contributions dominate in determining the ground-state (or equilibrium) crystal structure (or atomic configuration) at higher temperatures. However, the computational cost of calculations required to capture entropy contributions are significant, with most theoretical studies using the calculated ΔV at 0 K to screen for promising cathodes.

4.1.1.1 *Ground State Hull, Metastability and Average Voltages*

Reversible intercalation reactions are typically "topotactic" processes, where the structural framework at the intercalated and empty compositions remain identical, with the intercalant occupying well-defined vacant sites within the cathode. For example, when Mg reversibly (de-)intercalates into a MO_2 host, the cathode structure remains identical to that of either the $MgMO_2$ (fully intercalated) or MO_2 (fully empty) structures, while Mg is removed or added. The framework that the cathode adopts during topotactic intercalation depends on the structure with which the (de-)intercalation process is begun, *i.e.* a fully intercalated (empty) framework is adopted if $MgMO_2$ (MO_2) is the starting composition. Thus, the calculated intercalation voltage is heavily dependent on the cathode structure (or polymorph) of (de-)intercalation. Given eqn (4.6), the lower (higher) the energy of the intercalated (empty) cathode host, the higher the average voltage.

Since polymorphism plays an important role in determining (de-)intercalation voltages, it is crucial to determine the ground state structures as a function of various intercalant compositions within the cathode framework. Accurate (de-)intercalation voltages (eqn (4.6)) can be estimated from DFT calculations if accurate 0 K phase diagrams, also referred to as "ground state hulls", can be determined, as schematically illustrated for a hypothetical Mg-intercalation system in Figure 4.1a,b. The term *ground state hull* comes from

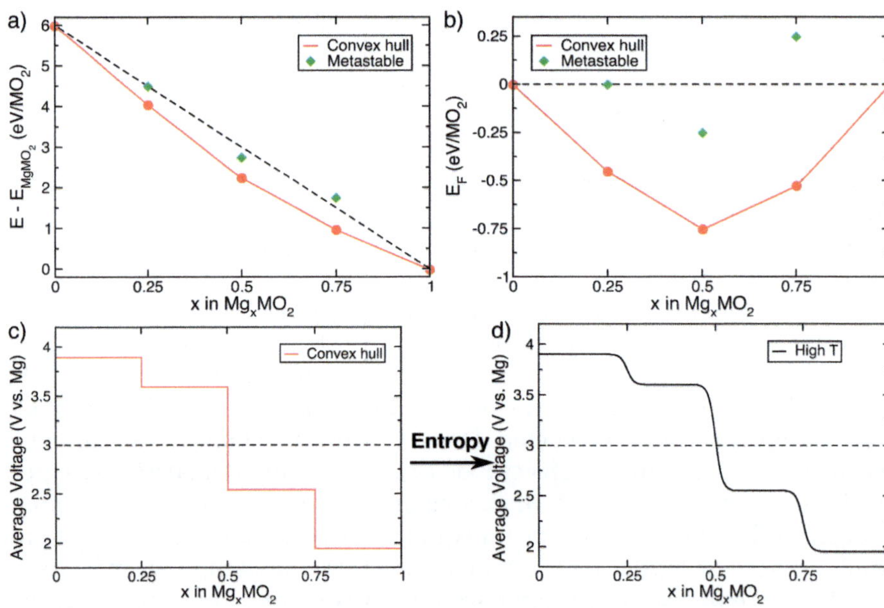

Figure 4.1 (a) The energy landscape in a hypothetical Mg intercalation material, where the energies are referenced to the intercalated composition, $MgMO_2$. (b) The ground state hull of the Mg intercalation system, with the energies referenced to both the intercalated and empty (MO_2) compositions. The red lines, red circles, and the green diamonds are the convex (or ground state) hull at 0 K, the ground states and metastable states, respectively. (c) The average voltage curve plotted as a function of Mg concentration (x) for Mg (de-)intercalation in Mg_xMO_2, obtained from the convex hull of panels (a) and (b), at 0 K. (d) Average voltage curve at a higher temperature including entropic contributions.

the fact that 0 K phase diagrams resemble convex hulls (see Figure 4.1b), since the ground state structure, at a given composition, always minimizes the Gibbs energy at that composition. Thus, given a set of compositions, structures, and Gibbs energies, the equilibrium phase diagram is mathematically obtained through a procedure of convex minimization. Once the ground state hull is obtained, eqn (4.6) is applied to calculate average voltages in a range of compositions, as shown in Figure 4.1c.

In Figure 4.1a, the internal energies are referenced with respect to the fully intercalated composition, *i.e.* E_{MgMO_2}, and the "uphill" nature of the energy landscape indicates the energy required (or the voltage to be applied) to de-intercalate Mg^{2+} ions from the $MgMO_2$ framework. For example, an uphill energy difference of ~6 eV between MO_2 and $MgMO_2$ signifies an average voltage of ~3 V *vs.* Mg (see eqn (4.6)) that will be required to charge $MgMO_2$ completely to MO_2. Visualizing ground state hulls for intercalating compositions is more intuitive if the energies of intermediate compositions, *i.e.* $0 < x < 1$ for x in Mg_xMO_2, are referenced to both the discharged (E_{MgMO_2}) and charged (E_{MO_2}) compositions, as shown in Figure 4.1b. Such a representation also allows an intuitive understanding of the metastability of various configurations, shown as the green diamonds in Figure 4.1b, where the metastability of each structure is quantified by the energy above the hull (E^{hull}). Mathematically, the E^{hull} at a given composition is defined as the energy released upon decomposition into the thermodynamic ground state, *i.e.* the distance to the ground state hull. For example, the energy of the metastable configuration at $x = 0.5$ in Figure 4.1b is ~−0.25 eV/MO_2, while the energy of the ground state configuration at the same composition is ~−0.75 eV/MO_2. Hence, the E^{hull} for the aforementioned configuration is ~0.5 eV/MO_2. Thus, $E^{hull} \geq 0$, with ground state configurations exhibiting $E^{hull} = 0$. Although metastable configurations at a given temperature should not exist, entropic or kinetic contributions can stabilize such configurations.

For the Mg_xMO_2 system shown in Figure 4.1, there are three thermodynamically stable configurations at $x_{Mg} = 0.25$, 0.5, and 0.75, as illustrated by the red circles in panels a and b. Since the Mg concentration is the only changing variable in the Mg_xMO_2 system, it can be treated as a (pseudo-)binary system. Note that the Gibbs phase rule in binary systems (at a given T,p) restricts the number of independent variables, which includes μ_{Mg}, to 1 in a single-phase (or ground state) region (red circles in Figure 4.1b), while there are no independent variables in a two-phase region (red lines in Figure 4.1b). Thus, μ_{Mg}, which is related to the average voltages *via* eqn (4.5), either exhibits a continuous range at ground state configurations ($x_{Mg} = 0.25$, 0.5, and 0.75) or a single value in two phase regions ($0.75 < x < 1$, $0.5 < x < 0.75$, $0.25 < x < 0.5$, and $0 < x < 0.25$). Consequently, the voltage profile in Figure 4.1c exhibits a "jump" across single-phase regions and a "plateau" across two-phase regions. Note that the voltage illustrations in Figure 4.1c are strictly applicable to 0 K, or for single-phases which do not exhibit any composition ranges.

Typically, by including entropy contributions, such as configurational, vibrational, *etc.*, (see Section 4.1.2) in intercalation solids extends the composition range over which a given phase is stable. For example, the Mg-vacancy configuration that forms the ground state at $x_{Mg} = 0.5$ in Figure 4.1b is stable only at $x = 0.5$ at 0 K. However, at a higher temperature, the $Mg_{0.5}MO_2$ ground state might be stable over a composition range around $x_{Mg} = 0.5$, say $0.45 \leq x \leq 0.55$. Consequently, the voltage profiles of Mg (de-)intercalation calculated at 0 K (Figure 4.1c) will change at higher temperatures (Figure 4.1d). While the voltage jumps are discrete at 0 K and will happen across distinct ground state compositions, the jumps are continuous and will occur across the range of stable compositions for each stable phase at higher temperatures. Thus, voltage profiles at higher temperatures will appear "smeared" compared to 0 K, with regions of smoothly sloped voltage curves indicating a "solid solution" behaviour.

4.1.1.2 Capturing Entropy Contributions and the Method of Cluster Expansion

Entropy contributions, including vibrational, configurational, magnetic, electronic, *etc.*, which are ignored in eqn (4.6), are important in determining the overall phase behaviour and the resultant voltage profiles. More importantly, entropy differences among solid phases determine which phase is stable at a given temperature. For example, consider a case where two metallic phases A and B in a given system exhibit relatively high but identical amounts of electronic entropy (*i.e.* low G). However, A will be preferentially stabilized at higher temperatures if A exhibits a marginally higher vibrational entropy than B, *i.e.* $S_A > S_B$. Among entropy contributions, vibrational and configurational are important since they can be appreciably different among solid phases, and can selectively stabilize a specific phase at higher temperatures. Capturing either vibrational or configurational entropy requires additional DFT calculations, requiring both significant computational and human time to obtain accurate estimates.

Vibrational entropy in solids can be estimated within the framework of phonons, *i.e.* by combining principles of statistical mechanics with phonon dynamics. In terms of DFT calculations, the phonon density of states (Figure 4.2a displays the density of states for a hypothetical cathode) and the phonon band structure of solids are calculated by obtaining the energies and forces corresponding to different symmetrically-distinct atomic displacements within a supercell of the solid. Subsequently, by calculating the force constants (weighed by the atomic masses) that correspond to various atomic displacements, the phonon properties and the resultant vibrational specific heat (C_v), vibrational entropy (S) and the vibrational Gibbs energies (G, plotted in Figure 4.2b) are estimated.[14–16] The contribution of phonons is almost insignificant to differences in the vibrational Gibbs energy between most solid phases at low temperatures,[17] *i.e.* $0 < T < 300$ K, justifying why DFT estimates of average voltages do not include phonon contributions. Note that the vibrational thermodynamic properties have to

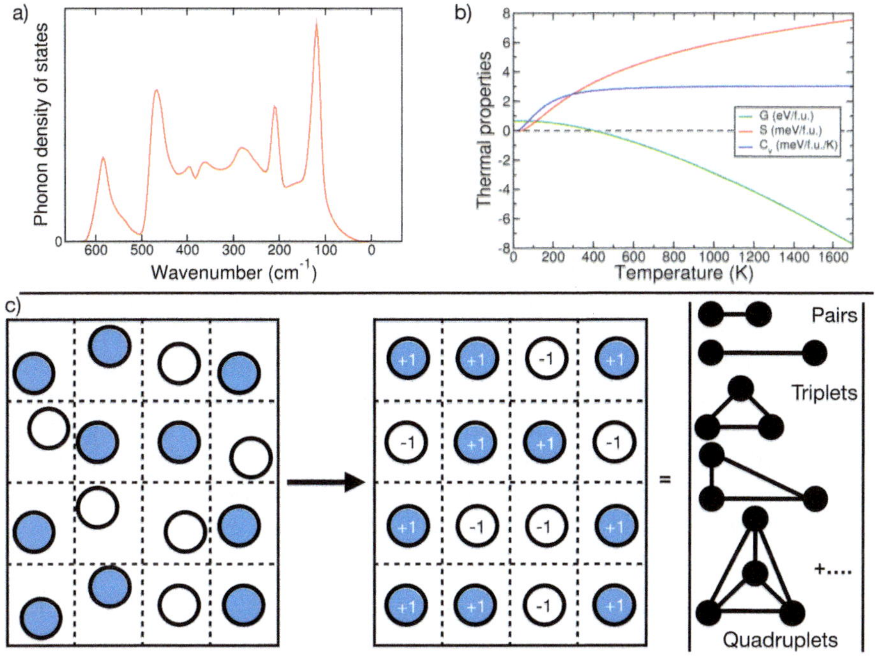

Figure 4.2 (a) Phonon density of states of a cathode material, and (b) computable thermodynamic properties, including the vibrational specific heat (blue curve), vibrational entropy (red), and vibrational Gibbs energy (green). (c) Schematic of the Ising model used to parameterize energies of a given configuration of Mg-ions (blue circles) and vacancies (white circles). Spin values of +1 (−1) are assigned to Mg-ions (vacancies). The energy of each configuration is decomposed to interactions of different clusters, such as pairs, triplets, quadruplets, *etc.*

be added to other entropy contributions to obtain overall thermodynamic quantities. In particular, the vibrational Gibbs energy has to be added to the internal energy (E), pressure–volume contributions ($p\Delta V$), configurational free energy (G_{config}), *etc.* to obtain the overall Gibbs energy of the solid phase.

Configurational entropy is important since the intercalant ions (and corresponding intercalant vacancies) can adopt various configurations within the cathode framework during (de-)intercalation, where several intercalant–vacancy configurations tend to be metastable at 0 K. A pathway to include the configurational entropy in solids is to use a sampling technique over an ensemble of microstates, such as Monte Carlo simulations to evaluate the energies of several possible configurations within the cathode, with the assumption that the cathode system is ergodic. Subsequently, we can use principles of statistical mechanics to obtain ensemble-averaged macroscopic quantities, such as the configurational Gibbs energy. However, performing Monte Carlo simulations on a large set of possible configurations requires the rapid evaluation of the energies of various configurations to ensure that the process is computationally tractable. Hence, the energy of different configurations within the intercalant systems can be parameterized by a generalized Ising model, often referred to as a "cluster expansion" (CE),[18,19] which in turn enables computationally feasible Monte Carlo simulations. Apart from energies, CE can be used to parameterize any configuration-dependent properties, such as forces, stresses, *etc.*[20]

A schematic of the Ising model is provided in Figure 4.2c, where a given configuration (σ) of intercalant atoms (blue circles) and vacancies (white circles) is mapped to an idealized cathode lattice, *i.e.* the energy of the configuration is coarse-grained over any microscopic atomic displacements. Each site is assigned a spin value, namely, +1 and −1 for the intercalant ion and the vacancy (centre panel in Figure 4.2c). Subsequently, the energy (or another property of interest) of the configuration ($E(\sigma)$), as evaluated using DFT, is decomposed into contributions from various clusters, such as pairs, triplets, quadruplets, *etc.* Thus, the CE is written as a summation of the interactions of symmetrically distinct clusters (α), as in eqn (4.7).

$$E(\sigma) = \sum_{\alpha} m_{\alpha} V_{\alpha} \left\langle \prod_{i \in \beta} \sigma_i \right\rangle. \tag{4.7}$$

$E(\sigma)$ in eqn (4.7) can be represented exactly if the summation over α is extended to an infinite number of clusters. In practice, the summation is truncated to a few clusters, with their corresponding effective cluster interactions (ECIs, V_{α} in eqn (4.7)) determined by suitably fitting the energies of multiple configurations ($E(\sigma^1)$, $E(\sigma^2)$, *etc.*). m_{α} is the multiplicity of a given cluster within the structure. σ_i is the occupation variable (±1) of each site within a cluster, which in turn is averaged over all clusters β that are symmetrically equivalent to α. For example, a pair term (β_1) containing a Mg^{2+} and a vacancy will have a value $\prod_{i \in \beta} \sigma_i = +1 \times -1 = -1$, while another pair ($\beta_2$) containing only Mg^{2+} ions will have a value of +1. If β_1 and β_2 are symmetrically equivalent and form a subset of α, then $\left\langle \prod_{i \in \beta} \sigma_i \right\rangle = \frac{-1+1}{2} = 0$, with m_{α} being 2 in this case. Finally, a CE fit to several possible configurations to obtain a reliable set of V_{α} can be used in conjunction with Monte Carlo simulations to obtain the configurational component of the Gibbs energy.

Note that configurational entropy contributions will not affect the overall average voltage normally (*i.e.* between MO_2 and $MgMO_2$, dashed black lines in Figure 4.1c,d), since the MO_2 and $MgMO_2$ ground states will not exhibit multiple Mg-vacancy configurations and will not exhibit a configurational entropy (there are exceptions, where $MgMO_2$ exhibits multiple possible configurations). However, significant vibrational entropy can change the Gibbs energies of both MO_2 and $MgMO_2$ and as a result, change the overall average voltage.

4.1.1.3 *Conversion* vs. *Intercalation*

When a MO_2 (or equivalently M_2O_4) cathode is reduced by Mg^{2+}, *i.e.* the cathode undergoes discharge, the host can either undergo a reversible intercalation that results in the formation of MgM_2O_4 or an irreversible conversion that results in the formation of decomposition products. In general, conversion reactions are driven by the thermodynamic driving force to form binary compounds, such as MgO and M_2O_3, MO, *etc.* Conversion reactions are especially exacerbated within Mg-based systems since Mg exhibits a large driving force to form MgO, as quantified by the MgO formation voltage of ~3.1 V *vs.* Mg. Furthermore, MgO formation stops any further electrochemical reactions due to its high stability and poor Mg^{2+} mobility.[5] Thus, intercalation reactions involving Mg^{2+} always have to compete against conversion reactions during discharge.

To quantify the ability of a cathode host to resist conversion reactions, a thermodynamic framework can be built upon the calculated ground state hull of Figure 4.3 (see Section 4.1.1).[21] Using DFT, one can calculate the energy ($G \approx E$) of multiple polymorphs of charged (M_2O_4) and discharged (MgM_2O_4) compositions and subsequently identify the lowest energy polymorphs in each scenario. For example, the lowest energy charged (discharged) polymorph in Figure 4.3 is labelled α (λ). The lowest energy polymorphs are the most relevant for calculating intercalation voltages as they exhibit the highest likelihood to be synthesized in experiments. Since intercalation reactions are ideally topotactic, we can define intercalation voltages for both the lowest energy charged and discharged polymorphs, labelled as $V_{\text{int}}^{(\text{charged})}$ and $V_{\text{int}}^{(\text{discharged})}$ (green arrows in Figure 4.3). Given the relationship of voltages to energies *via* eqn (4.6), $V_{\text{int}}^{(\text{discharged})}$ is always higher than (or at least equal to) $V_{\text{int}}^{(\text{charged})}$.

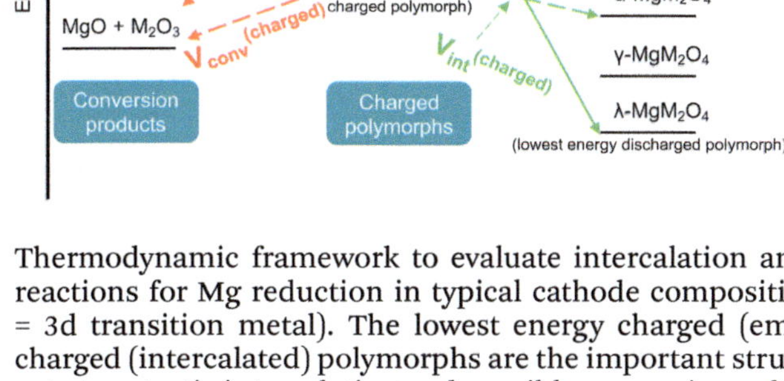

Figure 4.3 Thermodynamic framework to evaluate intercalation and conversion reactions for Mg reduction in typical cathode compositions, M_2O_4 (M = 3d transition metal). The lowest energy charged (empty) and discharged (intercalated) polymorphs are the important structures to evaluate topotactic intercalation and possible conversion voltages.

Similar to intercalation reactions starting with α and λ polymorphs, we can calculate the voltages for conversion reactions (red arrows in Figure 4.3) that lead to decomposition products, with different structures and compositions compared to either M_2O_4 or MgM_2O_4. The precise decomposition products that can form from a potential conversion reaction can be identified based on the ground state hull of the entire Mg–M–O phase space. For example, Mg reduction of M_2O_4 can lead to MgO and M_2O_3 during a conversion reaction, if they are the thermodynamically stable oxides that exist in the $Mg_xM_2O_4$ ($0 \le x \le 1$) region of the Mg–M–O ternary phase diagram.

Based on the calculated intercalation and conversion voltages for the lowest energy polymorphs, we can identify which reaction is thermodynamically favoured. Since we are considering scenarios of Mg reduction or discharge into a M_2O_4 cathode host, the reaction exhibiting the higher voltage *vs.* Mg will be favoured. For example, if $V_{\text{int}}^{(\text{discharged})} > V_{\text{conv}}^{(\text{discharged})}$, then the intercalation process is favoured and will likely occur. Similarly, if $V_{\text{conv}}^{(\text{charged})} > V_{\text{int}}^{(\text{charged})}$, the conversion reaction will likely occur. Notably, a given cathode composition (M_2O_4) can thermodynamically favour the intercalation process with a certain polymorph (say λ), while favouring the conversion reaction with another (α). Nevertheless, the framework displayed in Figure 4.3 is strictly thermodynamic ignoring any kinetic factors that may influence a given intercalation (or conversion) reaction.

4.1.1.4 *Solvent Co-intercalation*

Since poor Mg^{2+} mobility within bulk oxide frameworks is a major challenge in developing reversible Mg intercalation batteries, intercalated solvent molecules that can electrostatically "shield" the intercalating Mg^{2+} and increase the bulk Mg mobility may improve the performance of a cathode. Several experimental studies have demonstrated superior cycling performance with layered cathodes (*e.g.* xerogel–V_2O_5) that contain or that co-intercalate solvating water molecules alongside Mg^{2+}-ions. However, changes in the water (or an equivalent solvent) concentration within the cathode structure and the electrolyte can have a significant impact on the voltages measured, phase behaviour of the cathode, and the overall cycling performance. It is important to quantify the impact of solvent co-intercalation on measured voltages in cathode materials that can accommodate solvent molecules, which requires generalizing the voltage expressions of eqn (4.4) and (4.6).

Consider the process of Mg–H_2O co-intercalation within the xerogel–V_2O_5 cathode, which follows eqn (4.8).[22]

$$Mg_{x_1}V_2O_5{\cdot}n_1H_2O + (x_2 - x_1)Mg + (n_2 - n_1)H_2O \rightarrow Mg_{x_2}V_2O_5{\cdot}n_2H_2O \quad (4.8)$$

The voltage for the co-intercalation process can be written, similar to eqn (4.6), as

$$\Delta V = -\frac{\Phi_{Mg_{x_2}V_2O_5\cdot n_2H_2O} - \Phi_{Mg_{x_1}V_2O_5\cdot n_1H_2O} - (x_2 - x_1)G_{Mg}}{2(x_2 - x_1)F} \quad (4.9)$$

Φ in eqn (4.9) is the grand-potential of the cathode framework, which accounts for the chemical potential of the solvent molecule that undergoes intercalation. Φ for a given $Mg_xV_2O_5{\cdot}nH_2O$ composition is defined as,

$$\Phi = G_{Mg_xV_2O_5\cdot nH_2O} - n_{H_2O}\mu_{H_2O} \tag{4.10}$$

G, n_{H_2O}, and μ_{H_2O} in eqn (4.10) are the Gibbs energy of the cathode composition, the number of water (or solvent) molecules within the cathode framework, and the chemical potential of water, respectively. Thus, eqn (4.9) is identical to eqn (4.6) when $n_2 = n_1$. While G is obtained from DFT calculations ($G \approx E$), the electrolyte acts as the chemical reservoir for solvent molecules, *i.e.* sets μ_{H_2O}, similar to the anode being the reservoir for the intercalating Mg. If the activity of water (a_{H_2O}) within a given electrolyte is known *a priori*, then μ_{H_2O} can be obtained *via* eqn (4.11). For example, a_{H_2O} in a "wet" electrolyte (~1) should be several orders of magnitude higher than a "dry" electrolyte (~10^{-4}), signifying a_{H_2O} as a unique handle that spans various electrolytic conditions.

$$\mu_{H_2O} = \mu_{H_2O}^{O} + RT\ln a_{H_2O} \tag{4.11}$$

$\mu_{H_2O}^{O}$ in eqn (4.11) is the chemical potential of water in its standard state, which can be obtained by combining theory calculations and experimental data, as in eqn (4.12) and (4.13). For example, the enthalpy of water vapor (or ice) can be approximated from total energy DFT calculations ($H \approx E$) of an isolated vapor molecule (ice unit cell). Subsequently, the DFT data can be combined with the experimental enthalpy of vaporization (melting) and the entropy of liquid water to obtain an accurate $\mu_{H_2O}^{O}$ at a given temperature (T).

$$\mu_{H_2O}^{0} = E_{vapor}^{DFT} - \Delta H_{H_2O}^{vaporization}(\text{exp.}) - TS_{H_2O}^{liquid}(\text{exp.}) \tag{4.12}$$

$$\mu_{H_2O}^{0} = E_{Ice}^{DFT} + \Delta H_{H_2O}^{melting}(\text{exp.}) - TS_{H_2O}^{liquid}(\text{exp.}) \tag{4.13}$$

Combining eqn (4.9)–(4.13), one can estimate the impact of solvent co-intercalation on the average intercalating voltage within a co-intercalating cathode framework, at various Mg compositions and electrolytic conditions.

4.1.1.5 Stability Windows

An ideal electrolyte in a battery system has to remain electrochemically compatible with the anode and the cathode, *i.e.* it should not cause any performance-limiting decomposition products. However, being compatible with the electrodes requires the electrolyte to remain chemically inert, while conducting the redox species through it, over a large range of energies. For example, a Mg battery that exhibits an average voltage of ~2.5 V requires an electrolyte that can withstand a change in μ_{Mg} of ~5 eV across the electrodes. As demonstrated for Li-ion systems,[23] solid electrolytes often exhibit a lower range of energies over which they are stable compared to liquids. Thus, it is crucial to quantify (and subsequently predict) the electrochemical stability range of candidate (solid) electrolytes while designing a battery system.

Thermodynamically, a Mg-electrolyte can either be reduced by transfer of Mg atoms from the anode (*i.e.* electrochemical reduction) or to the cathode (electrochemical oxidation). In turn, the reduction and oxidation of the electrolyte will occur at well-defined voltages *vs.* Mg, nominally referred to as the reductive (or cathodic) and oxidative (or anodic) stability limits. Subsequently, the net voltage difference between the cathodic and anodic stability limits defines the electrochemical stability window (ESW) of an electrolyte. To a first order of approximation in liquid electrolytes, the stability window is given by the position of the lowest unoccupied molecular orbital (LUMO, relevant for reduction) and the highest occupied molecular orbital (HOMO, oxidation) levels, with respect to a Mg metal reference. More generally, the stability window depends on the range of Mg chemical potentials that the electrolyte is stable across.

To estimate the ESW of an electrolyte, we use eqn (4.14), where Φ, G, n, and μ are the grand-canonical potential, the Gibbs energy ($G \approx E$), number of Mg atoms within the electrolyte phase (c), and the Mg chemical potential, respectively.[23]

$$\Phi[c,\mu_{\mathrm{Mg}}] = G[c] - n_{\mathrm{Mg}}[c]\mu_{\mathrm{Mg}} \tag{4.14}$$

We can estimate the range of μ_{Mg} that the electrolyte is stable across, given the relevant 0 K phase diagram is calculated. For example, we can use eqn (4.14) to estimate the ESW of a solid Mg-ionic conductor, $MgSc_2Se_4$,[24] once we calculate the ternary Mg–Sc–Se phase diagram at 0 K. In case a specific electrode–electrolyte combination gives rise to complex interfacial reactions that may alter the stability limit of the electrolyte//electrode interface, eqn (4.14) has to be extended to include the entire electrode (e) and electrolyte (c) chemical space, as in eqn (4.15), where x is the degree of mixing between the electrode and electrolyte ($0 \leq x \leq 1$).

$$\Phi[c,e,\mu_{\mathrm{Mg}}] = \min(xG[c] + (1-x)G[e]) - n_{\mathrm{Mg}}[xc + (1-x)e]\mu_{\mathrm{Mg}} \tag{4.15}$$

where $\min(xG[c] + (1-x)G[e])$ and $n_{\mathrm{Mg}}[xc + (1-x)e]$ in eqn (4.15) refer to the lowest Gibbs energy within the electrode–electrolyte phase space, and the number of Mg atoms, respectively, at the mixing fraction, x. Thus, to estimate the oxidative stability of $MgSc_2Se_4$ conductor in contact with a $MgTi_2S_4$ cathode, we need to (i) calculate the entire Mg–Sc–Se–Ti–S phase-diagram with DFT at 0 K, (ii) consider possible interfacial products for different degrees of $MgSc_2Se_4$–$MgTi_2S_4$ mixing, and (iii) compute the oxidative stability for the resultant interfacial products, which will in turn determine the overall oxidative stability of the $MgSc_2Se_4$ electrolyte against $MgTi_2S_4$. Nevertheless, kinetic barriers involved in the oxidation or reduction processes can cause significant deviations from the thermodynamic estimates made using eqn (4.14) or (4.15).

4.1.2 Kinetics of Ionic Diffusion in Materials

In this section, we discuss the foundations of ion transport in solids and the derivation of the principal quantities that can be calculated from first-principles. We start by assuming that ionic motion in battery materials follows Fick's first law (Section 4.2.1) and arrive at the definition of the diffusion coefficient through the Green–Kubo relationship. We conclude by connecting the diffusion coefficient with the activation barrier for ionic migration, which is used in screening host materials for adequate ion transport.

4.1.2.1 *Fick's First Law and the Green–Kubo Model for Diffusion*

Under the assumption of steady state, Fick's first law of eqn (4.16) can be used to describe the flux (**J**) of ionic species in solids across a concentration gradient ($\vec{\nabla}C$).[25]

$$\mathbf{J} = -D_c \vec{\nabla} C \tag{4.16}$$

where D_c is the tensorial diffusion coefficient measuring the mobility of ions in the host structure, including any anisotropy in ionic motion. At thermodynamic equilibrium, D_c is defined by eqn (4.17).

$$D_c = \Theta D_J \tag{4.17}$$

Θ is the so-called thermodynamic factor, as in eqn (4.18).

$$\Theta = \left(\frac{\langle (\delta N)^2 \rangle}{\langle N \rangle} \right)^{-1} = \left(\frac{\partial \left(\frac{\mu}{kT} \right)}{\partial \ln x} \right) \tag{4.18}$$

where N is the number of diffusing species (*e.g.* Mg^{2+}), and μ is the chemical potential of the diffusing species at concentration x. Thus, eqn (4.18) indicates that the underlying driving force for ionic diffusion is not the concentration gradient as in eqn (4.16), but the gradient of μ. Under ideal conditions, μ is proportional to ln x, and diffusion is across a $\vec{\nabla}C$ (eqn (4.16)). Notably, $\langle (\delta N)^2 \rangle$ of eqn (4.18) is the fluctuation of the number of mobile species in an open system. For simplicity, we can assume that the system is closed to the exchange of species, and thus $\langle (\delta N)^2 \rangle$ is simply the fluctuation in a region of $\langle N \rangle$ species. Therefore, Θ becomes a function of the spatial configuration of the mobile ions at a specific concentration.

D_J of eqn (4.17) is the jump diffusion coefficient, which is,

$$D_J = \lim_{t\to\infty}\left[\frac{1}{2dt}\frac{1}{N}\left(\sum_{i=1}^{N}\vec{r}_i(t)\right)^2\right], \tag{4.19}$$

where d is the dimensionality of the host where ion i diffuses, while $\vec{r}_i(t)$ is its displacement after a period of time t. Therefore, as species i diffuses in the host material over time t, D_J measures the displacement of the centre of mass of the diffusing species. Altogether, eqn (4.17)–(4.19) form the Green–Kubo model of ionic diffusion. Theoretically, $\vec{r}_i(t)$ is estimated using either molecular dynamics (MD)[24] or kinetic Monte Carlo simulations.

Often ionic motion in solids is measured by a tracer species with a diffusion coefficient D^* (eqn (4.20)), which is different from D_J (eqn (4.19)).

$$D^* = \lim_{t\to\infty}\left[\frac{1}{2dt}\left(\frac{1}{N}\sum_{i=1}^{N}\left[\vec{r}_i(t)\right]^2\right)\right]. \tag{4.20}$$

D^* tracks the displacement of the individual tracer, as opposed to the centre of mass of the diffusing species in D_J. Importantly, D_J reduces to D^* if the displacements of the species are not correlated over time, *i.e.* diffusing atoms/ions move randomly.

4.1.2.2 *Diffusion Coefficients and Activation Barriers*

Eqn (4.21) defines the Arrhenius-type relationship between the diffusion coefficient (D_J or simply D, eqn (4.19)) and the activation or migration barrier, ΔE_a, in a crystalline material.

$$D = a^2 g\, f x_D \nu^* \exp\left(\frac{-\Delta E_a}{kT}\right). \tag{4.21}$$

where a is the hop distance of the migrating ion between two identical crystallographic sites, g is the geometric factor, f is a correlation factor and x_D is the concentration of diffusion carriers (*i.e.* vacancies or interstitials). The hopping frequency of eqn (4.21), ν^*, can be written in terms of the attempt frequency ν that is related to bond vibrations (or phonons) and the activation entropy (ΔS_a), which is the entropy difference between the stable and activated states. For simplicity, ν^* is often assumed to range between 10^{12} and 10^{13} Hz, while vibrational contributions dominate ΔS_a.

$$\nu^* = \nu \exp\left(\frac{\Delta S_a}{k}\right) \tag{4.22}$$

ΔE_a is defined as the energy difference between the initial stable state (E_i in Figure 4.4) and the activated state (E^*), as in eqn (4.23).

$$\Delta E_a = E^* - E_i \tag{4.23}$$

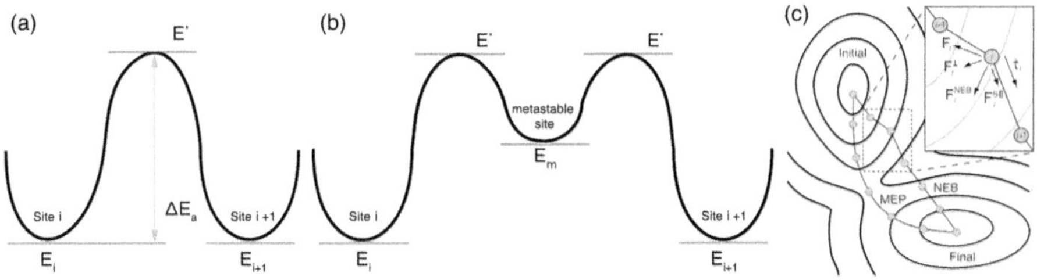

Figure 4.4 Minimum energy path (MEP) experienced by ions moving in a host material in two idealised scenarios (a) plateau-type and (b) valley-type. The initial site is identified by E_i and is separated by a ΔE_a barrier to its equivalent site E_{i+1} or to a metastable site with energy E_m. (c) The starting elastic band of images (grey circles), highlighted by the dashed black square, that converges to the MEP connecting two ground state geometries, *i.e.* initial and final, within the nudged elastic band (NEB) method. The inset indicates the forces that are relevant in NEB, namely, (i) the spring force $F^{S||}$ acting parallel to the NEB, and (ii) the force perpendicular to the NEB $F\perp$ set by the potential energy surface. F^{NEB} is the total force acting along the band. Reproduced from ref. 26 with permission from AIP Publishing, Copyright 2008.

Physically, ΔE_a represents a saddle-point in the energy landscape, *i.e.* it is the *lowest highest* energy state that migrating ions have to cross. The path connecting the stable sites *via* the saddle point is called the minimum energy path (MEP). Figure 4.4 illustrates two ideal scenarios of MEP (black lines) for ionic migration in solids, namely (a) plateau-like, whereby the diffusing ion crosses one activated state while migrating between two stable sites; (b) valley-type, where the ion traverses a metastable site as it migrates between stable sites. For example, a valley-type MEP is found for Li or Mg migration in a spinel-Mn_2O_4 structure,[1] while Li-MEP in a layered-NiO_2 structure is plateau-like.

4.1.2.3 *Estimating Migration Barriers*

ΔE_a in solids can be (directly or indirectly) measured from a number of experimental techniques, such as impedance spectroscopy, nuclear magnetic resonance (NMR), *etc.* Theoretically, ΔE_a can be estimated from utilizing saddle-point finding algorithms, such as the nudged elastic band (NEB) method,[1] or extrapolated from MD simulations over a range of temperatures.[24]

The NEB method[26–28] computes MEPs using an elastic band of geometric configurations, referred to as images (grey circles in Figure 4.4c), that approximate the MEP, *i.e.* each image represents a distinct configuration along the ionic migration. The starting band (dashed black square in Figure 4.4c) is eventually relaxed to the MEP through a force projection scheme, where the force due to the potential energy acts perpendicular ($F\perp$) to the band, and the spring force ($F^{S||}$) acts parallel to the band. $F^{S||}$ ensures (roughly) equal spacing among images, giving rise to the elastic nature of the band, and aids in obtaining a reliable MEP profile. Thus, the convergence of the starting band to the MEP is done by minimizing the net force, F^{NEB}. Since forces are straightforward to compute in DFT,[29–31] the NEB method has been extensively used in conjunction with DFT to compute the ΔE_a of ions in solids (see Section 4.4.6).[1,2,4,32,33]

Nevertheless, reliably predicting MEP is time consuming as each image in the band represents a distinct DFT calculation. Furthermore, a linear interpolation of images between initial and final states may not represent an ideal starting band which can decelerate NEB convergence. Rong *et al.*[34] have proposed two strategies to accelerate NEB calculations: (i) The *PathFinder* algorithm, based on the fact that ions in host structures migrate avoiding other atoms or bonds, *i.e.* avoiding regions with dramatic changes of electronic charge densities, and (ii) The *ApproxNEB*, which evaluates the energy of each image within the band using an inexpensive single-point DFT calculation. Thus, *PathFinder* (obtains approximate paths) and *ApproxNEB* (calculates approximate energies) can be used together to approximately predict ΔE_a. An alternative methodology to find approximate MEPs[35–37] is the bond valence site energy (BVSE) method, which can be thought of as an empirical force field that calculates the energies around various ionic sites. Nishitani *et al.*[37] applied the BVSE to Mg^{2+} migration in several hosts reproducing the barriers obtained using costly DFT-NEB calculations. To identify good Mg conductors from large datasets one can employ these inexpensive methods.[38]

4.1.2.4 Percolation Theory

The migration barriers extracted from structural models, as discussed in Sections 4.2.1 and 4.2.2, provide an atomic picture of ion dynamics. However, macroscopic diffusion of ions in materials relies on the existence of percolating networks of active migration channels, as displayed in Figure 4.5. Hence, in addition to recognizing facile microscopic hops from theory (*e.g.* DFT-NEB), it is important to identify whether a contiguous percolating network of low-barrier migration channels exist, which can be mathematically modelled and predicted by percolation theory. Understanding the existence of percolation networks is paramount to be able to quantify and engineer the extent of usable capacity in electrodes.

Theoretically, the solution to the site percolation problem[39–41] estimates the critical concentration $x = x_{crit}$ (also referred to as "percolation threshold") at which an infinite network of contiguous connected sites exists in a randomly occupied infinite lattice. For example, in Figure 4.5 a minimum of 5 yellow circles is required to form a percolating network (green line) that can span the 2D lattice of 45 blue circles, resulting in an $x_{crit} = 5/45 = 1/9$. The blue and yellow circles in Figure 4.5 can be considered to be the cathode framework and Mg atoms. Thus, two Mg atoms (yellow circles) are considered to be connected if there is an active Mg migration channel connecting the two sites. x_{crit} defines the critical concentration of Mg atoms

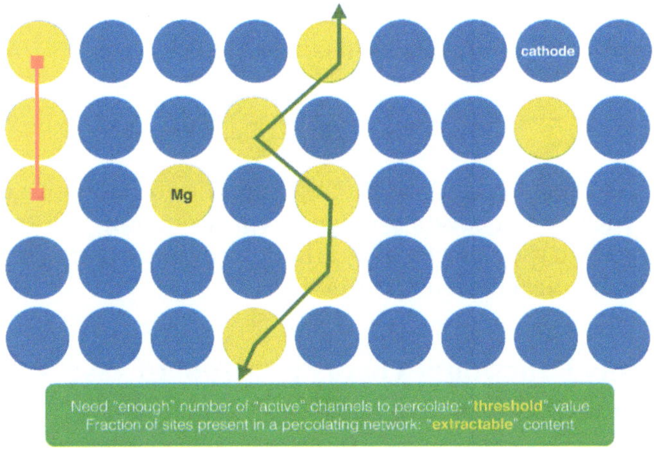

Figure 4.5 Schematic of a 2D lattice of blue circles with a percolating (green line) and a non-percolating (red line) network of contiguous connected yellow circles. The blue (yellow) circles can be considered to be cathode (Mg) atoms of an intercalation framework. The important quantities from percolation theory are listed in the green box at the bottom.

at which enough Mg sites are connected by active migration channels and macroscopic Mg transport is feasible. Thus, the probability that a percolating network exists within a lattice, $p(x)$, exhibits a step-function at x_{crit}, as in eqn (4.24).

$$p(x) = \begin{cases} 0 \forall x < x_{\text{crit}} \\ 1 \forall x \geq x_{\text{crit}} \end{cases} \tag{4.24}$$

Percolation thresholds can be analytically derived for 2D lattices.[39] However, in 3D lattices, numerical Monte Carlo simulations are used to assess x_{crit}.[4,41–43] Due to the use of finite supercells with periodic boundary conditions in Monte Carlo simulations (that result in "wrapping effects"), $p(x)$ becomes a sigmoidal functional form. However, Urban *et al.*[41] demonstrated that the inflection point of $p(x)$, defining x_{crit}, does not change significantly due to a change of functional form, and a well-converged x_{crit} can be obtained with computationally tractable 3D supercells. Thus, Monte Carlo simulations can accurately capture the percolation dynamics yielding reliable estimates of x_{crit} for practical cathodes.

Another important quantity accessible from Monte Carlo simulations is the fraction of sites, $F(x)$, participating in the percolating network. For example, although the concentration of yellow circles (x) in Figure 4.5 exceeds x_{crit}, not all yellow circles are connected to the percolating network (green line), with several isolated "clusters" that are connected (red line). Similarly, for $x > x_{\text{crit}}$ in a 3D cathode, each occupied Mg site is either part of a percolating network (and participates in macroscopic diffusion) or is only a part of a finite cluster in isolation. Subsequently, $F(x)$ can be defined as the fraction of the number of sites in the percolating cluster ($n_{\text{percolating}}(x)$) over the total number of sites in the 3D lattice ($n_{\text{sites}}(x)$), as in eqn (4.25).

$$F(x) = \frac{n_{\text{percoalting}}(x)}{n_{\text{sites}}} \approx \frac{1}{n_{\text{sites}} N_{\text{MC}}} \sum_{i}^{N_{\text{MC}}} n^{i}_{\text{percolating}}(x) \tag{4.25}$$

In practice, $F(x)$ is averaged over a large number of Monte Carlo sweeps (N_{MC}), where each sweep covers a range of concentrations (x). Importantly, $F(x)$ maps directly to the capacity of a cathode and can be compared immediately to electrochemical measurements.

4.1.3 Density Functional Theory as a Tool to Assess Thermodynamic and Kinetic Properties

While we introduced the thermodynamic and kinetic properties that govern battery materials in Sections 4.1 and 4.2, here we discuss the first-principles methodologies that provide reliable prediction of such properties. Our focus is limited to DFT,[12,13] which gives an accurate assessment of internal energies (total energies) that are used in the theoretical models presented in Sections 4.1 and 4.2.

In DFT, as in other quantum-mechanical methods, such as Hartree–Fock (HF), Møller–Plesset, coupled cluster, *etc.*,[44] the total energies are computed by numerically solving the time-independent Schrödinger equation ($H|\psi\rangle = E|\psi\rangle$), which only requires the atomic positions in space as input. The foundation of DFT relies firmly on the Hohenberg and Kohn theorem,[12] which proves that "the full many-particle ground state is a unique functional of $\rho(r)$" and that the only the true ground state $\rho(r)$ yields the lowest total energy, where $\rho(r)$ is the electronic charge density. Hence, the Schrödinger wavefunction ($|\psi\rangle$), which depends on the 3(N + M) spatial coordinates (N, M = number of electrons, nuclei), is recast in terms of $\rho(r)$ in DFT, which depends only on 3 spatial coordinates (x,y,z) and significantly reduces the computational cost.[31,44] To simplify the complex description of the charge density of materials with multiple correlated electrons, Kohn and Sham[13] introduced the concept of a non-interacting electron-gas, *i.e.* electrons do not interact with each other and instead individually interact with a mean-field $\rho(r)$. Thus, the many-electron wavefunction is mapped as a collection of Kohn–Sham (KS) one-electron orbitals. Within this approximation, the charge density of the non-interacting electron-gas is mapped to the ground state density ($\rho(r)$) of the actual interacting electron system, with the ground state energy (E) written as a functional of $\rho(r)$.

$$E[\rho(\vec{r})] = T_{\text{S}}[\rho] + J[\rho] + E_{\text{XC}}[\rho] + \int V_{\text{Ne}}\rho(\vec{r})\text{d}\vec{r}. \tag{4.26}$$

where $T_{\text{S}}[\rho]$ is the kinetic energy of the non-interacting electron gas, $J[\rho]$ is the classical Coulomb repulsion between electrons, and V_{Ne} is the electrostatic potential from the atomic nucleus. The exchange and correlation

(XC) functional, $E_{XC}[\rho]$, contains non-classical interactions, such as electronic exchange (which follows the Pauli exclusion principle), the difference in kinetic energy between an interacting and a non-interacting electron gas, and static and dynamic interactions of individual electrons that result in correlation effects. Thus, by approximating a many-body wavefunction as a linear combination of KS orbitals, and by solving eqn (4.26) self-consistently, one can obtain a good approximation for the ground state energy of most practical multi-electron systems.

The grand-challenge in DFT, as categorized by Jacobs' ladder,[45] is to accurately describe the electronic XC whose functional form is unknown. A number of XC functionals have been developed, including the linear density approximation (LDA), where E_{XC} depends on the local $\rho(r)$,[46] the generalized gradient approximation (GGA)[47] accounting for local $\rho(r)$ and $\vec{\nabla}\rho(r)$, and meta-GGA functionals that include higher-order gradients of ρ and/or kinetic energy densities. An example of recent meta-GGA functionals is the strongly constrained and appropriately normed (SCAN) functional,[48] satisfying the 17 known constraints for XC functionals.

*4.1.3.1 GGA+*U *and Hybrid Functionals*

The mean-field formulation of DFT does not penalize electrons enough from fictitiously interacting with themselves,[46] that results in the so-called self-interaction error (SIE), which affects total energy estimates. Particularly in oxides with open-shell 3d or 4f orbitals, which are relevant as intercalating cathodes (*e.g.* $LiCoO_2$ and $LiFePO_4$[49]) the SIE arising from insufficient localization of electrons in the 3d/4f orbitals can be sizable. A number of correction schemes have been proposed to reduce the SIE:

- **DFT+*U*–** The formalism by Anisimov *et al.*[50] lowers the SIE in 3d and 4f orbitals by adding an orbital-dependent on-site energy-penalty (*U*) for fractional occupation of electrons. DFT+*U* has been successfully applied to the describe the thermodynamics of intercalation in several cathodes.[4,33,49,51–53] Recently, it was reported that the SCAN XC functional also requires a *U* correction to accurately reproduce the ground state polymorph, electronic structure, and redox energetics of 3d and 4f oxides.[54]
- **Hybrid functionals–** SIE is reduced by adding a finite amount of the exact HF exchange in the XC functional. Two distinct classes of hybrid functionals exist:[55,56] (i) linear global hybrids, where a portion of the DFT exchange term is replaced by the HF exchange throughout the system, *e.g.* B3LYP[57–59] and PBE0;[60] and (ii) range-separated hybrids, where HF exchange is added within selective space domains, *e.g.* HSE06.[61,62]

The literature presents several examples where using the GGA+*U*[63,64] or HSE06[65] functional significantly improves the prediction of thermodynamic properties (*e.g.* formation energies and intercalation voltages) of oxides containing open-shell 3d and 4f transition metals compared to semi-local LDA and GGA XC functionals. In general, hybrid functionals are computationally more expensive than GGA+*U* calculations, especially when the KS orbitals are written in a plane-wave basis, since the HF exchange-integrals have to be explicitly computed. Therefore, GGA+*U* remains the strategy of choice for high-throughput DFT studies.[2,4,49,51,63,64]

Typically, GGA-based NEB (or molecular dynamics) calculations are used to estimate migration barriers (see Section 4.2.3),[1,2,4,66] which have been shown to agree satisfactorily with experimental measurements in Li-ion, and Na-ion solid electrolytes.[38] However, GGA may not accurately localize electrons within the 3d/4f orbitals within oxide cathodes. For example, Barnes *et al.*[67] recently reported that Mg^{2+} migration barriers in MnO_3 is significantly underestimated with GGA (~0.92 eV) compared to HSE06 (~1.53 eV). Although GGA+*U* can correct the SIE of GGA, the degree of electron localization is expected to vary between the images (or as the ion migrates) within a NEB calculation, which in turn will require different *U* values for different images. Hence DFT+*U* is clearly not suitable for NEB calculations. A plausible alternative is performing NEB calculations using expensive hybrid functionals, where more validation is required with experiments.

4.1.4 Application of First-principles Methods to Multivalent Ion Intercalation Hosts

Our attention turns to the application of the methodologies introduced in Sections 4.1–4.3 to the study of materials relevant for Mg batteries.

4.1.4.1 High-throughput Screening to Identify a Promising Intercalation Motif

The accuracy of DFT together with the availability of high-performance supercomputers, and materials informatics has fuelled a paradigm shift in materials design, thus allowing researchers to screen across material classes by systematic evaluation of properties. The combination of DFT, with pre/post-processing algorithms and databases is referred to as high-throughput screening (HTS), which has been applied[2] to identify candidate cathodes in a number of multivalent applications, including Al^{3+}, Ca^{2+}, Mg^{2+}, Y^{3+} and Zn^{2+}, as demonstrated in Figure 4.6.

Each point in Figure 4.6 represents the computed open circuit voltage of the intercalation MV ions using eqn (4.6), within M_2O_4 (M = 3d transition metal) cathodes that have the spinel structure (see Section 4.4.6). The voltages in Figure 4.6 were computed at the GGA+*U* level of theory (see Section 4.3.1). Expectedly, the computed voltages follow the electrochemical series across intercalant species (*i.e.* $Li^+/Li \sim -3.04$ V *vs.* SHE > Ca^{2+}/Ca −2.87 V > $Mg^{2+}/Mg \sim -2.37$ V > $Y^{3+}/Y \sim -2.37$ V > $Al^{3+}/Al \sim -1.67$ V > $Zn^{2+}/Zn \sim -0.76$ V), which provides further validation for calculating voltages using GGA+*U*. Notably, the multivalent intercalation voltages reported in Figure 4.6 are

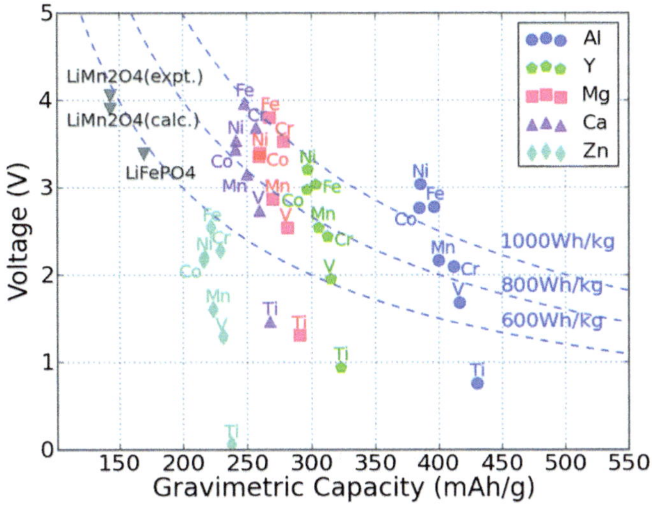

Figure 4.6 Predicted open circuit voltages *vs.* gravimetric capacities of multivalent ions Al, Ca, Mg, Y and Zn in M_2O_4 spinels (with M = Mn, Fe, Co, Ni, and Cr). Transition metals M are displayed next to each compound. Dashed blue lines are target energy densities, *i.e.* 600, 800, and 1000 W h kg^{-1}. Reproduced from ref. 2 with permission from the Royal Society of Chemistry.

always lower than 4 V (*vs.* the corresponding metal), with most Mg and Ca intercalation compounds providing voltages of between 2 and 4 V, which is lower than Li-intercalation within the same hosts. However, considering the additional charge carried by multivalent cations, a multivalent spinel cathode could exhibit a significantly higher energy density than the corresponding Li version.

Apart from computing intercalation voltages, it is important to verify the relative thermodynamic stability of the discharged and charged compositions, by calculating their E^{hull} (see Section 4.1.1), which is readily available on the "Battery Explorer" app[35,49] in the Materials Project website.[68] Also included in the app are theoretical (gravimetric and volumetric) capacities and energy densities. It has been estimated by Sun *et al.*[69] that structures with values of energy above the hull below ~70 meV/atom may be accessible *via* specific synthesis procedures even though they are metastable at 0 K.

4.1.4.2 *Voltage Curves as a Function of Temperature, the Case of TiS_2, CrO_2 and V_2O_5*

This section emphasizes the predictive power of DFT calculations when coupled with cluster expansion models and statistical thermodynamics. We present the simulated Mg intercalation voltage curves as a function of temperature for three prototypical cathode materials, including, TiS_2, CrO_2, and V_2O_5, which have been the subject of several experimental investigations.[70–74]

To date, Mg intercalation in spinel-TiS_2, against a Mg metal anode and an aluminium–chloride-based electrolyte has provided the highest energy density reversible Mg battery.[72] Figure 4.7 shows the computed phase diagram of Mg intercalation into various polymorphs of TiS_2 for Mg concentrations in the range of $0 < x < 0.5$,[66] where each point at a Mg composition represents a specific arrangement of Mg and vacancies, or Mg orderings, in TiS_2.

Experimentally, Mg_xTiS_2 exists in two phases, the layered-O1 (blue circles in Figure 4.7), and the spinel (orange). Other layered phases where Mg is coordinated by regular sulphur octahedra (O3) but following a different stacking sequence or prismatic structures (P3/P2) were also investigated in ref. 66. Figure 4.7 shows that Mg favourably intercalates in the TiS_2 framework up to a composition of $x_{Mg} = ½$. The layered-phase (blue), hitting the lowest formation energies, defines the ground state at Mg compositions

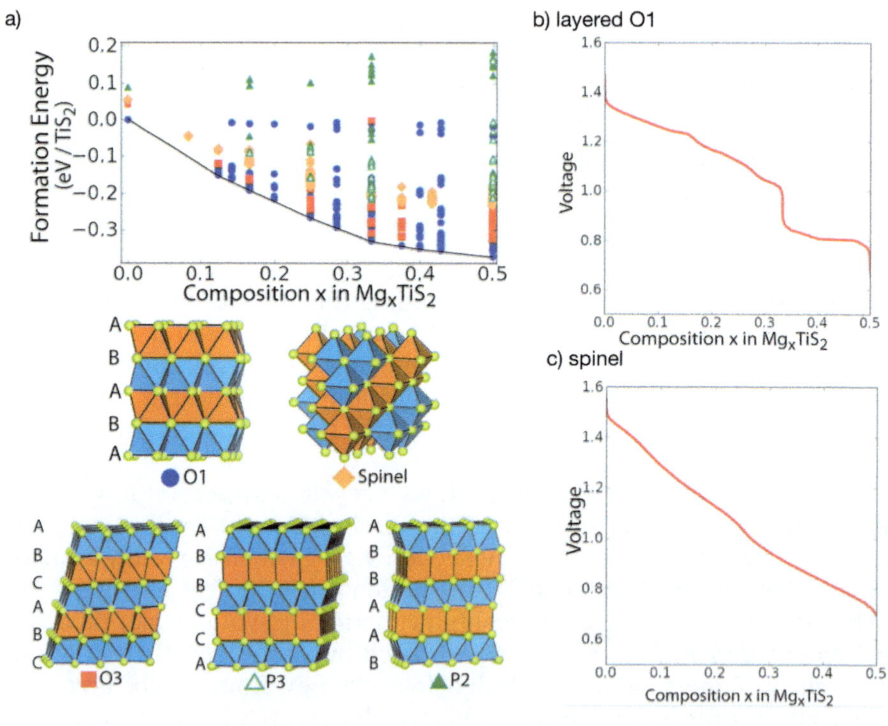

Figure 4.7 (a) Compositional phase diagram computed with DFT at 0 K of Mg intercalation in various phases of TiS_2.[66] The *x*-axis shows the Mg composition and *y*-axis the formation energy *vs.* empty TiS_2. The orange diamonds indicate the spinel-TiS_2. The red squares and blue circles represent the layered-TiS_2 with octahedrally coordinated Mg and two distinct Ti–S layer stacking. Similarly, empty and filled green triangles indicate layered-TiS_2 as well, where Mg is in a prismatic coordination. Computed voltage curves at 300 K for Mg intercalation in two TiS_2 phases, (b) layered O1 and (c) spinel. Reproduced from ref. 66 with permission from the American Chemical Society, Copyright 2015.

of $0 < x < 0.5$, and is followed in stability by the O3 phases. Experimental observations[70,72,73] verified that the spinel structure O3 is not ground-state at all. A certain range of Mg compositions and special care is required to intercalate Mg in this polymorph.[70,72,73] From a visual inspection of the O1 structures, several stable ordered phases can be identified, at $x_{Mg} = 1/6$, 1/3 and 1/2. The ordering at $x_{Mg} = 1/6$ displays every other Mg layer being completely empty while the remaining alternating layers are 1/3 occupied. In the case of $Mg_{1/3}TiS_2$, every Mg layer is 1/3 Mg occupied. At $x_{Mg} = 1/2$, Mg ions order into staggered arrangements of Mg rows (extending in each 2D plane) across different layers of TiS_2.

Using the energetics of the Mg orderings in TiS_2 (Figure 4.7a) the authors fitted a cluster expansion (see Section 4.1.2), which includes configurational contributions to the Gibbs energy of the Mg_xTiS_2 chemical space. Figure 4.7 shows the voltage profile at 300 K of the Mg_xTiS_2 system, calculated using Monte Carlo simulations based on the cluster expansion, in (b) the layered phase and (c) the spinel phase. The sloping nature of both voltage curves ($0 < x < 0.3$ in layered, and up to x ~0.5 in spinel) reflects the solid–solution character for Mg intercalation in the TiS_2 phases. In Figure 4.7b, three steps can be identified, corresponding to minima in panel a, which match the Mg orderings at $x_{Mg} = 1/6$, 1/3 and 1/2. In contrast, regions where the voltage curve is smooth reflect the solid–solution behaviour and correspond to regimes of disorder between Mg and vacancies. The spinel phase of Figure 4.7c shows a continuous solid–solution behaviour across Mg compositions, which the authors[66] attributed to a larger degree of electrostatic screening in the spinel compared to the O1 phase.

$Mg_xCr_2O_4$ is one of the highest voltage Mg intercalation cathodes as identified from the high-throughput computational screening of Liu *et al.* (Figure 4.6).[2] To understand the phase behaviour of Cr_2O_4 during Mg intercalation, Chen *et al.*[52] fitted a cluster expansion (see Section 4.1.2) of the Mg vacancy configurational space within $Mg_xCr_2O_4$. Subsequently, the authors performed Monte Carlo simulations to quantify changes in the voltage profile during Mg (de-)intercalation, as a function of temperature (Figure 4.8a). A fair agreement between the voltage profile computed by the cluster expansion at 0 K (yellow line of Figure 4.8a) to the DFT version (green line) was found. At room temperature (black line in Figure 4.8a) a fairly smeared voltage profile is obtained with distinct voltage jumps at $x_{Mg} = 0.5$ and 0.33. Thus, the Mg vacancy ground states forming at $x_{Mg} = 0.5$ and 0.33 are energetically stable compared to configurations at other Mg compositions and will exist during (de-)intercalation even under room temperature conditions. The authors demonstrated that ground states at $x_{Mg} = 0.5$ and 0.33 will curb Mg mobility during Mg (de-)intercalation, attributed to their corresponding stability. At $x_{Mg} < 0.25$ the steep voltage increase is due to the high instability of the empty spinel-Cr_2O_4 structure, which may potentially lead to decomposition of the spinel structure to a stable Cr_2O_4 polymorph.

In the case of V_2O_5, Sai Gautam *et al.*[53] identified δ-V_2O_5 as a promising Mg cathode owing to a combination of a reasonable average voltage (~2.5 V) and

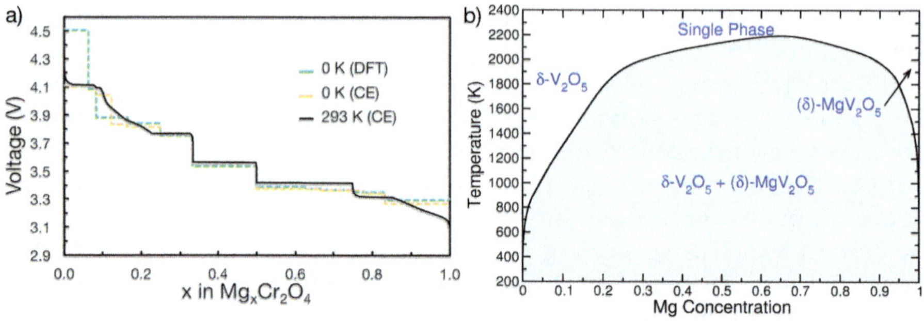

Figure 4.8 (a) 0 and 298 K voltage curves for Mg intercalation in spinel-Cr_2O_4. Reproduced from ref. 52 with permission from the American Chemical Society, Copyright 2018. (b) Temperature-composition phase diagram for Mg intercalation in δ-V_2O_5. Reproduced from ref. 53 with permission from the American Chemical Society, Copyright 2015.

migration barrier (~600–760 meV). Cluster expansion-based Monte Carlo simulations were performed to derive the temperature-composition phase-diagram (Figure 4.8b) of Mg intercalation, which indicated that δ-$Mg_xV_2O_5$ was a phase separating system, with the concomitant formation of Mg-rich and Mg-poor domains. The two-phase behaviour at intermediate Mg compositions ($0 < x < 1$) should exist up to high temperatures (melting point of V_2O_5 ~954 K). Given that the ground state of V_2O_5 is the α polymorph (α is more stable than δ by ~100 meV/f.u.) and the α and δ polymorphs are only separated by a shear transition of alternate V_2O_5 layers, it is crucial for the de-intercalated δ-V_2O_5 to remain metastable with Mg (de-)intercalation if δ-V_2O_5 is to be successfully used as a practical Mg cathode.

4.1.4.3 *Conversion* vs. *Intercalation During Mg Reduction*

Figure 4.9 plots the intercalation *vs.* conversion reaction voltages for a one electron reduction that involves the reaction of one mole of Mg per mole of M_2X_4, where M and X are a 3d transition metal and a chalcogen (*i.e.* O, S, Se), respectively.[21] The polymorph under consideration can exhibit the lowest energy at the intercalated/discharged (left panel in Figure 4.9) or empty/charged (right panel) state (see Section 4.1.3). Blue squares in Figure 4.9 indicate that intercalation is preferred, whereas red squares indicate a conversion preference. From Figure 4.9, it is clear that most 3d sulphides and selenides favour conversion reactions upon Mg reduction. For 3d oxides, when considering the lowest energy discharged polymorph, there are a few compounds that resist conversion reactions, such as oxides of V, Cr, Mn, Fe and Co that favour intercalation over conversion upon Mg reduction. Interestingly, Cr_2O_4 displays the highest level of intercalation preference (or resistance to conversion) among the 3d oxides. However, even among oxides, conversion reactions are favoured during Mg reduction if the lowest energy

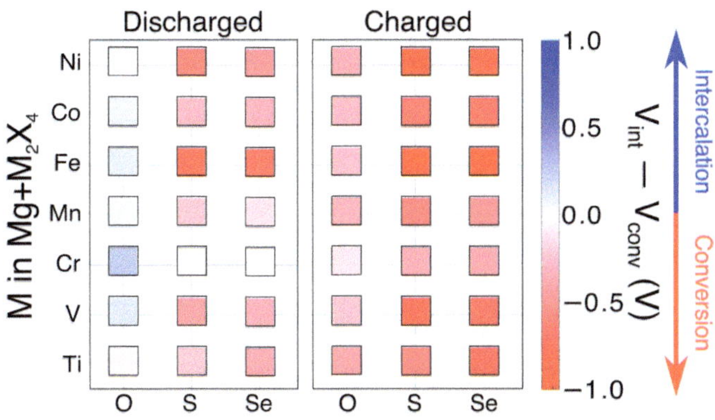

Figure 4.9 Difference between the intercalation and conversion voltages for Mg reduction in M_2X_4 cathodes, where M = 3d transition metal and X = O, S, or Se. Left and right panels correspond to reduction reactions using the lowest energy of the discharged and charged polymorphs. Adapted from ref. 21 with permission from John Wiley and Sons, © 2018 WILEY-VCH Verlag GmbH & Co. KGaA, Weinheim.

charged polymorph is considered (right panel). Hence, this analysis highlights three important trends: (i) oxides resist conversion better than sulphides and selenides, (ii) discharged polymorphs favour intercalation, (iii) Cr resists conversion to the largest degree among the 3d metals.

4.1.4.4 *Co-intercalation in Xerogel-V_2O_5*

Ref. 22 reports the impact of solvent co-intercalation under various electrolyte conditions for Mg discharge within xerogel-V_2O_5, which is plotted in Figure 4.10. Using a three-step computational strategy to resolve the xerogel-V_2O_5 structure, Sai Gautam *et al.*[22] calculated Mg intercalation voltages using the grand-potential (of Section 4.1.4) at various Mg concentrations and electrolyte conditions (*i.e.* a_{H_2O}). The authors reported that at low Mg concentrations ($x_{Mg} \leq 0.25$, red line in Figure 4.10), water co-intercalation is thermodynamically favoured in both wet (a_{H_2O}~1) and dry (a_{H_2O}~10^{-4}) electrolytes, as indicated by the linear dependence of voltages on a_{H_2O}. Water co-intercalation occurs only under wet conditions at higher Mg concentrations ($0.25 \leq x_{Mg} \leq 0.5$, blue line in Figure 4.10). Interestingly, a "superdry" electrolyte (a_{H_2O}~10^{-8}) suppresses water co-intercalation (no dependence of voltage on a_{H_2O}) and removes any existing water from the xerogel structure (see equation, $0 \leftrightarrow 0$). Superdry electrolytes also alter the phase behaviour of the xerogel, signified by the merging of the low and high Mg voltage curves in Figure 4.10. Therefore, water co-intercalation and the dependence of the intercalation voltage on electrolyte conditions have important consequences in the choice of electrode/electrolyte pairs, where theoretical studies, such as in ref. 22, can provide important insights for experiments.

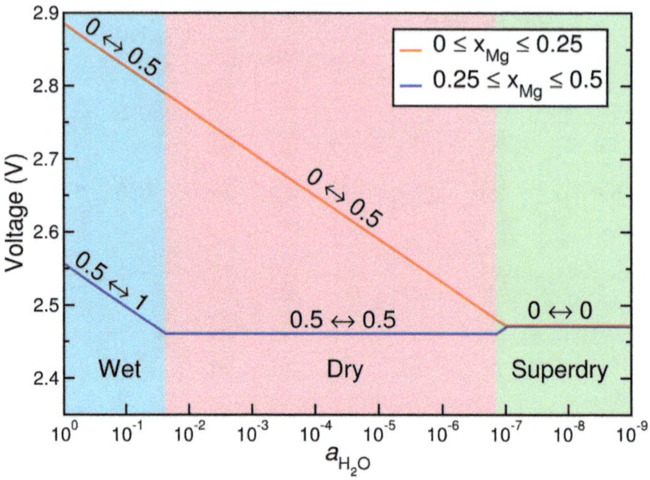

Figure 4.10 Average Mg intercalation voltages *vs.* Mg^{2+}/Mg at high (blue) and low (red) Mg concentrations as a function of water activity within the electrolyte. Equations along the voltage lines indicate the change in water content per V_2O_5 formula unit. Reproduced from ref. 22 with permission from the American Chemical Society, Copyright 2016.

4.1.4.5 Electrochemical Stability Windows of Coating Materials

A number of reports[6,7,75] have clearly demonstrated that Mg electrolytes display a limited ESW (~1.5 V–3.0 V *vs.* Mg) compared to Li-electrolytes (~1.5 V–5 V *vs.* Li).[76] Thus, using protective anodic/cathodic coatings is one strategy to mitigate the poor ESW of Mg electrolytes. Chen *et al.*[77] predicted the EWS of several Mg-containing compounds, which may form as a result of electrolyte decomposition at either the Mg metal anode or high-voltage cathodes. The authors presented an exhaustive study of binary, ternary, and quaternary chemical spaces that contain non-redox-active metals (or cations) and anions.

Figure 4.11 depicts ESWs (using the framework in Section 4.1.5) of Mg ternary and quaternary oxides. Left (right) end of each bar indicates the reductive (oxidative) stability, with the number near each bar signifying the ESW. The voltage scale is referenced to Mg^{2+}/Mg, *i.e.* 0 V is Mg metal while ~3.5 V is $Mg_xCr_2O_4$ (see Section 4.4.2). Interestingly, none of the Mg ternary (or quaternary) oxides are stable against Mg metal, as indicated by the lack of reductive stability of any compound up to 0 V, with $Mg(BH_4)_2$ (not shown in Figure 4.11) exhibiting the best reductive stability (0.01 V *vs.* Mg) among ternary Mg-compounds. Ternary oxides do not supersede the anodic stability of MgF_2 (~5.8 V,[77] not shown in Figure 4.11), with MgP_4O_{11} (~4.55 V), MgS_2O_7 (~4.45 V), and quaternary $Mg_{0.5}Ti_2(PO_4)_3$ (~3.82 V) displaying high enough oxidative stabilities (>3.5 V) that could be compatible with high voltage cathodes. As noted in ref. 77, Mg mobility in these compounds needs to be assessed (see the examples in Section 4.4.6) before practical coating strategies can be implemented.

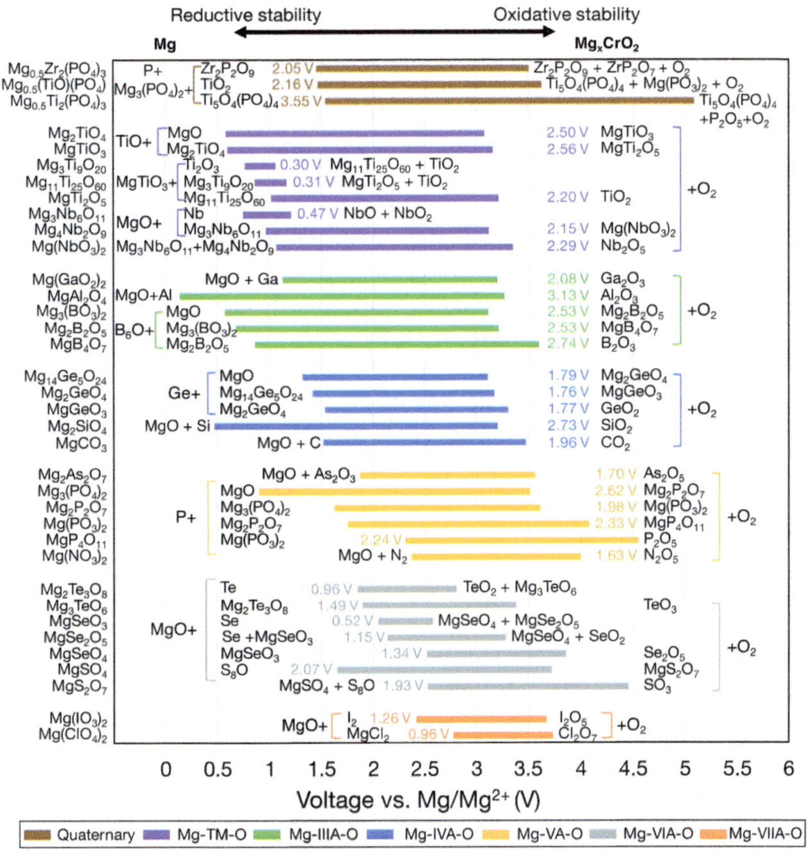

Figure 4.11 ESW of Mg-containing ternary and quaternary oxides as indicated next to each bar. The voltages (*vs.* Mg metal) across which the compound is stable is indicated by the width of each bar. Adapted from ref. 77, https://doi.org/10.3389/fchem.2019.00024, under the terms of the CC BY 4.0 licence, https://creativecommons.org/licenses/by/4.0/.

4.1.4.6 Assessing Mg Migration in a Spinel Structure

Here, we will examine the prediction of Mg migration barriers in a number of hosts exhibiting the spinel structure (Figure 4.12) using the methods in Section 4.2. Before delving into Mg migration, it is important to define an upper-limit of the migration barrier that electrodes can tolerate. As demonstrated previously,[1,2,5] by assuming reasonable battery performance, *e.g.* a 2 h (dis)charge time t for a particle of active materials of a size of 1 μm (diffusion length), one arrives to a minimum required Mg diffusivity, $D \sim 10^{-12}$ cm^2 s^{-1}, since the diffusion length scales as $\sqrt{Dt}$. If it assumed that a random-walk model for ion diffusion holds, $\nu \approx 10^{12}$ s^{-1} and $a \approx 3$ Å, using eqn (4.21), one defines a maximum ΔE_a ~525 meV. In nanoparticle cathodes operating at a higher temperature (~60 °C), the maximum ΔE_a can go up to ~750 meV.[5]

The accuracy of computed barriers is typically ± 50 meV, which represents ~1 order of magnitude in diffusivity.

In spinel-MgM_2X_4, the cation Mg and M order in a face-centred cubic (FCC) packing of anions X, as in Figure 4.12a. In "normal" spinels, ½ of the octahedral (oct) 16d sites are M atoms (*e.g.* Ti or Mn, blue polyhedra), while ⅛ of the tetrahedral (tet) 8a sites are Mg ions (orange polyhedra). Depending on the ionic radii of Mg^{2+} and $M^{2+/3+/4+}$, the spinel structure can exhibit "inversion", where a fraction of the 8a (16d) sites are occupied by M (Mg) atoms. An extensive account of the spinel structure is given in ref. 78.

In screening for high-voltage oxide spinel cathodes, Liu *et al.*[2] calculated the Mg migration barriers in $Mg_xM_2O_4$ (M = Mn, Co, Ni or Cr) with DFT-NEB calculations (see Section 4.2.3). The MEP in $Mg_xM_2O_4$ follows a tet(8a) → oct(16c) → tet(8a) topology (top left in Figure 4.13), with the barrier often determined by the size of the triangular face shared by the 8a and 16c sites (bottom left in Figure 4.13). The authors demonstrated that the specific choice of M does not affect the energy landscape during migration with the calculated barriers being (charged–discharged limits) 776–486 meV for $Mg_xMn_2O_4$, 698–520 meV for $Mg_xCo_2O_4$, 669–485 meV for $Mg_xNi_2O_4$, and 616–636 meV for $Mg_xCr_2O_4$. These barriers are larger in magnitude than for spinel-TiS_2 (~550 meV, Figure 4.12b), which can be attributed to the strong ionic bonding and electrostatic interactions between Mg^{2+} and O^{2-}.

Apart from calculating voltages, the authors in ref. 66 predicted the Mg migration barriers into spinel-TiS_2, which were later shown to reversibly intercalate Mg experimentally by Sun *et al.*[72] In spinel-TiS_2, experiments[72] have shown that a significant concentration of Mg (x ~0.6 per f.u.) resides in the usually-vacant oct sites, 16c (dashed square in Figure 4.12a). Therefore,

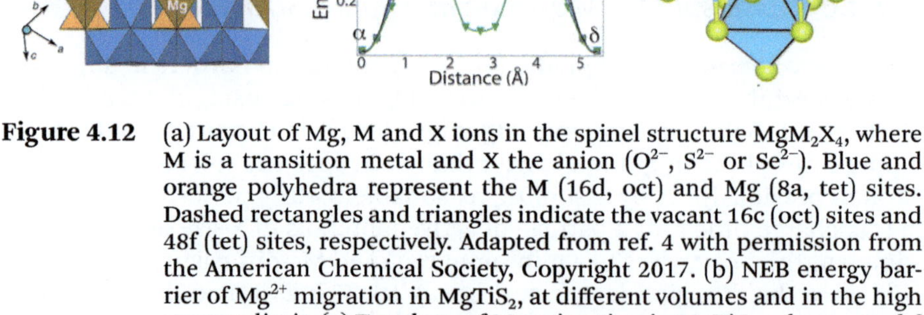

Figure 4.12 (a) Layout of Mg, M and X ions in the spinel structure MgM_2X_4, where M is a transition metal and X the anion (O^{2-}, S^{2-} or Se^{2-}). Blue and orange polyhedra represent the M (16d, oct) and Mg (8a, tet) sites. Dashed rectangles and triangles indicate the vacant 16c (oct) sites and 48f (tet) sites, respectively. Adapted from ref. 4 with permission from the American Chemical Society, Copyright 2017. (b) NEB energy barrier of Mg^{2+} migration in $MgTiS_2$, at different volumes and in the high vacancy limit. (c) Topology of Mg migration in $MgTiS_2$, where α and δ are the 16c sites, and γ is the 8a site. Adapted from ref. 66 with permission from the American Chemical Society, Copyright 2015.

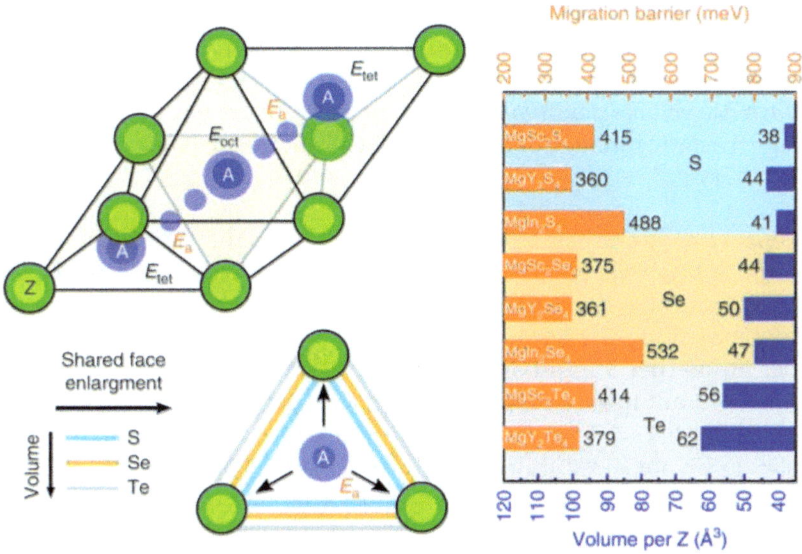

Figure 4.13 Right panel, calculated Mg migration barriers (orange bars) and the volume per anion (blue) in chalcogenides MgM_2Z_4 spinels. Top-left panel, typical tet → oct → tet migration path in spinels, with the energy of corresponding sites indicated by E_{tet}, and E_{oct}. E_a is the migration barrier. Bottom-left panel, representation of the effect of anion size on the triangular face shared between the 8a and 16c sites. Adapted from ref. 23, https://doi.org/10.1038/s41467-017-01772-1, under the terms of the CC BY 4.0 licence, https://creativecommons.org/licenses/by/4.0/.

the authors[66] assumed that the Mg^{2+} ions in spinel-TiS_2 migrate between 16c sites (α and δ in Figure 4.12c) *via* a local energy minima, the tet 8a site, where β in Figure 4.12c indicates the triangular face sharing site. The calculated barriers have been computed in the high vacancy limit – only one 16c amongst the 32 available in the lattice model is occupied by Mg. The calculated migration barriers decrease significantly from ~0.85 to ~0.55 eV as the volume increases by 10% from equilibrium (Figure 4.12b). These barriers are overestimated compared to values from galvanostatic intermittent titration experiments (~550 meV) and theory (500–600 meV) by Sun *et al.*,[72] which could be due to the specific XC functional used in the NEB calculations.

Based on Mg migration barriers of a variety of cathode hosts, three practical design rules were developed by Rong *et al.* to identify good Mg conductors:[1,24]

1. *Avoid materials with preferred Mg coordination.* A statistical analysis of the Inorganic Crystal Structural Database demonstrates that Mg highly prefers an octahedral coordination in oxides.[1,2,79] Hence, materials where Mg does not occupy octahedral sites should be preferred.[1,24]

2. *Limiting changes to ionic coordination.*[1] Smaller changes in coordination environment along the migration path flattens the energy landscape, and should reduce barriers.[1,80]
3. *Increase volume per anion.* Larger anion volumes correlate with weaker electrostatic interactions of Mg with the anion and consequently reduce barriers.[24,32]

Canepa *et al.*[24] applied these rules to identify the first class of room-temperature Mg ionic conductors, which are ternary spinel-MgM_2X_4, where M = In, Y or Sc and X = S or Se. Figure 4.13 depicts the computed barriers (orange bars) along the 8a → 16c → 8a (tet → oct → tet) pathway and volume per anion (per S^{2-} and Se^{2-}, blue bars) in these materials. The barriers in Figure 4.13 are lower compared to those of cathode sulphur spinels (*e.g.* ~550 meV in $MgTi_2S_4$), with MgY_2S_4 (~360 meV), MgY_2Se_4 (~361 meV), and $MgSc_2Se_4$ (~375 meV) exhibiting the lowest barriers among the spinels considered. For $MgSc_2Se_4$,[24] impedance and variable-temperature ^{25}Mg NMR measurements were used to validate the theoretical predictions. However, impedance experiments measured non-negligible electronic conductivity, detrimental to the utilization of $MgSc_2Se_4$ as a Mg solid electrolyte. Subsequent theory investigations revealed that the chalcogenide spinels are prone to *n*-type conductivity, especially when synthesized under anion-poor environments (or high temperatures).[81]

4.1.4.7 Probing Long-range Mg Transport with Percolation Theory

In Section 4.2.4 we introduced the percolation theory, which was used to predict conditions under which long-range Mg transport occurs in spinel-$Mg_xMn_2O_4$, a potential cathode for Mg batteries[2,71] but prone to inversion (see Section 4.4.6). A number of possible Mg migration pathways can be envisioned in inverted $Mg_xMn_2O_4$, and a complete description of these is given in ref. 4. A barrier of 750 meV was used as the defining criterion to classify active and inactive migration pathways, *i.e.* pathways that have barriers below (above) 750 meV facilitate or open (block or close) Mg migration. Within the percolation theory, sites in the spinel that are connected by open pathways (<750 meV barrier) are considered connected and can eventually form a contiguous percolating network.

Figure 4.14a displays the percolation threshold (x_{crit}, black lines), at various degrees of inversion (i) in $Mn_{3-x}O_4$ and various Mg vacancy concentrations (z). The x-axis starts at a Mn_3O_4 (*i.e.* 50% Mn-excess or 100% Mg-deficient) composition and covers Mg concentrations up to $Mg_{1.5}Mn_{1.5}O_4$ (*i.e.* 25% Mn-deficient, 50% Mg-excess). The dashed yellow line identifies the stoichiometric spinel (*i.e.* Mn : O = 2 : 4). The blue (red) coloured region signifies Mg concentrations which do (do not) facilitate percolation. In general, percolation thresholds in the Mn-excess domain (*i.e.* $x_{crit} < 1$) are desirable since they signify the presence of percolating networks in the stoichiometric spinel.

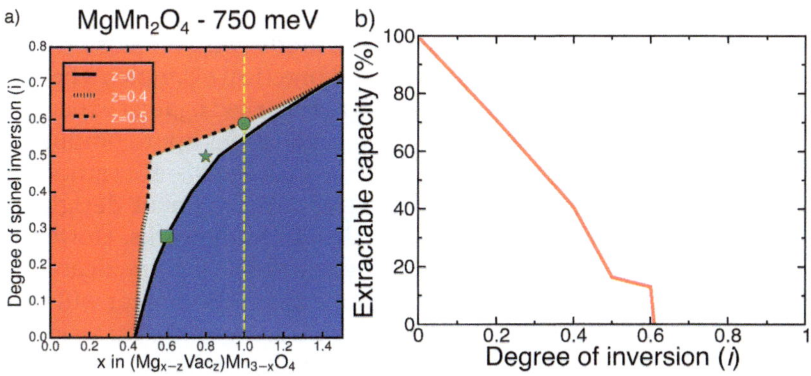

Figure 4.14 (a) Critical concentration (x_{crit}) of Mg percolation in ($Mg_{x-z}Vac_z$)$Mn_{3-x}O_4$. The stoichiometric concentration of ($Mg_{x-z}Vac_z$)$Mn_{3-x}O_4$ (Mn : O = 2 : 4) is illustrated by the yellow line. The zero on the *x*-axis indicates the Mn_3O_4 composition. *z* indicates the vacancy (Vac) concentration in ($Mg_{x-z}Vac_z$)$Mn_{3-x}O_4$. (b) The extractable Mg content (or capacity) for a stoichiometric Mn-oxide spinel as a function of the degree of inversion. Adapted from ref. 4 with permission from the American Chemical Society, Copyright 2017.

At $i = 0$, the stoichiometric spinel (yellow line) allows Mg percolation since the threshold ($x_{crit} \approx 0.44$) is in the Mn-excess domain. The stoichiometric discharged spinel ($z = 0$) can tolerate i ~0.55 before percolation stops, while a charged stoichiometric spinel ($z = 0.5$) enables Mg transport up to i ~0.59. Also, long-range Mg diffusion can be achieved at higher degrees of inversion ($0.59 < i < 0.77$) only under Mn-deficient concentrations (*i.e.* $x > 1$). Furthermore, the extractable Mg content (or the measured electrochemical capacity) from the Mn-oxide spinel is plotted as a function of i in Figure 4.14b for a stoichiometric spinel. Note that the extractable capacity decreases continuously with i, signifying the increasing number of Mg atoms that form isolated clusters within the spinel and hence do not participate in long-range diffusion. Finally, the extractable capacity reaches zero at i ~0.6, which corresponds to the highest degree of inversion that a stoichiometric spinel can tolerate before Mg percolation stops (Figure 4.14a). Thus, percolation theory can act as a robust theoretical framework to translate calculated microscopic migration barriers into long-range macroscopic transport properties.

4.2 Conclusions

The development of high energy density Mg batteries has been constrained by the lack of high voltage cathodes with good enough Mg mobility and reliable electrolytes that are stable against Mg metal and a high voltage cathode. Materials discovery, specifically using theory and computations to screen a variety of chemical spaces is required to make intercalation Mg batteries a reality. Here, we have introduced concepts of thermodynamics and kinetics that are relevant for battery materials, briefly overviewed the levels of approximations used in the electronic structure methods (with a focus on DFT) to assess material properties and demonstrated practical examples where accurate properties have been calculated and benchmarked against experiments. We have also discussed examples where a high-throughput computational infrastructure has been used to screen for conversion-resistant cathodes, stable coating materials (and Mg ionic conductors) and formulate design principles that govern Mg mobility within typical solid frameworks. However, the hunt for reliable cathodes, coatings, and electrolytes for practical Mg and multivalent batteries is still ongoing. We hope that this chapter will provide sufficient guidance to the entire scientific community on the materials properties that can be accurately predicted theoretically, which should help in the screening and/or understanding of novel materials for high energy density Mg batteries.

Acknowledgement

P. C. acknowledges support from the Singapore Ministry of Education Academic Fund Tier 1 (R-284-000-186-133).

References

1. Z. Rong, R. Malik, P. Canepa, G. Sai Gautam, M. Liu, A. Jain, K. Persson and G. Ceder, *Chem. Mater.*, 2015, **27**, 6016.
2. M. Liu, Z. Rong, R. Malik, P. Canepa, A. Jain, G. Ceder and K. A. Persson, *Energy Environ. Sci.*, 2015, **8**, 964.
3. D. Aurbach, Z. Lu, A. Schechter, Y. Gofer, H. Gizbar, R. Turgeman, Y. Cohen, M. Moshkovich and E. Levi, *Nature*, 2000, **407**, 724.
4. G. Sai Gautam, P. Canepa, A. Urban, S.-H. Bo and G. Ceder, *Chem. Mater.*, 2017, **29**, 7918.
5. P. Canepa, G. Sai Gautam, D. C. Hannah, R. Malik, M. Liu, K. G. Gallagher, K. A. Persson and G. Ceder, *Chem. Rev.*, 2017, **117**, 4287.
6. H. D. Yoo, I. Shterenberg, Y. Gofer, G. Gershinsky, N. Pour and D. Aurbach, *Energy Environ. Sci.*, 2013, **6**, 2265.
7. J. Muldoon, C. B. Bucur and T. Gregory, *Chem. Rev.*, 2014, **114**, 11683.
8. P. Canepa, S. Jayaraman, L. Cheng, N. N. Rajput, W. D. Richards, G. S. Gautam, L. A. Curtiss, K. A. Persson and G. Ceder, *Energy Environ. Sci.*, 2015, **8**, 3718.
9. A. Ponrouch, C. Frontera, F. Bardé and M. R. Palacín, *Nat. Mater.*, 2016, **15**, 169.
10. P. Canepa, G. S. Gautam, R. Malik, S. Jayaraman, Z. Rong, K. R. Zavadil, K. Persson and G. Ceder, *Chem. Mater.*, 2015, **27**, 3317.
11. M. K. Aydinol, A. F. Kohan and G. Ceder, *J. Power Sources*, 1997, **68**, 664.
12. P. Hohenberg and W. Kohn, *Phys. Rev.*, 1964, **136**, B864.
13. W. Kohn and L. J. Sham, *Phys. Rev.*, 1965, **140**, A1133.

14. N. W. Ashcroft and N. D. Mermin, *Solid State Physics*, Cengage Learning, Inc, CA, United States, 1st edn, 1976.
15. C. Kittel, *Introduction to Solid State Physics*, John Wiley & Sons Inc, New York, United States, 8th edn, 2004.
16. D. A. McQuarrie and J. D. Simon, *Physical Chemistry: A Molecular Approach*, University Science Books, Sausalito, United States, 1st edn, 2011.
17. A. van de Walle and G. Ceder, *Rev. Mod. Phys.*, 2002, **74**, 11.
18. J. M. Sanchez, F. Ducastelle and D. Gratias, *Phys. A*, 1984, **128**, 334.
19. G. Ceder, *Comput. Mater. Sci.*, 1993, **1**, 144.
20. S. K. Kolli and A. Van der Ven, *Chem. Mater.*, 2018, **30**, 2436.
21. D. C. Hannah, G. Sai Gautam, P. Canepa and G. Ceder, *Adv. Energy Mater.*, 2018, **8**, 1800379.
22. G. Sai Gautam, P. Canepa, W. D. Richards, R. Malik and G. Ceder, *Nano Lett.*, 2016, **16**, 2426.
23. W. D. Richards, L. J. Miara, Y. Wang, J. C. Kim and G. Ceder, *Chem. Mater.*, 2016, **28**, 266.
24. P. Canepa, S.-H. Bo, G. Sai Gautam, B. Key, W. D. Richards, T. Shi, Y. Tian, Y. Wang, J. Li and G. Ceder, *Nat. Commun.*, 2017, **8**, 1759.
25. A. Van der Ven, G. Ceder, M. Asta and P. D. Tepesch, *Phys. Rev. B*, 2001, **64**, 184307.
26. D. Sheppard, R. Terrell and G. Henkelman, *J. Chem. Phys.*, 2008, **128**, 134106.
27. G. Henkelman, B. P. Uberuaga and H. Jónsson, *J. Chem. Phys.*, 2000, **113**, 9901–9904.
28. G. Henkelman and H. Jónsson, *J. Chem. Phys.*, 2000, **113**, 9978.
29. *Hans Hellmann: Einführung in die Quantenchemie*, ed. D. Andrae, Springer Berlin Heidelberg, Berlin, Heidelberg, 2015.
30. R. P. Feynman, *Phys. Rev.*, 1939, **56**, 340.
31. R. M. Martin, *Electronic Structure*, Cambridge University Press, Cambridge, 2004.
32. Y. Wang, W. D. Richards, S. P. Ong, L. J. Miara, J. C. Kim, Y. Mo and G. Ceder, *Nat. Mater.*, 2015, **14**, 1026.
33. S. P. Ong, V. L. Chevrier, G. Hautier, A. Jain, C. Moore, S. Kim, X. Ma and G. Ceder, *Energy Environ. Sci.*, 2011, **4**, 3680.
34. Z. Rong, D. Kitchaev, P. Canepa, W. Huang and G. Ceder, *J. Chem. Phys.*, 2016, **145**, 074112.
35. S. Adams and R. P. Rao, *Phys. Status Solidi*, 2011, **208**, 1746.
36. S. Adams and R. P. Rao, *Phys. Chem. Chem. Phys.*, 2009, **11**, 3210.
37. Y. Nishitani, S. Adams, K. Ichikawa and T. Tsujita, *Solid State Ionics*, 2018, **315**, 111.
38. F. Meutzner, T. Nestler, M. Zschornak, P. Canepa, G. S. Gautam, S. Leoni, S. Adams, T. Leisegang, V. A. Blatov and D. C. Meyer, *Phys. Sci. Rev.*, 2018, **4**, 20180044.
39. M. B. Isichenko, *Rev. Mod. Phys.*, 1992, **64**, 961.
40. J. W. Essam, *Rep. Prog. Phys.*, 1980, **43**, 833.
41. A. Urban, J. Lee and G. Ceder, *Adv. Energy Mater.*, 2014, **4**, 1400478.
42. S. C. van der Marck, *Int. J. Mod. Phys. C*, 1998, **09**, 529.

43. C. D. Lorenz and R. M. Ziff, *Phys. Rev. E*, 1998, **57**, 230.
44. W. Koch and M. C. Holthausen, *A Chemist's Guide to Density Functional Theory*, Wiley-VCH Verlag GmbH, Weinheim, FRG, 2001.
45. J. P. Perdew, in *AIP Conference Proceedings*, AIP, 2001, vol. 577, p. 1.
46. J. P. Perdew and A. Zunger, *Phys. Rev. B*, 1981, **23**, 5048.
47. J. P. Perdew, K. Burke and M. Ernzerhof, *Phys. Rev. Lett.*, 1996, **77**, 3865.
48. J. Sun, A. Ruzsinszky and J. Perdew, *Phys. Rev. Lett.*, 2015, **115**, 1.
49. F. Zhou, M. Cococcioni, C. A. Marianetti, D. Morgan and G. Ceder, *Phys. Rev. B*, 2004, **70**, 235121.
50. V. I. Anisimov, J. Zaanen and O. K. Andersen, *Phys. Rev. B*, 1991, **44**, 943.
51. S. P. Ong, L. Wang, B. Kang and G. Ceder, *Chem. Mater.*, 2008, **20**, 1798.
52. T. Chen, G. Sai Gautam, W. Huang and G. Ceder, *Chem. Mater.*, 2018, **30**, 153.
53. G. Sai Gautam, P. Canepa, A. Abdellahi, A. Urban, R. Malik and G. Ceder, *Chem. Mater.*, 2015, **27**, 3733.
54. G. Sai Gautam and E. A. Carter, *Phys. Rev. Mater.*, 2018, **2**, 095401.
55. P. Pernot, B. Civalleri, D. Presti and A. Savin, *J. Phys. Chem. A*, 2015, **119**, 5288.
56. F. Corà, M. Alfredsson, G. Mallia, D. S. Middlemiss, W. C. Mackrodt, R. Dovesi and R. Orlando, The Performance of Hybrid Density Functionals in Solid State Chemistry, in *Principles and Applications of Density Functional Theory in Inorganic Chemistry II*, Springer, Berlin, Heidelberg, 2004, pp. 171.
57. A. D. Becke, *J. Chem. Phys.*, 1993, **98**, 5648.
58. C. Lee, C. Hill and N. Carolina, *Chem. Phys. Lett.*, 1989, **162**, 165.
59. P. J. Stephens, F. J. Devlin, C. F. Chabalowski and M. J. Frisch, *J. Phys. Chem.*, 1994, **98**, 11623.
60. C. Adamo and V. Barone, *J. Chem. Phys.*, 1999, **110**, 6158.
61. J. Heyd, G. E. Scuseria and M. Ernzerhof, *J. Chem. Phys.*, 2003, **118**, 8207.
62. J. Heyd and G. E. Scuseria, *J. Chem. Phys.*, 2004, **121**, 1187.
63. A. Jain, G. Hautier, S. P. Ong, C. J. Moore, C. C. Fischer, K. a. Persson and G. Ceder, *Phys. Rev. B*, 2011, **84**, 045115.
64. L. Wang, T. Maxisch and G. Ceder, *Phys. Rev. B*, 2006, **73**, 195107.
65. V. L. Chevrier, S. P. Ong, R. Armiento, M. K. Y. Chan and G. Ceder, *Phys. Rev. B*, 2010, **82**, 075122.
66. A. Emly and A. Van der Ven, *Inorg. Chem.*, 2015, **54**, 4394.
67. T. A. Barnes, L. F. Wan, P. R. C. Kent and D. Prendergast, *Phys. Chem. Chem. Phys.*, 2018, **20**, 24877.
68. A. Jain, S. P. Ong, G. Hautier, W. Chen, W. D. Richards, S. Dacek, S. Cholia, D. Gunter, D. Skinner, G. Ceder and K. A. Persson, *APL Mater.*, 2013, **1**, 011002.
69. W. Sun, S. T. Dacek, S. P. Ong, G. Hautier, A. Jain, W. D. Richards, A. C. Gamst, K. A. Persson and G. Ceder, *Sci. Adv.*, 2016, **2**, e1600225.
70. N. Amir, Y. Vestfrid, O. Chusid, Y. Gofer and D. Aurbach, *J. Power Sources*, 2007, **174**, 1234.

71. C. Kim, P. J. Phillips, B. Key, T. Yi, D. Nordlund, Y.-S. Yu, R. D. Bayliss, S.-D. Han, M. He, Z. Zhang, A. K. Burrell, R. F. Klie and J. Cabana, *Adv. Mater.*, 2015, **27**, 3377.
72. X. Sun, P. Bonnick, V. Duffort, M. Liu, Z. Rong, K. A. Persson, G. Ceder and L. F. Nazar, *Energy Environ. Sci.*, 2016, **9**, 2273.
73. X. Sun, P. Bonnick and L. F. Nazar, *ACS Energy Lett.*, 2016, **1**, 297.
74. G. Gershinsky, H. D. Yoo, Y. Gofer and D. Aurbach, *Langmuir*, 2013, **29**, 10964.
75. A. L. Lipson, S.-D. Han, B. Pan, K. A. See, A. A. Gewirth, C. Liao, J. T. Vaughey and B. J. Ingram, *J. Electrochem. Soc.*, 2016, **163**, A2253.
76. J. B. Goodenough and Y. Kim, *Chem. Mater.*, 2010, **22**, 587.
77. T. Chen, G. Ceder, G. Sai Gautam and P. Canepa, *Front. Chem.*, 2019, **7**, 24.
78. K. E. Sickafus, J. M. Wills and N. W. Grimes, *J. Am. Ceram. Soc.*, 1999, **82**, 3279.
79. I. D. Brown, *Acta Crystallogr., Sect. B: Struct. Sci.*, 1988, **44**, 545.
80. G. S. Gautam, P. Canepa, R. Malik, M. Liu, K. Persson and G. Ceder, *Chem. Commun.*, 2015, **51**, 13619.
81. P. Canepa, G. Sai Gautam, D. Broberg, S.-H. Bo and G. Ceder, *Chem. Mater.*, 2017, **29**, 9657.

CHAPTER 5

Anode Materials for Rechargeable Mg Batteries

K. JAYASAYEE*[a], R. BERTHELOT[b,c], K. C. LETHESH[d] AND E. M. SHERIDAN[a]

[a]New Energy Solutions, SINTEF Industry, 7034 Trondheim, Norway; [b]ICGM, Université de Montpellier, CNRS, Montpellier 34095, France; [c]Réseau sur le Stockage Electrochimique de l'Energie (RS2E), CNRS, Amiens, France; [d]Department of Chemistry, Norwegian University of Science and Technology (NTNU), Trondheim, Norway
*E-mail: kaushik.jayasayee@sintef.no

5.1 Introduction

Rechargeable magnesium batteries (RMBs) are promising candidates for post-lithium ion battery technologies. RMBs have the potential to advance the development of affordable and efficient electric energy storage (EES) systems due to their advantages in cost and safety. Their high theoretical volumetric capacity (3832 mA h cm^{-3}), theoretical specific capacity (2205 mA h g^{-1}), favorable negative reduction potential (−2.37 V *vs.* SHE), abundance and good chemical stability compared to Li makes Mg an ideal anode material for the development of non-aqueous rechargeable RMBs.[1–3]

The electrodeposition of Mg has been carried out in complex non-aqueous electrolyte systems based on, *e.g.* Grignard reagents in ethereal solvents such as tetrahydrofuran (THF) and glymes. Most of the other polar aprotic solvents such as pyridine, aniline, dimethylaniline, esters, acetonitriles and

Energy and Environment Series No. 23
Magnesium Batteries: Research and Applications
Edited by Maximilian Fichtner

Published by the Royal Society of Chemistry, www.rsc.org

alkyl carbonates are not suitable as they are unstable due to the nucleophilic nature of Grignard reagents.[4–9] Electrodeposition of Mg from ethereal Grignard electrolytes shows high reversibility, coulombic efficiency and does not form dentrites during the cycling process, which is very important for the safe operation of batteries.

Metallic Mg is known to form a dense passive layer in electrolytes other than ethers, similar to the solid electrolyte interphase (SEI) in LIBs. The passive layer contains insoluble organic and inorganic salts of Mg and its formation is driven by the instability of the electrolyte at the low reduction potential of Mg metal. Contrary to LIBs, where the SEI layer transports Li^+, the Mg surface film does not conduct Mg^{2+} ions. Therefore, the formation of the Mg passivation layer makes the electrodeposition/stripping process irreversible.[6] However, ethereal solvents such as THF and glymes are intolerant towards even trace amounts of water, oxygen and other contaminants, which reduce to form passivation films near the Mg deposition potential. In addition, the corrosive nature of these electrolytes towards the current collectors, poor oxidation stability at potentials of >1.5 V *vs.* Mg and poor oxidation stability essentially rule out the use of high voltage insertion cathodes for rechargeable RMBs.[2,10,11]

To achieve a high energy density and simple fabrication method, it is advantageous to use Mg as a metal sheet, but there are many factors which necessitate the development of alternative Mg alloy materials. First, incompatibility of elemental Mg with many electrolytes (other than salts based on highly corrosive Grignard reagents and hydride anions dissolved in ethereal solvents) irreversibly passivates pure Mg electrodes. In addition, elemental Mg exhibits low ductility at temperatures below 200 °C arising from its simple hexagonal close packed (hcp) structure, which is not sufficient for forming and workability.[12,13] To improve the mechanical properties of elemental Mg, Gollas *et al.* investigated the performance of Mg alloys with Mg in different ratios with Zn and/or Gd (Mg–1.63 wt% Zn, Mg–1.55 wt% Gd and Mg–1.02 wt% Zn–1.01 wt% Gd) by melting the metals to produce the respective saturated solid solutions. It was found that the workability of Mg was vastly improved by alloying, with the possibility to extrude 100 μm thick foils to be used as anodes for RMBs. The electrochemical studies carried out in $(PhMgCl)_2$–$AlCl_3$/THF showed that the alloying elements dissolve and redeposit during galvanostatic cycling without adversely affecting the electrochemical properties of Mg.[12] Similarly, Mg alloy foils of type AZ31 (3% Al and 1% Zn) have been used by Aurbach *et al.* towards their development of solid-state RMBs owing to their improved mechanical properties.[14] Nevertheless, the critical issue related to the stability at high operational voltage is not solved with this method.

One way to improve the electrochemical properties of elemental Mg is through the development of Mg nano- and meso-structures.[15–17] Mg sea-urchin-like nanostructures developed through vapor deposition for aqueous primary Mg/air batteries were found to show much higher gravimetric energy density and rate capability than commercial Mg powder.[15] It was concluded that by reducing the particle size and creating a porous particle structure, they were able to improve the material utilization as well as increase the number of particles at the vicinity of the surface making them more active. Chen *et al.* synthesized ultrasmall Mg nanoparticles with a narrowly distributed spherical structure of around 2.5 nm in diameter for rechargeable RMBs using an ion-exchange route.[17] When tested in a coin cell with a graphene-like MoS_2 cathode and $Mg(AlCl_3Bu)_2$ in THF as an electrolyte, nano-sized Mg was found to show better cycling capabilities than bulk Mg without any changes in morphology or particle growth, even after cycling. It was found that the high specific surface area of the nanoparticles allowed the formation of a comparatively thinner surface passivation layer that is conductive enough to support Mg^{2+} migration during cycling.

Coating the Mg surface with an artificial Mg^{2+} conducting polymeric interphase was reported recently by Ban *et al.* to enable reversible stripping and deposition Mg in a high-voltage electrolyte composed of $Mg(TFSI)_2$ dissolved in an oxidation resistant carbonate solvent.[18] The polymeric coating of about 100 nm thick with an ionic conductivity of 1.19 μS^{-1} was made of thermal-cyclized polyacrylonitrile (cPAN, $(C_3H_3N)_n$) and Mg trifluoromethanesulfonate ($Mg(CF_3SO_3)_2$). Figure 5.1 compares the deposition/stripping cycles of both uncoated and coated Mg particles in a conventional ethereal APC (all phenyl complex, PhMgCl/THF–$AlCl_3$) electrolyte and one with non-corrosive carbonate ester. The red lines in the figure indicate that the coated Mg particles show enhanced reversibility with excellent control of the overpotential for at least 1000 h in the carbonate-based electrolyte compared to the non-coated Mg particles. The electronically-insulating polymeric interphase is considered to resist reduction of the electrolyte components including water traces on the Mg surface at the Mg^{2+} deposition potential when tested with V_2O_5 cathodes in a full cell configuration, without hindering the ionic transport (Figure 5.2).

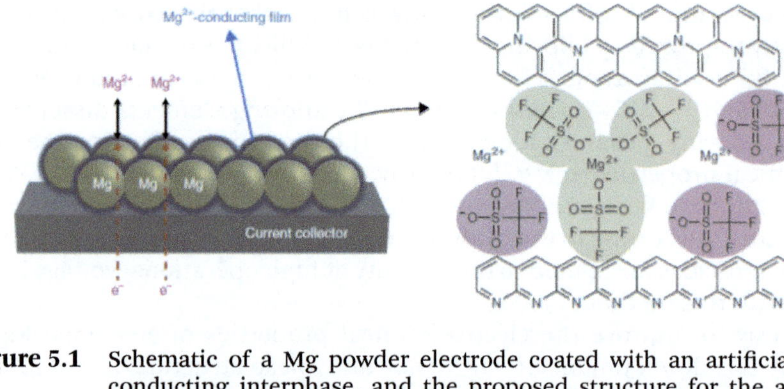

Figure 5.1 Schematic of a Mg powder electrode coated with an artificial Mg^{2+}-conducting interphase, and the proposed structure for the artificial Mg^{2+}-conducting interphase based on XPS, TOF-SIMS and TGA analysis. Reproduced from ref. 18 with permission from Springer Nature, Copyright 2018.

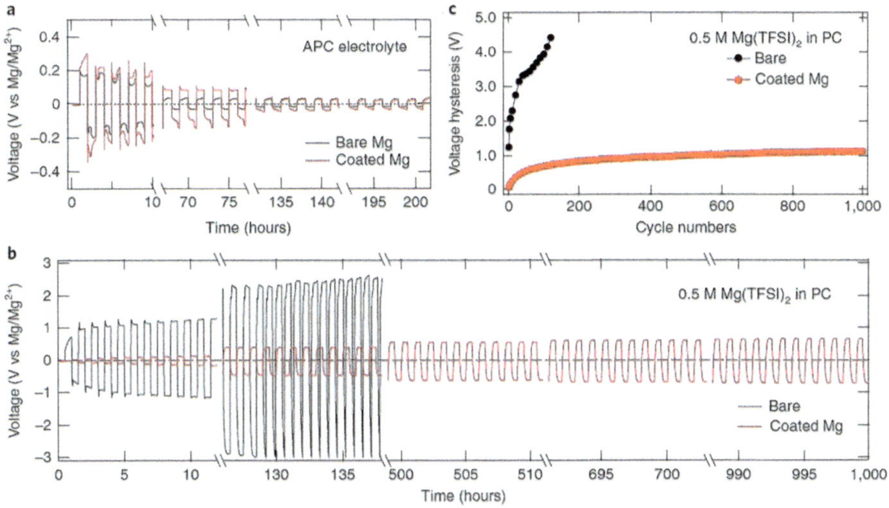

Figure 5.2 Voltage responses of symmetric Mg cells under repeated polarization with and without an artificial interphase in different electrolyte systems at a current density of 0.01 $mAcm^{-2}$. Reversible Mg deposition/stripping in APC electrolyte (a), where each deposition/stripping cycle lasts for an hour, and in 0.5 M $Mg(TFSI)_2$/PC electrolyte (b), where each deposition/stripping cycle lasts for half an hour. The cell made with pristine Mg electrodes shows a huge overpotential at the beginning and fails after 135 cycles, whereas the cell made with the Mg^{2+}-conducting interphase-protected Mg electrodes performs over prolonged cycles in carbonate-based electrolytes. The reversibility is conspicuous in the latter as proven by up to 1000 cycles. (c) Voltage hysteresis *versus* cycle numbers for symmetric Mg electrodes with 0.5 M $Mg(TFSI)_2$/PC electrolyte. Lower voltage hysteresis is observed on an interphase protected Mg electrode, where reductive decomposition of PC is prevented because of the artificial interphase. Reproduced from ref. 18 with permission from Springer Nature, Copyright 2018.

While several pioneering works have been reported over the years by the groups of Gregory, Aurbach, Muldoon and Fichtner on the development of alternate electrolytes, it is also important to continue with the search for new anode materials that are compatible with different classes of electrolytes without compromising the operating voltage window, specific capacity, rate capability and cycling stability. Similar to LIBs and sodium-ion batteries (SIBs), negative electrode materials for RMBs can be classified into insertion-type materials or alloying-type materials depending on the mechanisms involved in the magnesiation/demagnesiation reactions. For the first type, Mg^{2+} ions are inserted and extracted out of the structure during electrochemical cycling, whereas for alloy-type materials an electrochemical conversion occurs with the magnesium ions. This chapter summarizes the electrochemical behavior of these materials and discusses the latest trends to provide guidelines for designing better anode materials in the near future.

5.2 Insertion-type Anodes

5.2.1 Graphite

Graphite is a promising anode material that can reversibly intercalate both Li^+ and K^+ ions.[19–21] In addition, graphite is already a commercialized anode material for LIBs. However, it is reported that graphite is unable to store Na^+ and Mg^{2+} ions.[22] The effective ionic radii of Li^+ and K^+ are 0.076 and 0.138 nm, respectively, which are higher than the ionic radii of Mg^{2+} 0.072 nm and Na^+ 0.102 nm, respectively. The authors claimed that the size of the ions is not the major factor that determines the storage capacity of graphite. The low capacity of graphite for Mg^{2+} and Na^+ was attributed to the weak chemical binding nature of the Mg^{2+} and Na^+ to carbon substrates such as graphene and its derivatives.[23] However, Kim *et al.* reported that Mg^{2+} can be reversibly co-intercalated into graphite in combination with ether solvents such as dimethoxyethane (DME) and diethylene glycol dimethyl ether (DEGDME). Initially, they explored the possibility of the co-intercalation of Mg^{2+} into graphite with linear ether solvents by density functional theory (DFT) calculations.[24] DFT calculations indicated that the intercalation Mg^{2+} into graphite is favourable through a co-intercalation mechanism. They calculated the binding energy between Mg^{2+} ions and various ethereal solvents to study the co-intercalation of Mg^{2+} ions into graphite. Among the solvents studied, a Mg^{2+}–DEGDME complex has the highest binding energy (10.05 eV). This indicates that DEGDME is the most promising solvent for the successful co-intercalation of Mg^{2+} ions. The same authors studied single layer and double layer graphite structures co-intercalated with the Mg^{2+}–DEGDME complex. The calculation of the co-intercalation energies of the Mg^{2+}–DEGDME complex in graphite indicated that their intercalation in graphite is thermodynamically favourable. In addition, the authors pointed out that the double layer structure was more favourable for co-intercalation compared to the single layer structure.

To validate the DFT prediction about the co-intercalation of Mg^{2+} ions into graphite, galvanostatic cycling was performed with 0.3 M $Mg(TFSI)_2$ in various linear ether solvents. Graphite and Mg metal were used as the working and counter electrodes for DME and DEGDME. As mentioned in the DFT calculations, the graphite electrode in combination with 0.3 M $Mg(TFSI)_2$ in DEGDME or DME showed a reversible voltage profile. X-ray diffraction (XRD) analysis at various stages of charging and discharging was used to study the intercalation mechanism of Mg^{2+} with graphite and the results confirmed the reversible intercalation of Mg^{2+} into graphite. Fourier-transform infrared (FTIR) analysis also supported the co-intercalation of Mg^{2+}-DEGDME as a double layer structure. However, graphite did not manage to retain its original structure after the reversible intercalation and deintercalation of Mg^{2+} ions.

In another study, Er and co-workers showed the potential of 2D defective carbon-based (porous graphene systems) materials as high capacity anode materials for Mg-ion rechargeable batteries by first-principles DFT calculations.[25] They predicted the Mg storage capacity of graphene with a varying degree of Stone–Wales and divalent defects. High concentrations of vacancies and topological defects in graphene can marginally improve its Mg storage capacity. The defect rich regions in the 2D graphene favour the adsorption of Mg.

5.2.2 Phospherenes

Although phosphorous-based materials have been studied as anode materials for LIBs and SIBs, not much has been reported for RMBs.[26–29] Recently, Han *et al.* investigated the possibility of using black phosphorus and phosphorene (exfoliated from black phosphorous) as potential anode materials for rechargeable RMBs by studying both the adsorption and diffusion of Mg using DFT.[30] A high theoretical capacity of 865 mA h g^{-1} was estimated for phosphorene, which is almost twice that for LIBs and SIBs.

By exploring the surface of phosphorene for Mg adsorption, the authors found that the most stable binding site for the Mg atom on the phosphorene surface is the top of the channel created between two P atoms, as shown in Figure 5.3a and b. The binding energy (E_b) calculation of Mg on the phosphorene and black phosphorous surfaces showed that the binding energy of Mg on monolayer phosphorene (−0.716 eV) is significantly higher than the binding energy of the bulk material (−1.366 eV) suggesting preferential affinity of Mg^{2+} for the monolayers. Calculation of the spatial charge distribution between Mg and phosphorene revealed that Mg possesses a unit positive charge (a cationic state) due to the complete transfer of its 3 s^2 electrons to phosphorene. Phosphorene possess structural anisotropy (puckered structure) because of the arrangement of P atoms in a honeycomb lattice and this structural feature plays an important role in the migration of Mg atoms on the surface of phosphorene. According to calculations, Mg^{2+} diffuses along

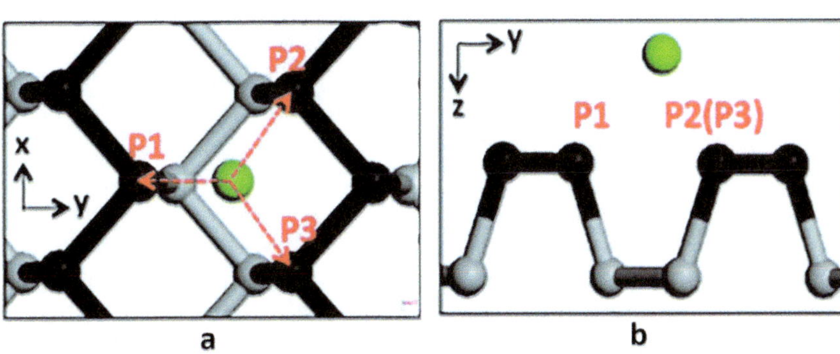

Figure 5.3 (a) Top and (b) side view of the most stable binding position of Mg on monolayer phosphorene. Reproduced from ref. 30 with permission from the Royal Society of Chemistry.

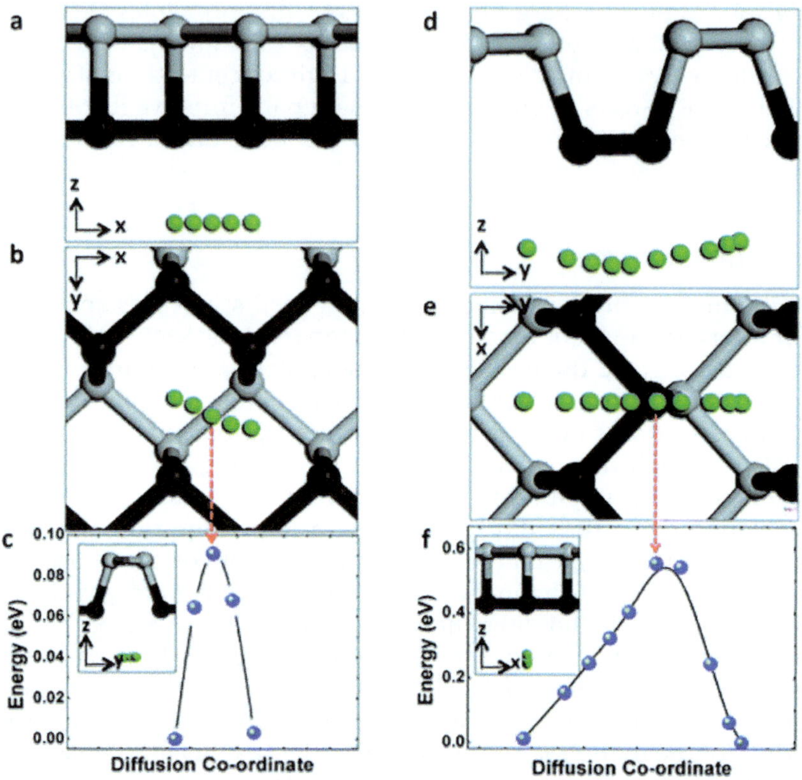

Figure 5.4 (a) and (b) schematic representation of Mg along the zigzag direction and (c) energy profile of the diffusion of Mg in the zigzag direction; (d), (e) and (f) schematic and energy profiles of Mg diffusion along the armchair direction in phosphorene. Reproduced from ref. 30 with permission from the Royal Society of Chemistry.

the zigzag direction (Figure 5.4a–c) with a lower migration barrier of 0.09 eV and in the armchair direction (Figure 5.4d–e) with a comparatively higher migration barrier of 0.56 eV at room temperature. Similar mechanisms were observed by Jin *et al.* for monolayer black phosphorous as an anode material for RMBs, again by DFT calculations.[31]

5.2.3 Borophenes

Boron-based materials, due to their light weight, have a distinct advantage for being employing as anode materials for rechargeable batteries. Recently, the synthesis of three types of borophenes (exfoliated from bulk boron) named S1, S2 and S3 were reported.[32,33] Borophenes were investigated as anode materials for LIBs.[34,35] Xiang *et al.* studied borophenes as anode materials for Mg-ion rechargeable batteries using DFT calculations.[36] The optimized structures of the borophenes S1, S2 and S3 used in their study are shown in Figure 5.5a–c.

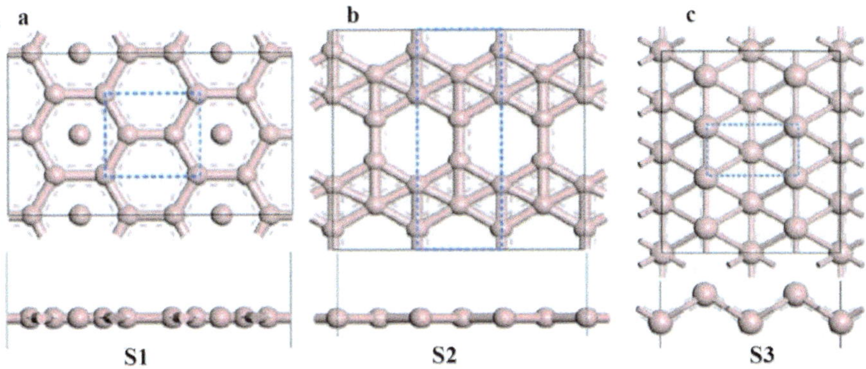

Figure 5.5 (a), (b) and (c) Top and side views of the optimized structure of S1, S2 and S3 respectively. Reproduced from ref. 36 with permission from the PCCP owner societies.

The adsorption, diffusion, theoretical specific capacity and open circuit voltage (OCV) of a Mg metal ion on the abovementioned (S1, S2 and S3) borophenes were investigated. S1 and S2 have flat structures such as that of graphene and S3 has a W-shaped corrugated structure. Nevertheless, the 2p orbital of boron crossing the Fermi level ensures the favorable electrical conductivity required for anode materials. The determination of the energy of adsorption (E_{ad}) of a Mg metal atom on the borophene monolayer showed that the E_{ad} value on S2 has a positive value (0.10 eV) making it incompatible for applications as a Mg anode material. On the contrary, the E_{ad} value of Mg on S1 and S3 have negative values (exothermic), which makes them suitable for Mg anode materials. In other words, the configuration of borophene has a major influence on the adsorption of Mg metal ions.

The investigation of the diffusion barrier of metal ions on borophene revealed that structural anisotropy played a major role in the diffusion of metal ions in borophene. As in the case of phosphorene, there are two diffusion pathways in borophene. One pathway is along the zigzag direction and the second one along the armchair direction. In the case of S1 and S2, the diffusion of Mg metal ions along the zigzag direction is faster than in the armchair direction at room temperature. However, in the case of S3, contrasting results were observed. The energy barrier for the diffusion of Mg metal ions along the armchair direction in S3 is considerably lower compared to those in S1 and S2. Determination of the capacity of adsorption of Mg ion on single layer borophene revealed that the capacity mainly depends on the configuration of borophene. For instance, a capacity of 684 mA h g^{-1} was observed for S1, while the capacity of S3 was only 300 mA h g^{-1} and no capacity was obtained for S2.

5.2.4 Transition Metal Carbides

The mechanism of non-lithium metal ion storage in two dimensional transition metal carbides (MXene nanosheets) has been derived by investigating the potential of O-terminated, OH-terminated and bare MXenes through first-principles simulations by Xie *et al.*[37] The calculated adsorption energies of Mg^{2+} on OH-terminated MXenes were very positive, which indicated that the Mg metal ions were not well adsorbed on OH-terminated MXenes. This is because of the Coulombic repulsion between the Ti atoms in MXene and Mg ions. However, Mg ions undergo physisorption on OH-terminated MXenes. The adsorption energies of O-terminated MXene show negative values, revealing the possibility of the formation of a full adsorption layer on O-terminated MXene. Mg transfers about 0.76 e/atom to the oxygen atoms on the Ti_2CO_2 surface. The capacity of Mg for O-terminated MXenes (Ti_2CO_2) is 570 mA h g^{-1}, which is very similar to the theoretical capacities of its alloy anodes.[38,39] The possibility of metal multi-layer adsorption on O-terminated MXenes was studied by calculating the second layer adsorption energy for various atoms. It is interesting to note that the formation of second layer Mg is thermodynamically favored on top of the magnesiated MXene. Adsorbed Mg exhibits bulk behavior when more than three layers are formed. The capacity increases with the increase in the number of layers. However, the increase in capacity becomes smaller with the addition of each layer because of the increase in weight. The storage mechanism of Mg involves normal insertion/extraction (in the first layer) and a plating/stripping mechanism from the second layer as in the case of simple Mg metal anodes.[40,41] A detailed study of the electronic structure of Ti_2CO_2 with two Mg metal layers revealed the formation of metallic bonds (formation of a negative electron cloud, NEC). In addition, only a weak repulsive interaction between Ti and Mg was observed, which contributed to the lower adsorption energy.

5.2.5 $Li_4Ti_5O_{12}$

$Li_4Ti_5O_{12}$ (LTO), with its zero-strain spinel framework structure and a theoretical capacity of 175 mA h g^{-1}, is successfully commercialized as an insertion-type anode material for LIBs. The advantages of LTO for LIBs include an ultralong cycle life, high power capability, excellent safety characteristics, negligible volume change over cycling, a wide temperature window of operation and lower cost of production.[42–44]

Spinel LTO nanoparticles synthesized through a facile sol–gel method have been reported by Y-G Guo in 2014 as the first insertion-type anode material for RMBs.[45] This material was found to achieve a reversible capacity of 175 mA h g^{-1} corresponding to 1.5 Mg atom per formula unit and a cycle life stability of about 5% capacity decay after 500 cycles with an average coulombic efficiency of close to 100%, as shown in Figure 5.6. The authors observed an initial capacity loss in the first cycle due to the decomposition of electrolyte at the electrode surface. An activation process was proposed during the initial discharge/charge cycles where Mg insertion and extraction processes happen with the gradual replacement of Li by Mg ions. This, over consecutive cycles, led to the formation of highly reversible Mg-rich spinel magnesium titanate phases such as Mg_4Ti, $Mg_{3.25}Li$ and $Mg_{2.5}Li$ through a 3-step phase separation and transition mechanism (Figure 5.7).

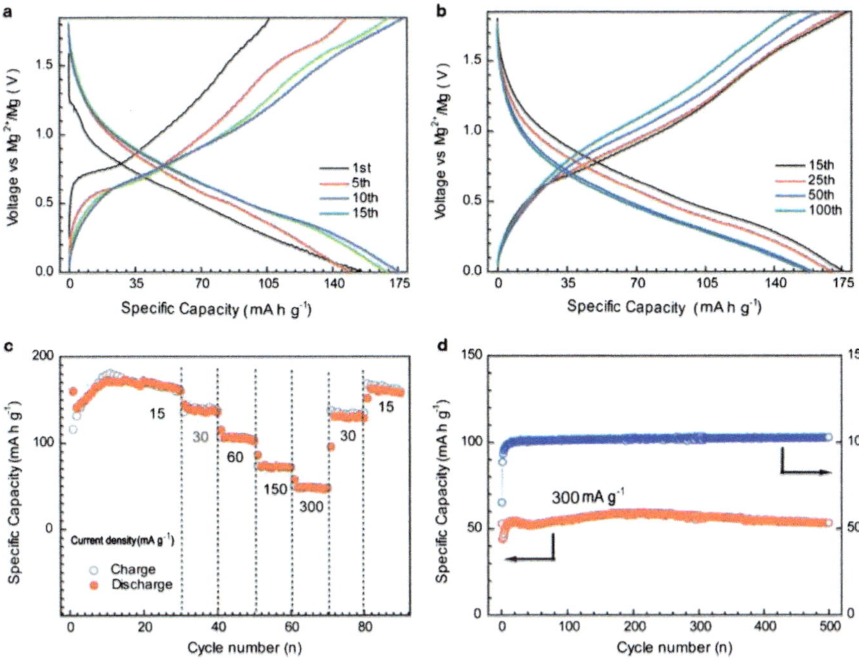

Figure 5.6 Charge–discharge characteristics of LTO electrodes in rechargeable Mg batteries (a and b). Typical galvanostatic discharge–charge voltage profiles of the cell cycled before (a) and after (b) 15 cycles at a current density of 15 mAg^{-1}. (c) Comparison of the rate capabilities of the cell cycled at different current densities. (d) Cycling performance of the cells using LTO electrodes cycled at a current density of 300 mAg^{-1}. Reproduced from ref. 45 with permission from Springer Nature, Copyright 2014.

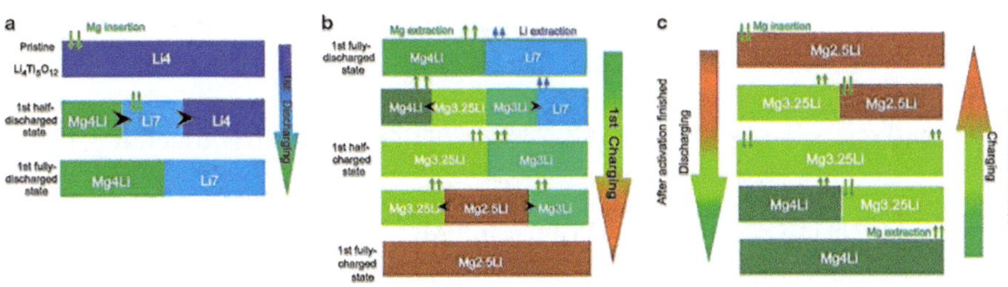

Figure 5.7 (a) The 1st cycle of the discharging and (b) charging processes in a Mg battery. (c) The discharge–charge processes after activation has finished in the Mg battery. Directions of phase boundary movement are marked by colored arrows. Reproduced from ref. 45 with permission from Springer Nature, Copyright 2014.

However, like other insertion-type anode materials, LTO suffers from the sluggish kinetics of Mg diffusion into the structure, probably due to the high polarization of the Mg^{2+} divalent cation. The same authors found that by effectively limiting the particle size of LTO to below 10 nm, a significant improvement in the electrochemical Mg^{2+} insertion could be achieved.[46] The superior diffusion kinetics and electronic conductivity in nanomaterials for an insertion mechanism is generally attributed to better inter-particle contact in the active material as well as between the active materials and the electrolyte created by the larger surface area.

Another proposal to improve the charge storage capability of the LTO electrode is to introduce Li^+ in to the RMB system to synergistically exploit the advantages of both magnesiation and lithiation by employing a suitable hybrid electrolyte comprising both Mg^{2+} and Li^+.[47] By tuning the concentration of Li^+ in 0.25 M $Mg(AlCl_2BuEt_2)_2$/THF through the addition of 0.25 M LiCl, a high reversible capacity and cycling stability is achieved for LTO with particle sizes of larger than 100 nm.

5.2.6 $Na_2Ti_3O_7$

$Na_2Ti_nO_{2n+1}$ ($2 \leq n \leq 9$) compounds, with representative materials such as $Na_2Ti_3O_7$, $Na_2Ti_6O_{13}$ and $Na_2Ti_7O_{15}$ are extensively studied as insertion-anode materials for sodium ion batteries (SIBs) due to their ease of synthesis, low redox potentials of Ti^{4+}/Ti^{3+}, non-toxicity, low cost and their favourable tunnel or layered structure for diffusion of Na^+ insertion/extraction. For example, layered $Na_2Ti_3O_7$ was found to have a promising reversible capacity of 200 mA h g^{-1} at average voltages of between 0.5 and 1.0 V. However, the rate capability and structural stability of $Na_2Ti_3O_7$ was found to be poor in SIBs.[48–50]

The smaller ionic radii of Mg^{2+} when compared to Na^+ (0.102 nm) and the lower solubility of Mg^{2+} in Ti-based compounds led Chen *et al.* to investigate $Na_2Ti_3O_7$ as an anode material for RMBs.[51] A facile hydrothermal procedure was used to synthesize $Na_2Ti_3O_7$ nanoribbons (NTO–NR) with a special layered structure where Na^+ was 7-coordinated to oxygen atoms between the layers and 9-coordinated to oxygen atoms on the corner. A magnesium insertion/extraction process through this material was demonstrated, as shown in Figure 5.8, to happen in two steps:

$$[Na]^{VII}[Na]^{IX}Ti_3O_7 + Mg^{2+} + e^- \rightarrow [Mg]^{VII}[Na]^{IX}Ti_3O_7 + Na^+ \text{ (first discharge)} \quad (5.1)$$

$$MgNaTi_3O_7 \leftrightarrow Mg_{0.5}NaTi_3O_7 + 0.5\,Mg^{2+} + e^- \text{ (cycle process)} \quad (5.2)$$

The first step is the activation step where the as-synthesized $Na_2Ti_3O_7$ is electrochemically activated and structurally transformed to a highly stable and reversible $MgNaTi_3O_7$ phase during the first discharge. During this process, Mg^{2+} occupies Na^{VII} sites with an irreversible layer shrinkage. In the subsequent cycles, 0.5 M Mg^{2+} would then reversibly insert and extract from the structure without any major transformation, resulting in a theoretical capacity of 88 mA h g^{-1}. The $Na_2Ti_3O_7$ nanoribbons investigated in this work were found to deliver a reversible capacity of 78 mA h g^{-1} with a remarkable

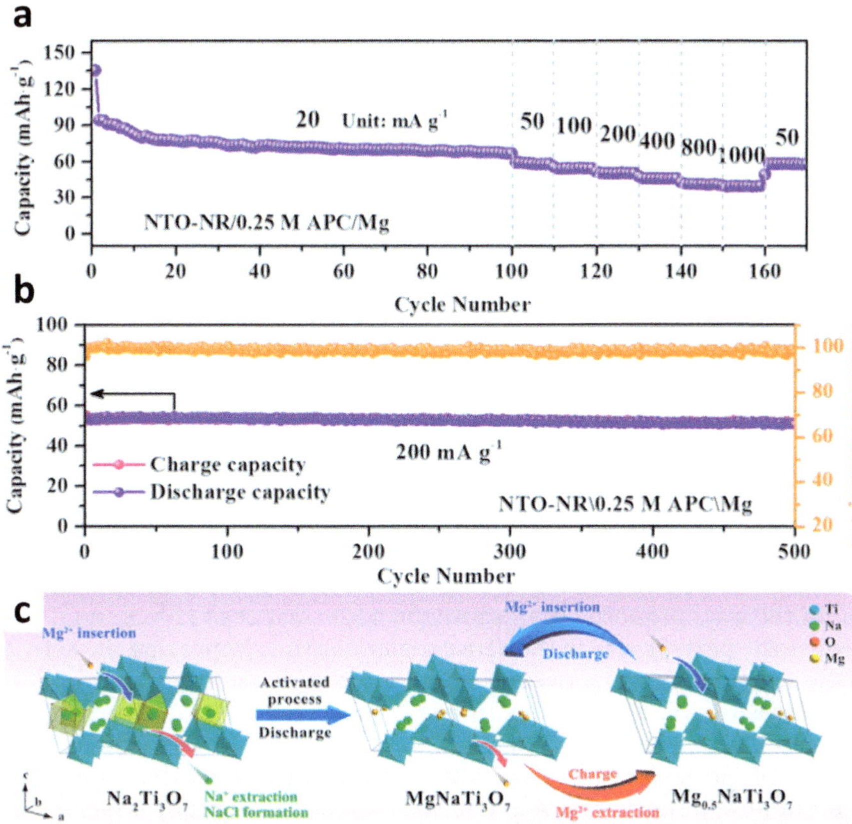

Figure 5.8 (a) Rate capability and cycling performance of the NTO–NR electrode in the potential window of 0.01–2.00 V. (b) Long-term cycling performance of the NTO–NR electrode at a high current density of 200 mAg^{-1}. (c) Schematic diagram of the Mg^{2+} insertion–extraction mechanism in the NTO structure. Reproduced from ref. 51 with permission from the American Chemical Society, Copyright 2016.

cycle life and rate capability. It was also concluded that improved contact points and shorter transport paths for the Mg^{2+} diffusion provided by 1D structures such as nanoribbons and nanotubes show better electrochemical performance than other morphologies such as bulk and nanosheet $Na_2Ti_3O_7$.

5.2.7 Li_3VO_4

Transition metal oxides such as Li_3VO_4 (LVO) were first developed for LIBs as low-cost anode materials with high chemical stability, theoretical capacity (592 mA g^{-1}), ionic conductivity, suitable electrode potential (0.5–1.0 V *vs.* Li/Li$^+$) and safety. These materials were reported to show an initial reversible capacity and energy density more than twice than that of $Li_5Ti_5O_{12}$. However,

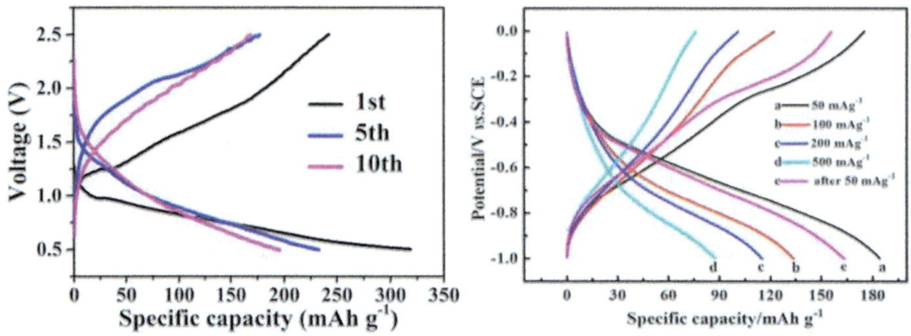

Figure 5.9 Left: Typical charge/discharge curves at the 1st, 5th and 10th cycle of LVO/C at 20 mAg^{-1}. Reproduced from ref. 56 with permission from Elsevier, Copyright 2017. Right: the rate performance of $FeVO_4$/C. Reproduced from ref. 61 with permission from John Wiley and Sons, © 2017 Wiley-VCH Verlag GmbH & Co. KGaA, Weinheim.

these materials suffer from poor rate capability, capacity retention on cycling and coulombic efficiency arising from their low electronic conductivity. Some of the ways proposed to improve the performance of Li_3VO_4 in LIBs are to reduce the particle size of the active materials, thus improving the activity, and the addition of conductive carbon-based materials and surface modification of the active materials by coating to enhance the electronic conductivity and coulombic efficiency.[52–55] Reports on employing LVO as an anode material for RMBs are so far very limited. Zhao *et al.* synthesized mesoporous LVO/carbon hollow spheres with a diameter of about 0.5–5 μm by a facile spray drying method resulting in an orthorhombic LVO phase.[56] Electrochemical studies (Figure 5.9, left) revealed that the Mg transport in the material occurs through an insertion mechanism, however yielding a low coulombic efficiency of about 76% at the first cycle. The authors suggested that the efficiency loss could be due to the entrapment of inserted Mg^{2+} in the host material. Due to the strong coulombic interaction between the host LVO and Mg^{2+} this material achieved a low capacity utilization for LVO of about 54% of its theoretical capacity. At the end of 15 cycles the LVO/carbon hollow spheres delivered an encouraging specific capacity of about 195 mA h g^{-1}, which was attributed to the shortened Mg^{2+} diffusion path and improved Mg^{2+} insertion kinetics created by the high surface area mesoporous structures.

5.2.8 $FeVO_4$

$FeVO_4$ in the form of bulk, nanosheets and graphene composites were reported to show good electrochemical performance in LIBs with a reversible capacity ranging from 400–1200 mA h g^{-1} depending on the morphologies. Generally, the enhanced performance showed by $FeVO_4$ nanostructures and composites with carbon-based materials compared to bulk materials is attributed to the high electronic conductivity, low volume expansion and high Li^+ transport rates resulting from the unique structures.[57–60]

However, reports on $FeVO_4$ as an anode material for RMBs are very rare. Cao *et al.* synthesized porous $FeVO_4$ through a hydrothermal method for their application as an anode material for aqueous rechargeable RMBs.[61] The as-synthesized $FeVO_4$ was modified by coating with carbon to enhance electronic conductivity and surface area by forming a $FeVO_4$/C composite with a hierarchical pore structure. Cyclic voltammetry (CV) studies carried out in 1 M $MgSO_4$ electrolyte at 1 mVs^{-1} showed good reversibility of the two redox couples with $FeVO_4$/C exhibiting decreased electrode polarization and supressed side reactions (oxygen evolution) compared to the uncoated material. The authors were also able to achieve a reasonable reversible capacity of 184 mA h g^{-1} at a current density of 50 mA h g^{-1} and a capacity retention rate of about 63% for $FeVO_4$/C after 50 cycles at 100 mA h g^{-1} (Figure 5.9, right).

5.3 Alloying-type Negative Electrode Materials

In comparison with the insertion-type mechanism, alloying-type electrodes work by a different electrochemical process. Instead of a topotactic intercalation for which the crystallographic structure is not deeply changed, electrochemical alloying involves a conversion process that creates compounds with a new atomic organization (eqn (5.3)). During the dealloying process, the pristine material is formed again (eqn (5.4))

$$M + xA^{n+} + xne^{-} \rightarrow A_xM \tag{5.3}$$

$$A_xM \rightarrow M + xA^{n+} + xne^{-} \tag{5.4}$$

When the active element M is combined with a non-active one, M′, in the pristine material, the electrochemical conversion results in the formation of the alloy (A_xM) and nanoparticles M′, according to eqn (5.5). There are two pathways during oxidation. In general, the as-created nanoparticles M′ remain in the electrode composite, and fresh particles of the element M are formed again (eqn (5.6)). In specific cases, the fresh particles M can chemically react with the nanoparticles M′ to again form the pristine material M–M′ (eqn (5.7))

$$M\text{–}M' + xA^{n+} + xne^{-} \rightarrow A_xM + M'^{0} \tag{5.5}$$

$$A_xM + M' \rightarrow M + xA^{n+} + xne^{-} + M' \tag{5.6}$$

$$A_xM + M' \rightarrow M\text{–}M' + xA^{n+} + xne^{-} \tag{5.7}$$

In comparison with classical insertion-type electrodes (*e.g.* $LiCoO_2$, $LiFePO_4$) for which only one cation is exchanged, here more than one ion could electrochemically react. Therefore, an alloy-type electrode usually exhibits high specific and/or volumetric capacities. However, substantial volume changes occur during electrochemical cycling, and successive expansion and contraction alter the electrochemical performance and can result in a loss of electrical contacts at the electrode level. At the particle scale, the dramatic volume expansion can locally induce the process of destruction and reformation of SEI layers, consequently impacting upon the cycling efficiency.[62] In order to compensate or at least mitigate these issues, strict control of the electrode formulation is mandatory in order to buffer the volume expansion.

Alloy-type electrode materials have been studied for long as anode materials in LIBs. There, silicon and tin are particularly interesting. The formation of $Li_{22}Si_5$ and $Li_{22}Sn_5$ corresponds to very high specific capacities of 4200 and 990 mA h g^{-1}, respectively. However, when compared to cycling-stable graphite-based electrodes, the progressive alteration of the composite electrode due to successive volume changes is critical.[63,64] Emerging SIBs show the same trend: despite a high specific capacity, for example 600 mA h g^{-1} with antimony-based electrodes,[65] alloy-type materials cannot compete with hard carbons.[66] However, the case is quite different for RMBs, where carbonaceous materials exhibit very low electrochemical Mg insertion consequently allowing alloy-type materials to play a role in designing the negative electrode.

5.3.1 Electrochemical Behavior of Single Metal Alloy Electrodes

Antimony and bismuth were the first investigated elements for alloy anodes, studied by the Toyota Research Institute of North America.[67] Metallic Sb and Bi, as well as the intermediate alloy compositions of $Sb_{1-x}Bi_x$, have been electrodeposited on titanium/platinum-coated copper substrates and tested in a half-cell configuration with EtMgCl–Et_2AlCl/THF solution. Galvanostatic profiles recorded at a very low current rate (C/100) show flat plateaus at 0.32 and 0.25 V for Sb and Bi, respectively, indicating a two-phase transition. Owing to the binary phase diagrams and the corresponding specific capacities, the formation of alloys, α-Mg_3Sb_2 and α-Mg_3Bi_2, was proposed and confirmed by X-ray diffraction (XRD) at the end of discharge (for Mg_3Bi_2 only). While the Bi-based electrode exhibits good cycling performance with more than 200 mA h g^{-1} for 100 cycles at 1C rate, the reversible capacity for the Sb-based electrode quickly drops to 20 mA h g^{-1}. Shao and coworkers compared the influence of particle grain size on the electrochemical performance and reported Bi nanotubes (Figures 5.10 and 5.11, left) to show a very low capacity fade of <8% and a reversible capacity of above 300 mA h g^{-1} for 200 cycles at 2C in 0.1 M $Mg(BH_4)_2$–1.5 M $LiBH_4$/diglyme.[38] In addition, a high-rate performance and cycle life have been achieved by fabricating anodes using a bismuth/reduced graphene oxide nanocomposite.[68]

Bi-based electrodes have been investigated by various characterization techniques to understand the mechanisms of reversible alloying and phase transitions.[69–71] Among them, *operando* X-ray diffraction, shown in Figure

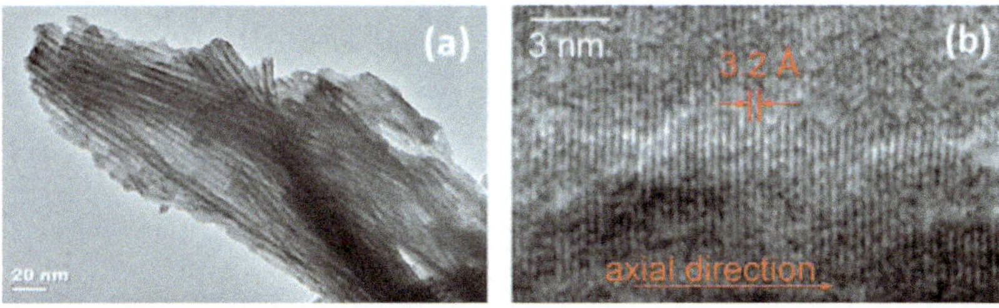

Figure 5.10 (a) TEM image of a bismuth nanotube bundle and (b) high-resolution TEM image of Bi nanotubes. Reproduced from ref. 38 with permission from the American Chemical Society, Copyright 2014.

5.11, right confirmed a two-phase reaction from Bi to α-Mg_3Bi_2, without any amorphization step. Interestingly, the group of Kovalenko investigated colloidal bismuth nanoparticles and reported that the partial formation of a high-temperature polymorph β-Mg_3Bi_2 might also be possible.[72] The formation of high-temperature and/or high-pressure products during electrochemical alloying has already been reported in LIBs, proving that electrochemistry could be an extremely strong driving force at the atomic scale. With solid state[25] Mg nuclear magnetic resonance spectroscopy (NMR) a fast exchange ionic exchange by hopping mechanism between the two Mg sites of α-Mg_3Bi_2 structure has been proposed to explain the distinct behavior of Bi toward magnesium electrochemistry. In parallel, high Mg^{2+} diffusion into Bi has been calculated by DFT.[73,74]

The reasons for the very weak cycling properties of antimony are still unclear, although some conclusions have been drawn by investigating intermediate compositions of $Sb_{1-x}Bi_x$ alongside the complete Sb–Bi solid solution. An almost complete alloying mechanism during the first discharge is observed for intermediate compositions of $Sb_{1-x}Bi_x$. However, two distinct plateaus are noticed with electrodeposited electrodes, whereas a unique plateau is clearly seen with micrometric-sized compositions obtained by high-energy ball milling.[67,69] The explanation might be found by comparing the alloys Mg_3Sb_2 and Mg_3Bi_2. Arthur *et al.* highlighted the higher ionicity of the Mg–Sb bond in Mg_3Sb_2 (compared to the Mg–Bi bond in Mg_3Bi_2) to explain the weakly reversible process. Operando X-ray diffraction confirmed this trend, with a progressive release of Bi particles and formation of Mg_3Sb_2 during the dealloying process of $Mg_3\ (Sb_{1-x}Bi_x)_2$. In order to support this observation, preliminary DFT calculations showed that the electronic band involved in oxidation displays a dominant Bi contribution.[75]

The electrochemical magnesiation of tin was also shown for the first time by the Toyota Research Institute of North America.[39] The formation of crystalline Mg_2Sn was confirmed by *ex situ* X-ray diffraction. The high corresponding specific capacity approaching 900 mA h g^{-1} is unfortunately not retained during cycling (Figure 5.12, left). As for lithium-ion batteries, tin particles are well known to suffer from the volume expansion which results in pulverization and loss of electrical contact inside the electrode.[64] Thanks

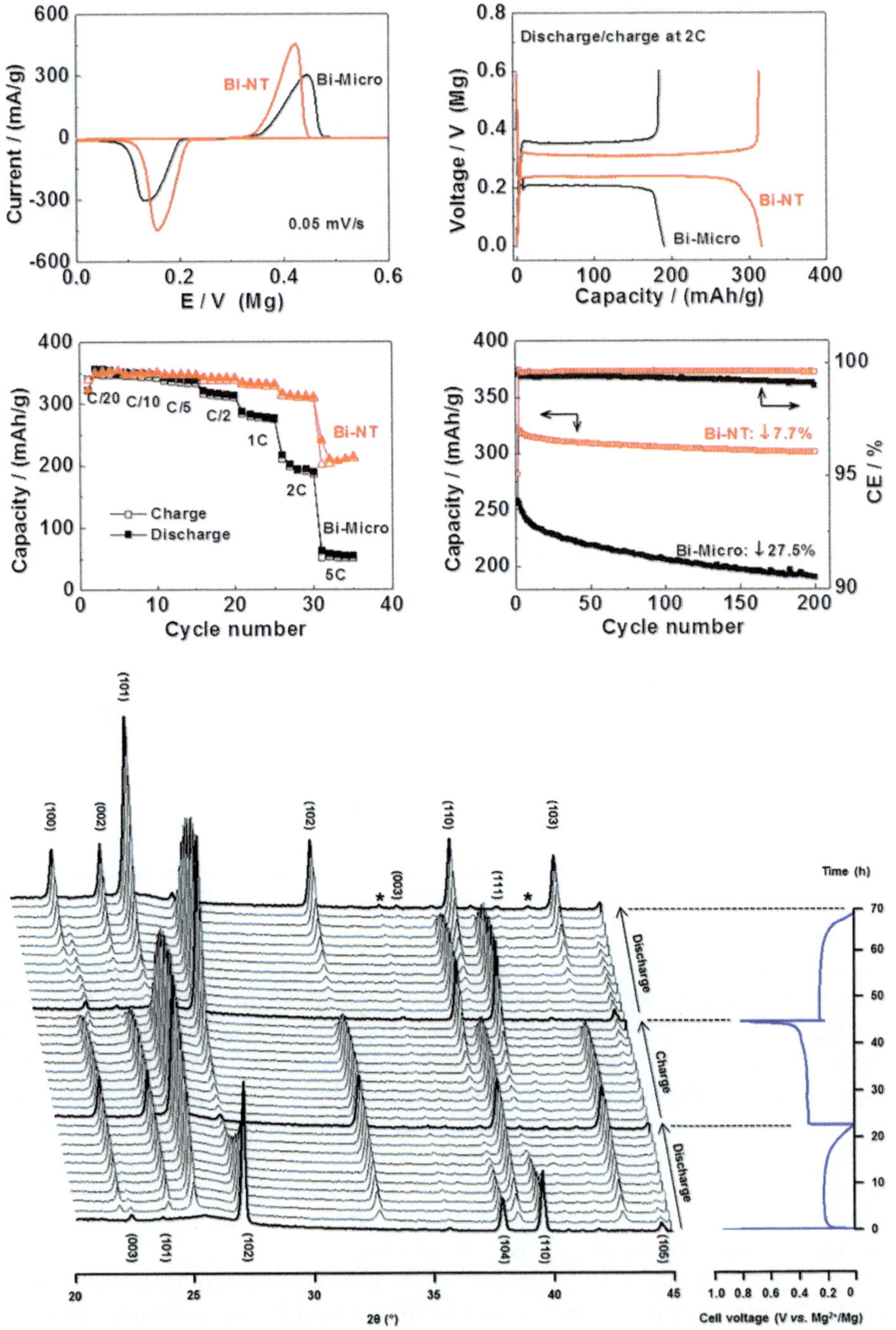

Figure 5.11 Top: Electrochemical behavior of Bi micrometric particles and Bi nanotubes. Reproduced from ref. 38 with permission from the American Chemical Society, Copyright 2014. Bottom: Operando X-ray diffraction highlighting the reversible 2-phase alloying from Bi to Mg_3Bi_2. Reproduced from ref. 69 with permission from the Royal Society of Chemistry.

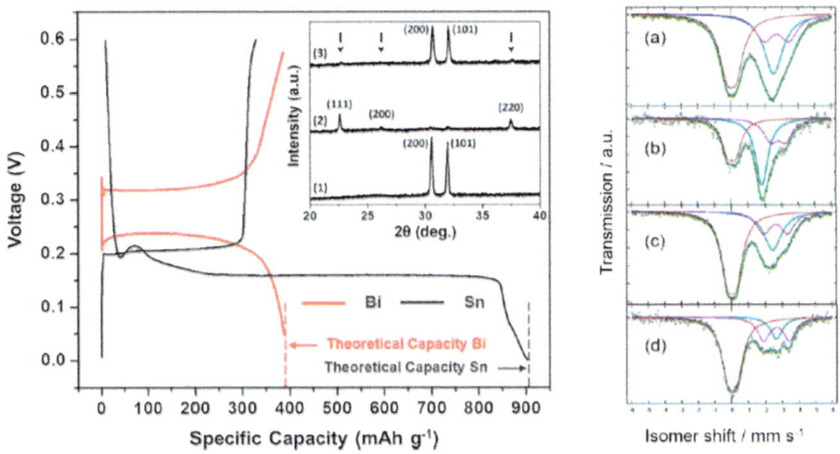

Figure 5.12 Left: first cycle of a Sn-based electrode in a Mg battery, showing the very high irreversible process during the first charge. The inset proves the formation of a Mg_2Sn alloy during the electrochemical reduction. Reproduced from ref. 39 with permission from the Royal Society of Chemistry. Right: Mössbauer spectrum following the reversible magnesiation of a Sn-based electrode. Reproduced from ref. 76, https://doi.org/10.3390/nano8070501, under the terms of the CC BY 4.0 licence, https://creativecommons.org/licenses/by/4.0/.

to Mössbauer spectroscopy, the group of Tirado followed the progressive magnesiation of tin and underlined a multiphase mechanism involving the coexistence of c-Mg_2Sn, distorted $Mg_{2-\delta}Sn$, and Mg-doped-Sn (Figure 5.13, right).[76] Nguyen *et al.* investigated the evolution of a tin electrode surface during electrochemical cycling in a RMB with an APC-type electrode.[77] Their XPS analysis revealed the interfacial relationship between Sn and PhMgCl, as well as THF decomposition at a very low voltage. The resulting film slows down the cation diffusion – with a Mg^{2+} diffusivity of 10–11 $cm^2\ s^{-1}$ – explaining the overall sluggish kinetics.

The alloying behavior of lead (Figure 5.13, left) was studied by Obrovac and coworkers on both sputtered and composite electrodes with the classical Grignard-based electrolyte (EtMgCl:–$AlCl_3$ in THF).[78] Interestingly, all the electrochemical analysis was performed at 60 °C. A first discharge down to 0 V shows a continuous plateau at low voltage; however with a corresponding capacity way beyond the theoretical one. The authors suggested a catalytic reaction between the electrode surface and the electrolyte, as has also been reported on a Pb-surface in sodium- and potassium-ion batteries.[79,80] An initial applied voltage of 5 mV appeared necessary to inhibit the catalytic reaction and to allow magnesiation. At C/40, the galvanostatic reduction shows a very flat plateau at 125 mV, and the formation of Mg_2Pb at the end of discharge was confirmed by *ex situ* X-ray diffraction. This two-phase reaction offers theoretically a very high volumetric capacity of 2300 Ah L^{-1} thanks to the high density of lead.

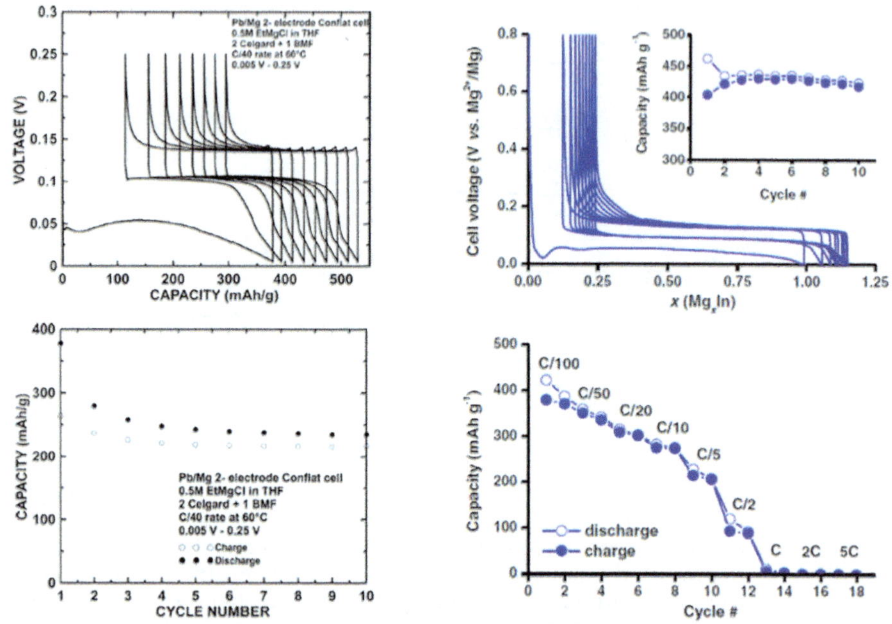

Figure 5.13 Electrochemical behavior of lead- and indium-based electrodes (left and right columns, respectively). Left: Reproduced from ref. 78 with permission from Elsevier, Copyright 2016. Right: Reproduced from ref. 81 with permission from Elsevier, Copyright 2015.

The electrochemical alloying of magnesium with indium has been also investigated following the above mentioned strategies.[81] Electrochemical reduction leads to the formation of crystalline MgIn, with a high theoretical capacity of 467 mA h g^{-1}. At C/100, the lowest alloying potential ever reported *versus* magnesium was measured (0.06 V) (Figure 5.13, right). It is important to mention that MgIn is not the most magnesiated alloy reported in the Mg–In phase diagram. Indeed, Mg_3In also exists, and would lead to an even greater specific capacity. Electrochemical tests performed at a higher current density showed that the kinetic parameters of this alloying reaction are important.

5.3.2 Electrochemical Behavior of Bimetallic Alloy Electrodes

The investigation of intermetallic compounds as electrode materials mainly follows two strategies. The first one is to seek a synergetic effect between two electrochemically active elements. The second one is to expect a positive effect given by another metallic element that does not participate to the electrochemical reaction. For the specific case of the negative electrode in RMBs this concept was patented in 2014 with two different metal couples, Sn–Bi and In–Sn.[82,83]

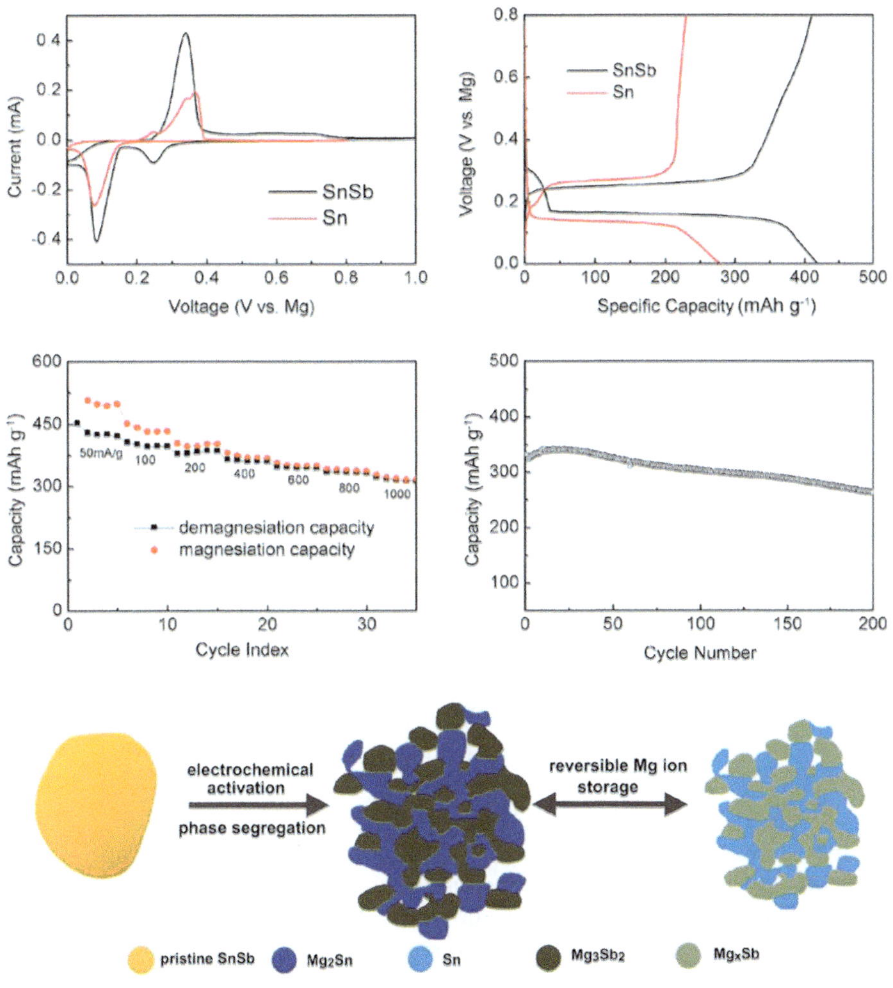

Figure 5.14 Electrochemical comparison between Sn and SnSb electrodes, showing the positive impact of antimony, and corresponding schematic picture of the electrochemical reaction of SnSb. Reproduced from ref. 87 with permission from John Wiley and Sons, Copyright 2015.

A SnSb alloy has already been investigated as a negative electrode material for Li-ion batteries.[84,85] In the context of magnesium systems, the use of this alloy is interesting to promote the activity of tin and to then improve the performance. From SnSb nanoparticles directly obtained on graphene sheets, a high reversible capacity of 420 mA h g^{-1} was obtained, with good rate capability and cyclic stability with an APC-based electrolyte (Figure 5.14). This very good behavior was explained by transmission electronic microscopy when revealing the formation of well-dispersed and stable tin particles during electrochemical conditioning in the presence of antimony.[86,87]

In 2016, Murgia and coworkers tried to combine the high specific capacity of indium and the good cycling behavior of bismuth.[88] The intermetallic

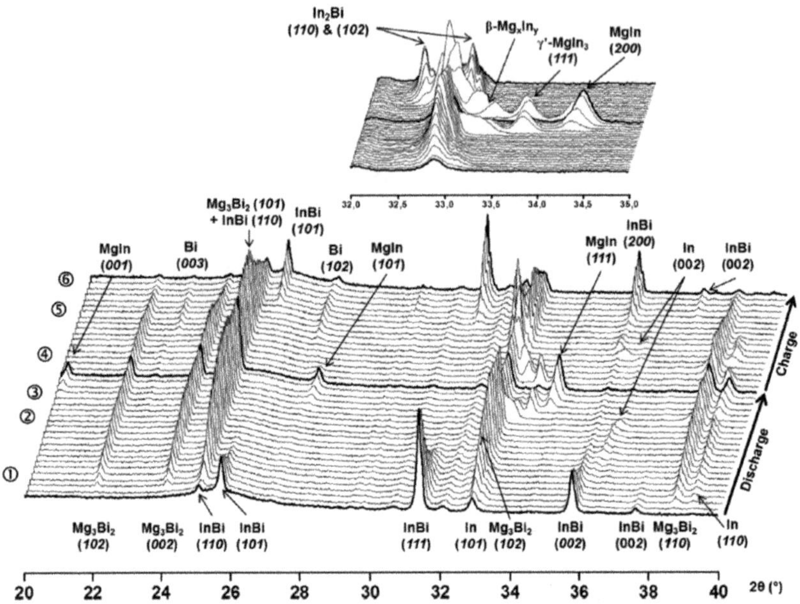

Figure 5.15 Operando XRD showing the complex magnesiation pathway of an intermetallic InBi electrode. The strong difference between this alloy and pure indium indicates the influence of the particle size in the reactivity towards electrochemical magnesium alloying. Reproduced from ref. 88 with permission from Elsevier, Copyright 2016.

compound InBi can be easily obtained by high-energy ball milling of metal powders. The first electrochemical discharge enables the formation of two alloys Mg_3Bi_2 and MgIn. The subsequent charge shows two distinct voltage plateaus in agreement with the de-alloying of MgIn and Mg_3Bi_2. Surprisingly, the pristine compound InBi is formed again at the end of charging, instead of a mixture of bismuth and indium. This result makes InBi the first example of a conversion-type electrode material in RMBs. Interestingly, an operando XRD follow-up of the reversible magnesiation demonstrated a completely different reaction pathway in comparison with the pure indium composite. Indeed, the first magnesiation of bismuth in the intermetallic compound InBi leads to very reactive indium particles. Then, instead of a simple two-phase process from In to MgIn, the reaction goes through the formation of the intermediate phases Mg_xIn_y and $MgIn_3$ (Figure 5.15).

The expected synergy between the two metals unfortunately does not resist higher current rates, in which only bismuth particles remain electrochemically active.

5.3.3 Interest in the Direct Use of Mg_xM Alloys

All of the previous examples demonstrate that tin, antimony, indium, bismuth and lead can undergo electrochemical alloying with magnesium. Magnesium removal from the as-formed alloys was also shown. As these alloys can also be synthesized by other means (high-temperature treatment, mechanochemical alloying), they can also be directly used as negative electrodes in RMBs. From a practical point of view, working with alloy powder could sometimes be easier than with soft and sticky metal particles (such as tin or indium). In addition, alloy powders enable easy electrode slurry-casted shaping. More importantly, they are compatible with more conventional magnesium electrolytes and thus open the door to the investigation of a high-voltage and Mg-free positive electrode material. This concept was patented in 2016 by Toyota when they highlighted the composite electrode Mg_xM/C prepared by high-energy mechanical alloying.[89]

Regarding the particularly good behavior of bismuth-based electrodes, Mg_3Bi_2 is now also widely investigated. Murgia and coworkers showed the first proof-of-concept of RMBs with Mg_3Bi_2 and Chevrel phase Mo_6S_8 as positive electrode materials.[69] With a simple electrode – $Mg(TFSI)_2$ dissolved in acetonitrile – a few galvanostatic cycles were reported. Following that, Tan and coworkers used a Mg_3Bi_2-based negative electrode to revisit several cathode materials (Prussian blue analogs (PBA), V_2O_5 nanosheets and Birnessite-MnO_2) with a $Mg(TFSI)_2$–LiTFSI/ACN electrolyte. An attractive PB–Mg_3Bi_2 full cell achieved a high discharge voltage of 2 V and delivered a specific capacity of 92 mA h g^{-1} with a low capacity fade rate of less than 0.06% per cycle at 200 mA g^{-1}. The corresponding energy density of 81 Wh kg^{-1} might indicate a potential application for grid-scale energy storage (Figure 5.16, top).[90]

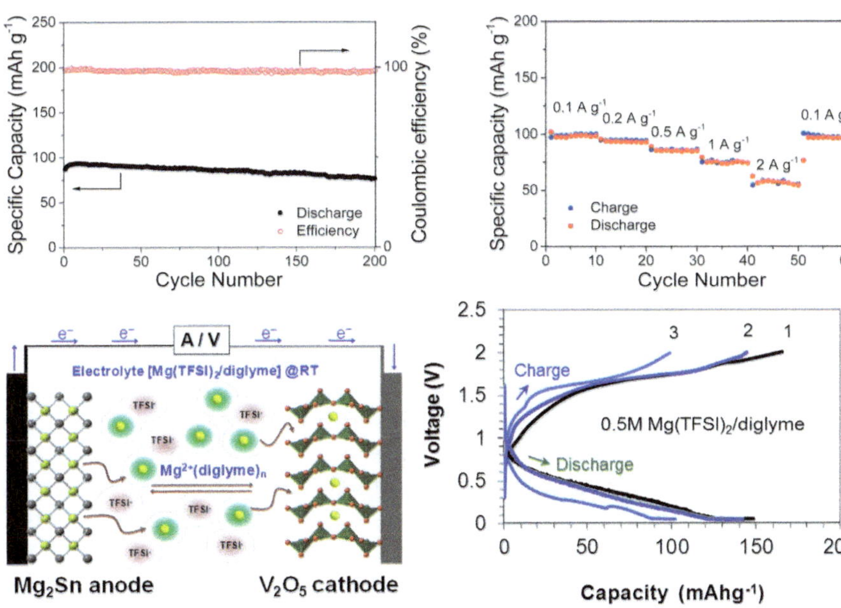

Figure 5.16 Prototype full-cells with Mg_3Bi_2 and Mg_2Sn as negative electrode materials. Top: Cycling performance of Mg_3Bi_2/PBA with 1 M LiTFSI–$Mg(TFSI)_2$/ACN as an electrolyte. Reproduced from ref. 90 with permission from the American Chemical Society, Copyright 2018. Bottom: Cycling performance of Mg_2Sn/V_2O_5 with 0.5 M $Mg(TFSI)_2$/diglyme as an electrolyte. Reproduced from ref. 92 with permission from Elsevier, Copyright 2017.

By investigating a Mg_2Sn alloy prepared though high-temperature melting, the group of Detsi demonstrated that the stress induced during the magnesium removal induces pulverization and formation of nanostructured Sn with a characteristic feature size in the sub-100 nm range.[91] With these tin particles, a reversible capacity of 300 mA h g^{-1} was demonstrated over 150 cycles at a rate of C/5 after the application of a combined sequence of regular galvanostatic cycling with an oxidative pulse to control passive film formation. However, the tests were only performed on half-cells with Grignard-based APC electrolytes. The use of Mg_2Sn with non-Grignard electrolytes was demonstrated by Nguyen and Son.[92] With Mg_2Sn powder prepared by a mechanochemical process, composite materials were prepared and then evaluated as a negative electrode in a complete Mg-ion battery with V_2O_5-based cathode and $Mg(TFSI)_2$ dissolved in diglyme as an electrolyte.

5.4 Conclusions and Perspective

The emerging research on RMBs is inspired by the high theoretical volumetric capacity and low toxicity of Mg as well as the sustainability issues that concern LIB materials and technology. The low cost and inherent safety of Mg advances RMBs as one of the viable technologies for large-scale stand-alone batteries for grid services and commercial deployment where the gravimetric energy density is not a crucial factor. However, RMB technology is not yet commercialized owing to the perennial challenges concerning the materials associated with both the electrodes and electrolytes with respect to performance, stability and cyclability. Among the components of RMBs, the anode is the least investigated, which is a concern considering the limitations it has on the choice of electrolytes and cathodes. This is also quite evident when comparing the volume of work reported towards the development of anodes for LIBs and SIBs to that for RMBs.

Although important breakthroughs in the development of high-voltage, non-chloride-based electrolytes have been reported in recent years, the issues with respect to the passivation of magnesium metal in the electrolytes are still persistent. Different approaches that are advocated in the development of new anode materials and structures for RMBs were discussed in the above sections. Unlike LIBs, graphite or other carbon-based compounds are thermodynamically unfavorable towards insertion of Mg^{2+} into the structure, which makes them a poor choice for anode materials. Other insertion-type materials based on oxides of titanium and vanadium were found to show moderate specific capacity with good reversibility. However, these materials suffer from sluggish diffusion of Mg^{2+} in the oxide hosts, poor initial coulombic efficiency and low rate capability. Some of the challenges related to diffusion kinetics and electronic conductivity of the oxide materials could be addressed through nanostructuring, tailoring of the morphology and particle size as well as conductive coatings. No significant work has yet been reported on these type of materials towards long term cyclability and issues related to volume expansion. The lack of availability of stable cathodes and electrolytes also limits investigation in these directions.

Alloy-type electrodes made from p-block elements offer an interesting playground for those who want to reach high capacities with RMBs. Today, the information on the type of magnesiated alloy that could possibly form through the electrochemical process is available in the literature for each alloying element. From a theoretical point of view, lead and tin are the most promising electrode materials. In particular, lead offers a high volumetric capacity exceeding 5800 mA h cm^{-3} with the formation of Mg_2Pb, whereas tin exhibits a remarkable specific capacity of 900 mA h g^{-1} suggesting that the theoretical capacity could be practically achieved. However, bismuth-based anode materials have exhibited the best electrochemical behavior so far.

Given the limited specific capacity of RMBs (which often arises from the cathode), alloying-type anodes are still a promising research direction. Alloying-type anodes offer the best possibility so far to be combined with high-voltage electrolytes and offer reversible capacities significantly greater than those of insertion-type materials. Nevertheless, the challenges related to volume expansion, fast capacity fading and poor rate capability of these type of anodes are yet to be addressed. Upcoming trends in negative electrode materials should take a composite approach to seek a synergetic effect between various materials. Whereas alloying-type materials can provide a high specific capacity, oxide insertion-type materials could be of interest as additives to strengthen particular properties. In addition, the novel work by Ban *et al.* regarding the development of an artificial Mg^{2+} conducting layer directly over Mg particles is worth mentioning as it addresses the compatibility issues between the anode and electrolyte.[18] Another interesting approach is the capacitive storage of Mg^{2+} on the surface and interface of high surface area anodes.[93]

References

1. M. Fichtner, *Beilstein J. Nanotechnol.*, 2015, **6**, 1601.
2. J. Muldoon, C. B. Bucur and T. Gregory, *Chem. Rev.*, 2014, **114**, 11683.
3. J. Song, E. Sahadeo, M. Noked and S. B. Lee, *J. Phys. Chem. Lett.*, 2016, 7, 1736.
4. D. M. Overcash and F. Mathers, *Trans. Electrochem. Soc.*, 1933, **64**, 305.
5. T. D. Gregory, R. J. Hoffman and R. C. Winterton, *J. Electrochem. Soc.*, 1990, **137**, 775.
6. Z. Lu, A. Schechter and M. Moshkovich, *J. Electroanal. Chem.*, 1999, **466**, 203.
7. L. Gaddum and H. French, *J. Am. Chem. Soc.*, 1927, **49**, 1295.
8. J. D. Genders and D. Pletcher, *J. Electroanal. Chem. Interfacial Electrochem.*, 1986, **199**, 93.

9. C. Liebenow, *J. Appl. Electrochem.*, 1997, **27**, 221.
10. D. Aurbach, I. Weissman, Y. Gofer and E. Levi, *Chem. Rec.*, 2003, **3**, 61.
11. P. Saha, M. K. Datta, O. I. Velikokhatnyi, A. Manivannan, D. Alman and P. N. Kumta, *Prog. Mater. Sci.*, 2014, **66**, 1.
12. D. Schloffer, S. Bozorgi, P. Sherstnev, C. Lenardt and B. Gollas, *J. Power Sources*, 2017, **367**, 138.
13. Z. Wu and W. Curtin, *Nature*, 2015, **526**, 62.
14. O. Chusid, Y. Gofer, H. Gizbar, Y. Vestfrid, E. Levi, D. Aurbach and I. Riech, *Adv. Mater.*, 2003, **15**, 627.
15. W. Li, C. Li, C. Zhou, H. Ma and J. Chen, *Angew. Chem., Int. Ed.*, 2006, **45**, 6009.
16. B. Peng, J. Liang, Z. Tao and J. Chen, *J. Mater. Chem.*, 2009, **19**, 2877.
17. Y. Liang, R. Feng, S. Yang, H. Ma, J. Liang and J. Chen, *Adv. Mater.*, 2011, **23**, 640.
18. S.-B. Son, T. Gao, S. P. Harvey, K. X. Steirer, A. Stokes, A. Norman, C. Wang, A. Cresce, K. Xu and C. Ban, *Nat. Chem.*, 2018, **10**, 532.
19. J. Dahn, *Phys. Rev. B*, 1991, **44**, 9170.
20. Z. Jian, W. Luo and X. Ji, *J. Am. Chem. Soc.*, 2015, **137**, 11566.
21. A. P. Cohn, N. Muralidharan, R. Carter, K. Share, L. Oakes and C. L. Pint, *J. Mater. Chem. A*, 2016, **4**, 14954.
22. P. Novák, R. Imhof and O. Haas, *Electrochim. Acta*, 1999, **45**, 351.
23. Y. Liu, B. V. Merinov and W. A. Goddard, *Proc. Natl. Acad. Sci. U. S. A.*, 2016, **113**, 3735.
24. D.-M. Kim, S. C. Jung, S. Ha, Y. Kim, Y. Park, J. H. Ryu, Y.-K. Han and K. T. Lee, *Chem. Mater.*, 2018, **30**, 3199.
25. D. Er, E. Detsi, H. Kumar and V. B. Shenoy, *ACS Energy Lett.*, 2016, **1**, 638.
26. J. Sun, G. Zheng, H.-W. Lee, N. Liu, H. Wang, H. Yao, W. Yang and Y. Cui, *Nano Lett.*, 2014, **14**, 4573.
27. C. M. Park and H. J. Sohn, *Adv. Mater.*, 2007, **19**, 2465.
28. Y. Kim, Y. Park, A. Choi, N. S. Choi, J. Kim, J. Lee, J. H. Ryu, S. M. Oh and K. T. Lee, *Adv. Mater.*, 2013, **25**, 3045.
29. W.-J. Li, S.-L. Chou, J.-Z. Wang, H.-K. Liu and S.-X. Dou, *Nano Lett.*, 2013, **13**, 5480.
30. X. Han, C. Liu, J. Sun, A. D. Sendek and W. Yang, *RSC Adv.*, 2018, **8**, 7196.
31. W. Jin, Z. Wang and Y. Q. Fu, *J. Mater. Sci.*, 2016, **51**, 7355.
32. A. J. Mannix, X.-F. Zhou, B. Kiraly, J. D. Wood, D. Alducin, B. D. Myers, X. Liu, B. L. Fisher, U. Santiago and J. R. Guest, *Science*, 2015, **350**, 1513.
33. B. Feng, J. Zhang, Q. Zhong, W. Li, S. Li, H. Li, P. Cheng, S. Meng, L. Chen and K. Wu, *Nat. Chem.*, 2016, **8**, 563.
34. H. Jiang, Z. Lu, M. Wu, F. Ciucci and T. Zhao, *Nano Energy*, 2016, **23**, 97.
35. Y. Zhang, Z.-F. Wu, P.-F. Gao, S.-L. Zhang and Y.-H. Wen, *ACS Appl. Mater. Interfaces*, 2016, **8**, 22175.
36. P. Xiang, X. Chen, W. Zhang, J. Li, B. Xiao, L. Li and K. Deng, *Phys. Chem. Chem. Phys.*, 2017, **19**, 24945.
37. Y. Xie, Y. Dall'Agnese, M. Naguib, Y. Gogotsi, M. W. Barsoum, H. L. Zhuang and P. R. Kent, *ACS Nano*, 2014, **8**, 9606.

38. Y. Shao, M. Gu, X. Li, Z. Nie, P. Zuo, G. Li, T. Liu, J. Xiao, Y. Cheng, C. Wang, J. G. Zhang and J. Liu, *Nano Lett.*, 2014, **14**, 255.
39. N. Singh, T. S. Arthur, C. Ling, M. Matsui and F. Mizuno, *Chem. Commun.*, 2013, **49**, 149.
40. H. D. Yoo, I. Shterenberg, Y. Gofer, G. Gershinsky, N. Pour and D. Aurbach, *Energy Environ. Sci.*, 2013, **6**, 2265.
41. D. Aurbach, Z. Lu, A. Schechter, Y. Gofer, H. Gizbar, R. Turgeman, Y. Cohen, M. Moshkovich and E. Levi, *Nature*, 2000, **407**, 724.
42. T.-F. Yi, L.-J. Jiang, J. Shu, C.-B. Yue, R.-S. Zhu and H.-B. Qiao, *J. Phys. Chem. Solids*, 2010, **71**, 1236.
43. X. Lin, F. Pan and H. J. Wang, *Mater. Technol.*, 2014, **29**, A82.
44. X. Sun, P. V. Radovanovic and B. Cui, *New J. Chem.*, 2015, **39**, 38.
45. N. Wu, Y.-C. Lyu, R.-J. Xiao, X. Yu, Y.-X. Yin, X.-Q. Yang, H. Li, L. Gu and Y.-G. Guo, *NPG Asia Mater.*, 2014, **6**, e120.
46. N. Wu, Y. X. Yin and Y. G. Guo, *Chem. – Asian J.*, 2014, **9**, 2099.
47. N. Wu, Z. Z. Yang, H. R. Yao, Y. X. Yin, L. Gu and Y. G. Guo, *Angew. Chem., Int. Ed.*, 2015, **54**, 5757.
48. H. Kang, Y. Liu, K. Cao, Y. Zhao, L. Jiao, Y. Wang and H. Yuan, *J. Mater. Chem. A*, 2015, **3**, 17899.
49. Q. Wang, C. Zhao, Y. Lu, Y. Li, Y. Zheng, Y. Qi, X. Rong, L. Jiang, X. Qi and Y. J. S. Shao, *Small*, 2017, **13**, 1701835.
50. W. Wang, W. Li, S. Wang, Z. Miao, H. K. Liu and S. Chou, *J. Mater. Chem. A*, 2018, **6**, 6183.
51. C. Chen, J. Wang, Q. Zhao, Y. Wang and J. Chen, *ACS Energy Lett.*, 2016, **1**, 1165.
52. H. Li, X. Liu, T. Zhai, D. Li and H. Zhou, *Adv. Energy Mater.*, 2013, **3**, 428.
53. C. Zhang, C. Liu, X. Nan, H. Song, Y. Liu, C. Zhang and G. Cao, *ACS Appl. Mater. Interfaces*, 2015, **8**, 680.
54. C. Liao, Q. Zhang, T. Zhai, H. Li and H. Zhou, *Energy Storage Mater.*, 2017, **7**, 17.
55. Q. Li, Q. Wei, J. Sheng, M. Yan, L. Zhou, W. Luo, R. Sun and L. Mai, *Adv. Sci.*, 2015, **2**, 1500284.
56. J. Zeng, Y. Yang, C. Li, J. Li, J. Huang, J. Wang and J. Zhao, *Electrochim. Acta*, 2017, **247**, 265.
57. P. Poizot, E. Baudrin, S. Laruelle, L. Dupont, M. Touboul and J.-M. Tarascon, *Solid State Ionics*, 2000, **138**, 31.
58. N. Yan, Y. Xu, H. Li and W. J. Chen, *Mater. Lett.*, 2016, **165**, 223.
59. X. Liu, Y. Cao, H. Zheng, X. Chen and C. Feng, *Appl. Surf. Sci.*, 2017, **394**, 183.
60. H. Zheng, Y. Yang, X. Liu, Z. Guo and C. Feng, *J. Nanopart. Res.*, 2017, **19**, 243.
61. H. Zhang, K. Ye, K. Zhu, R. Cang, J. Yan, K. Cheng, G. Wang and D. Cao, *Chem. - Eur. J.*, 2017, **23**, 17118.
62. M. Obrovac and V. Chevrier, *Chem. Rev.*, 2014, **114**, 11444.
63. A. Casimir, H. Zhang, O. Ogoke, J. C. Amine, J. Lu and G. Wu, *Nano Energy*, 2016, **27**, 359.

64. M. Zhao, Q. Zhao, J. Qiu, H. Xue and H. Pang, *RSC Adv.*, 2016, **6**, 95449.
65. A. Darwiche, C. Marino, M. T. Sougrati, B. Fraisse, L. Stievano and L. Monconduit, *J. Am. Chem. Soc.*, 2012, **134**, 20805.
66. E. Irisarri, A. Ponrouch and M. Palacin, *J. Electrochem. Soc.*, 2015, **162**, A2476.
67. T. S. Arthur, N. Singh and M. Matsui, *Electrochem. Commun.*, 2012, **16**, 103.
68. T. R. Penki, G. Valurouthu, S. Shivakumara, V. A. Sethuraman and N. Munichandraiah, *New J. Chem.*, 2018, **42**, 5996.
69. F. Murgia, L. Stievano, L. Monconduit and R. Berthelot, *J. Mater. Chem. A*, 2015, **3**, 16478.
70. M. Hattori, K. Yamamoto, M. Matsui, K. Nakanishi, T. Mandai, A. Choudhary, Y. Tateyama, K. Sodeyama, T. Uchiyama, Y. Orikasa, Y. Tamenori, T. Takeguchi, K. Kanamura and Y. Uchimoto, *J. Phys. Chem. C*, 2018, **122**, 25204.
71. Z. Liu, J. Lee, G. Xiang, H. F. Glass, E. N. Keyzer, S. E. Dutton and C. P. Grey, *Chem. Commun.*, 2017, **53**, 743.
72. K. V. Kravchyk, L. Piveteau, R. Caputo, M. He, N. P. Stadie, M. I. Bodnarchuk, R. T. Lechner and M. V. Kovalenko, *ACS Nano*, 2018, **12**, 8297.
73. W. Jin, Z. Li, Z. Wang and Y. Q. Fu, *Mater. Chem. Phys.*, 2016, **182**, 167.
74. S. C. Jung and Y.-K. Han, *J. Phys. Chem. C*, 2018, **122**, 17643.
75. F. Murgia, D. Laurencin, E. T. Weldekidan, L. Stievano, L. Monconduit, M.-L. Doublet and R. Berthelot, *Electrochim. Acta*, 2018, **259**, 276.
76. F. Nacimiento, M. Cabello, C. Pérez-Vicente, R. Alcántara, P. Lavela, G. Ortiz and J. J. N. Tirado, *Nanomaterials*, 2018, **8**, 501.
77. D. T. Nguyen, X. M. Tran, J. Kang and S. W. Song, *ChemElectroChem*, 2016, **3**, 1813.
78. K. Periyapperuma, T. T. Tran, M. Purcell and M. Obrovac, *Electrochim. Acta*, 2015, **165**, 162.
79. L. Ellis, B. Wilkes, T. Hatchard and M. Obrovac, *J. Electrochem. Soc.*, 2014, **161**, A416.
80. V. Gabaudan, R. Berthelot, L. Stievano and L. Monconduit, *ACS Omega*, 2018, **3**, 12195.
81. F. Murgia, E. T. Weldekidan, L. Stievano, L. Monconduit and R. Berthelot, *Electrochem. Commun.*, 2015, **60**, 56.
82. N. Singh and M. Matsui, *US Pat.*, 8647770 B2, 2014.
83. N. Singh and M. Matsui, *US Pat.*, 8673493 B2, 2014.
84. F. J. Fernández-Madrigal, P. Lavela, C. P. Vicente, J. L. Tirado, J. C. Jumas and J. Olivier-Fourcade, *Chem. Mater.*, 2002, **14**, 2962.
85. P. Antitomaso, B. Fraisse, L. Stievano, S. Biscaglia, D. Aymé-Perrot, P. Girard, M. T. Sougrati and L. Monconduit, *J. Mater. Chem. A*, 2017, **5**, 6546.
86. L. R. Parent, Y. Cheng, P. V. Sushko, Y. Shao, J. Liu, C. M. Wang and N. D. Browning, *Nano Lett.*, 2015, **15**, 1177.

87. Y. Cheng, Y. Shao, L. R. Parent, M. L. Sushko, G. Li, P. V. Sushko, N. D. Browning, C. Wang and J. Liu, *Adv. Mater.*, 2015, **27**, 6598.
88. F. Murgia, L. Monconduit, L. Stievano and R. J. E. A. Berthelot, *Electrochim. Acta*, 2016, **209**, 730.
89. R. Zhang and F. Mizuno, *US Pat.*, 9444094 B2, 2016.
90. Y. H. Tan, W. T. Yao, T. Zhang, T. Ma, L. L. Lu, F. Zhou, H. B. Yao and S. H. Yu, *ACS Nano*, 2018, **12**, 5856.
91. H. Yaghoobnejad Asl, J. Fu, H. Kumar, S. S. Welborn, V. B. Shenoy and E. Detsi, *Chem. Mater.*, 2018, **30**, 1815.
92. D.-T. Nguyen and S.-W. Song, *J. Power Sources*, 2017, **368**, 11.
93. Y. Meng, D. Wang, Y. Zhao, R. Lian, Y. Wei, X. Bian, Y. Gao, F. Du, B. Liu and G. Chen, *Nanoscale*, 2017, **9**, 12934.

CHAPTER 6

Mg Stripping and Plating at Magnesium Metal and Intermetallic Anodes

M. MATSUI*

Department of Chemical Science and Engineering, Kobe University, 1-1 Rokkodai-cho, Nada-ku, Kobe, 657-8501, Japan
*E-mail: matsui@godzilla.kobe-u.ac.jp

6.1 Introduction

Magnesium metal is an ideal negative electrode material because of its high volumetric capacity (3833 mA h cm^{-3}), low equilibrium potential (−2.3 V *vs.* SHE), and non-dendritic growth during the deposition process.

The electrodeposition of magnesium metal in high-temperature molten salts has been studied since the early 19th century and still remains as a common production process of magnesium metal at this time.[1] On the other hand, the electrodeposition of magnesium metal for rechargeable battery applications has only been studied since the late 20th century. Early work by Gregory *et al.* was based on organomagnesium compounds such as Grignard reagents.[2] Aurbach *et al.* designed various electrolyte solutions, so-called "magnesium organohaloaluminates", which show excellent reversibility and a relatively wide electrochemical window.[3] After these studies, various new classes of electrolyte solutions were developed over the last decade. In

Energy and Environment Series No. 23
Magnesium Batteries: Research and Applications
Edited by Maximilian Fichtner

Published by the Royal Society of Chemistry, www.rsc.org

addition, a series of analytical or theoretical studies were also carried out to understand the electrochemical deposition/dissolution process of magnesium metal. In the present chapter, we review the recent progress of the magnesium metal anode and magnesium-based intermetallic compounds. We start with an overview of the electrolyte solutions, because the choice of the electrolyte solution is very limited compared with lithium or sodium systems. In this part, we focus on the classification of the various electrolyte solutions to understand the electrochemically active species. Secondly, we dig into some of the details in the deposition/dissolution process of the magnesium metal. We review recent studies on the magnesium anode carried out by electrochemical, spectroscopic and theoretical approaches. In addition, two of the key challenges of magnesium metal anodes, dendritic growth and passivation layer, are reviewed for further studies. Finally, we discuss the recent updates concerning magnesium-based intermetallic compounds as alternate negative electrode materials, because intermetallic anodes have advantages in the development of a battery system, due to their compatibilities with a wide variety of electrolyte solutions.

6.2 Overview of the Electrolyte Solutions

In this section, we give a brief overall review of electrolyte solutions for the electrodeposition of magnesium. As described in the introduction, the electrodeposition of magnesium metal in high temperature molten salt is a common process for the production of magnesium metal. Typically molten halide electrolytes such as $MgCl_2$–KCl binary or $MgCl_2$–KCl–NaCl ternary systems are used as the electrolysis bath at an operating temperature of 700–750 °C. The electrochemically active species in the molten halide electrolytes are complex anions such as $MgCl_6^{4-}$, $MgCl_4^{2-}$ and $MgCl_3^{-}$.

The electrodeposition of magnesium metal at room temperature was originally studied using electrolyte solutions containing organomagnesium compounds such as Grignard reagents, because magnesium metal is not deposited in the electrolysis bath containing simple magnesium salts such as magnesium perchlorate. Jolibois firstly reported the room temperature electrodeposition of magnesium metal using a solution of diethylmagnesium (Et_2Mg) and magnesium iodide (MgI_2) in diethyl ether (Et_2O).[4] The electrolysis of a Grignard reagent was reported by Gaddum *et al.*[5] They reported that benzylmagnesium chloride shows better coulombic efficiency than phenylmagnesium bromide, and proposed a possible deposition mechanism. 25 years later, Connor *et al.* successfully deposited magnesium metal using organomagnesium-free electrolysis baths.[6] There, the magnesium is electrodeposited from *in situ* formed $Mg(BH_4)_2$ solution from a mixture of $MgBr_2$ and $LiBH_4$ solutions. In this work, the coulombic efficiency of the deposition process was significantly improved up to 90% compared with previous works. The reversible deposition/dissolution process of the magnesium metal in EtMgBr in THF solution was investigated by Genders *et al.*[7] In this paper, they also reported that the magnesium metal cannot be deposited

Table 6.1 Magnesium plating from organomagnesium chloride/$AlCl_3$/THF baths at 1.25 mA cm^{-2}. Reproduced from ref. 2 with permission from The Electrochemical Society, Copyright 1990.

R Group	[RMgCl] (mol L^{-1})	[$AlCl_3$] (mol L^{-1})>	Current efficiency, % Anodic	Cathodic	Mg deposition quality
Methyl	1.5	0.10	100	101	Dendritic
Ethyl	0.8	0.10	98.9	100	Excellent, small grain
Ethyl	1.5	0.10	99.5	100	Excellent, large grain
Ethyl	2.0	0.20	90.1	—	Very dendritic
Butyl	1.0	0.1	104	98.4	Excellent, small grain

from conventional electrolyte solutions using $MgBr_2$ and $Mg(ClO_4)_2$. They concluded that the simple Mg^{2+} cation is unlikely to be the electrochemically active species.

Based on these works, Gregory *et al.* performed a comprehensive study on rechargeable magnesium batteries in 1990.[2] Their electrolyte solutions, Grignard reagents (MeMgCl and EtMgCl) with a small amount of aluminum chloride in THF, showed an excellent coulombic efficiency of 100%, as shown in Table 6.1. Interestingly, they claimed that dendritic growth of magnesium occurred in some of the electrolyte solutions. 10 years after this study, Aurbach *et al.* finally succeeded in the development of a rechargeable magnesium battery prototype using Bu_2Mg–$EtAlCl_2$ complex salt (dichloro complex: DCC) in THF as the electrolyte solution.[8] The electrolyte solution showed improved anodic stability >2.0 V *vs.* Mg while normal Grignard reagents such as BuMgCl solution are only stable below 1.5 V *vs.* Mg. The electrolyte solution coupled with a Mo_6S_8 Chevrel phase allowed stable charge/discharge cycling of >500 cycles. The excellent cycling performance suggests no dendritic growth of the magnesium metal anodes. They also confirmed the smooth surface morphology of the electrodeposited magnesium using *in situ* STM in the following works, as shown in Figure 6.1.[9,10] They also made continuous efforts towards the improvement of the magnesium organohaloaluminate electrolytes and discovered an all phenyl complex (APC) electrolyte that shows a high anodic stability of >3.0 V *vs.* Mg, as shown in Figure 6.2.[11] Despite the improvement of the electrolyte properties, there still remains a question concerning the electrochemically active species in the electrolyte solution. Kim *et al.* developed a non-nucleophilic electrolyte [$Mg_2(\mu\text{-}Cl)_3\cdot 6THF$][$HMDSAlCl_3$] for Mg–S batteries.[12] In this paper, they discussed the formation process of the electrochemically active [$Mg_2(\mu\text{-}Cl)_3\cdot 6THF$]$^+$ cation. They also claim that this "dimer" magnesium cation is probably the electrochemically active species in the organohaloaluminate electrolytes as well. The anodic stability of the [$Mg_2(\mu\text{-}Cl)_3\cdot 6THF$][$HMDSAlCl_3$] electrolyte solution was improved *via* the removal of a [$HMDS_2Mg$] species during the recrystallization process.

Despite the improvement in the anodic stability of the electrolyte solutions, they reported corrosive properties of the electrolyte solutions containing

Figure 6.1 *In situ* STM images of a gold electrode treated in a 1.0 mol L^{-1} BuMgCl in THF solution. (a) A 3D image of the as-prepared electrode at OCV, (b) 2D and 3D images after galvanostatic Mg deposition (0.9 C cm^{-2}), a magnified image of (b) is removed from the original copy. Reproduced from ref. 9 with permission from The Electrochemical Society, Copyright 2001.

halides towards various metals at high electrode potential in their following paper. Although the oxidative stability of the electrolyte solution was improved to >3.0 V, the practical voltage window of the electrolyte solutions containing halides still remains <2.5 V due to their severe corrosive properties, as shown in Figure 6.3.[13,14]

In order to avoid corrosion problems, halide-free electrolyte solutions have been reported by several groups. The groups used $Mg(TFSA)_2$ dissolved in glymes as the electrolyte solutions, and successfully confirmed the electrodeposition of magnesium metal by XRD.[15–17] Figure 6.4 shows a cyclic voltammogram of a $Mg(TFSA)_2$-based electrolyte solution reproduced by our group. The cyclic voltammogram (CV) shows a very high overpotential and poor coulombic efficiencies during the deposition/dissolution process. The reversibility of the magnesium metal anode in the $Mg(TFSA)_2$-based electrolyte solution was improved by the addition of $MgCl_2$, as shown in Figure 6.5.[18] In this case, the dimer $[Mg_2(\mu\text{-}Cl)_3]^+$ cation analogues are probably the electrochemically active species.

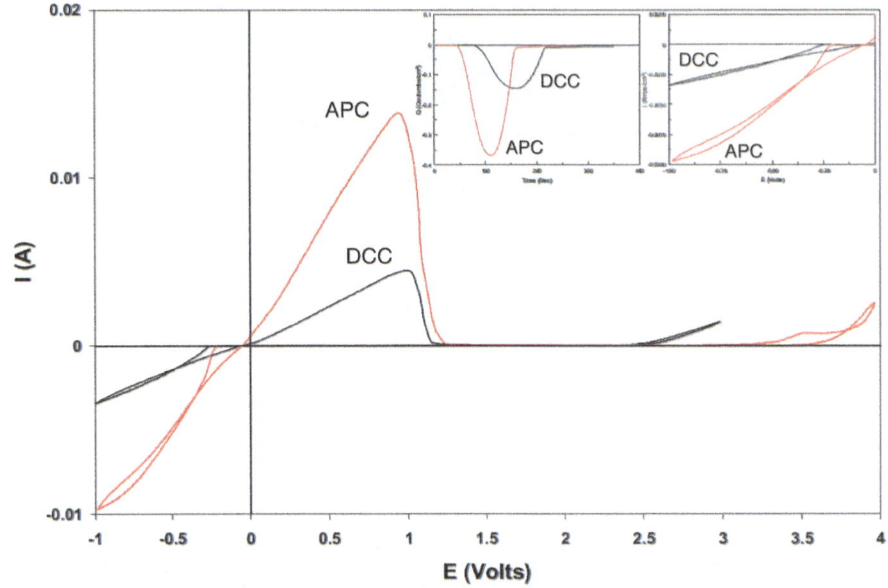

Figure 6.2 Comparison between the voltammetric behaviour of the APC and DCC electrolytes. The APC electrolyte shows higher deposition/dissolution current and higher oxidation stability. Right inset: enlargement of the CV at around 0 V *vs.* Mg, left inset: the charge balance upon Mg deposition/dissolution cycling. Reproduced from ref. 11 with permission from John Wiley and Sons, Copyright © 2007 WILEY-VCH Verlag GmbH & Co. KGaA, Weinheim.

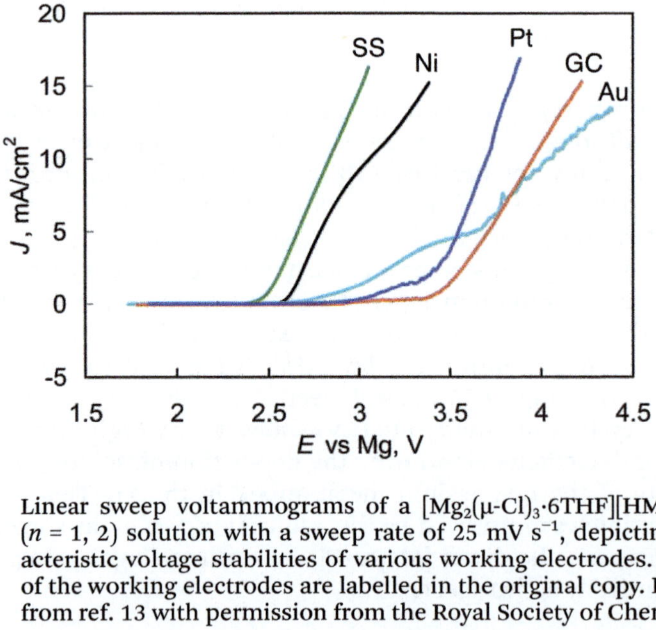

Figure 6.3 Linear sweep voltammograms of a $[Mg_2(\mu\text{-}Cl)_3{\cdot}6THF][HMDS_nAlCl_{4-n}]$ (n = 1, 2) solution with a sweep rate of 25 mV s^{-1}, depicting the characteristic voltage stabilities of various working electrodes. The names of the working electrodes are labelled in the original copy. Reproduced from ref. 13 with permission from the Royal Society of Chemistry.

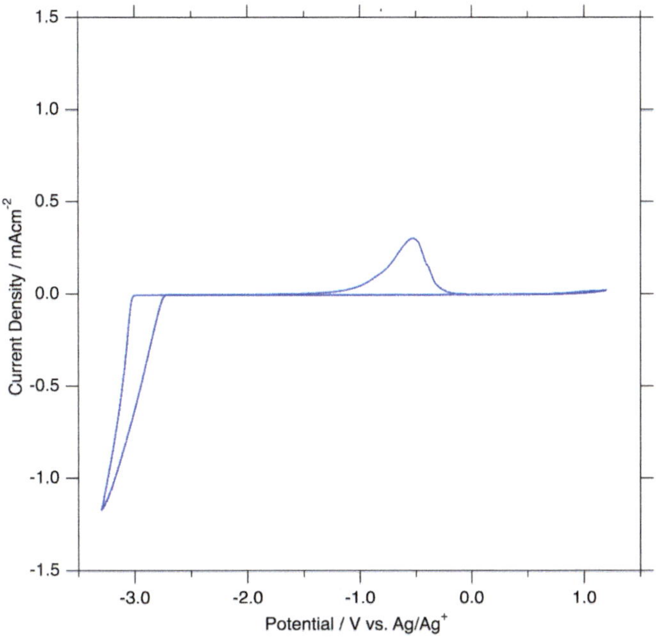

Figure 6.4 A cyclic voltammogram of 0.5 mol L^{-1} $Mg(TFSA)_2$ in BuMeG3 with a sweep rate of 1.0 mV s^{-1} using a Pt electrode taken by the authors' group.

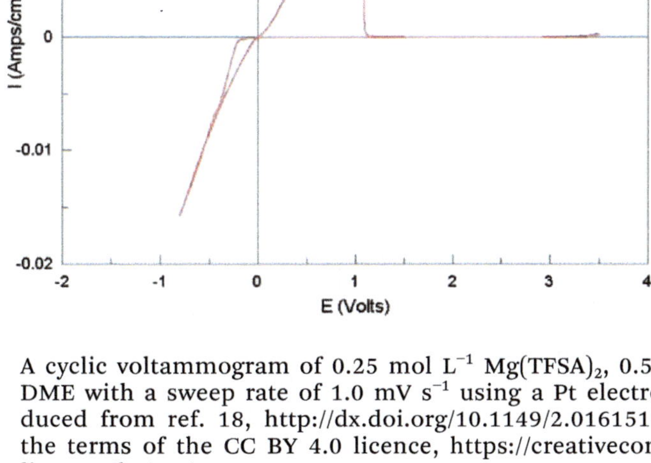

Figure 6.5 A cyclic voltammogram of 0.25 mol L^{-1} $Mg(TFSA)_2$, 0.5 mol L^{-1} in DME with a sweep rate of 1.0 mV s^{-1} using a Pt electrode. Reproduced from ref. 18, http://dx.doi.org/10.1149/2.0161513jes, under the terms of the CC BY 4.0 licence, https://creativecommons.org/licenses/by/4.0/.

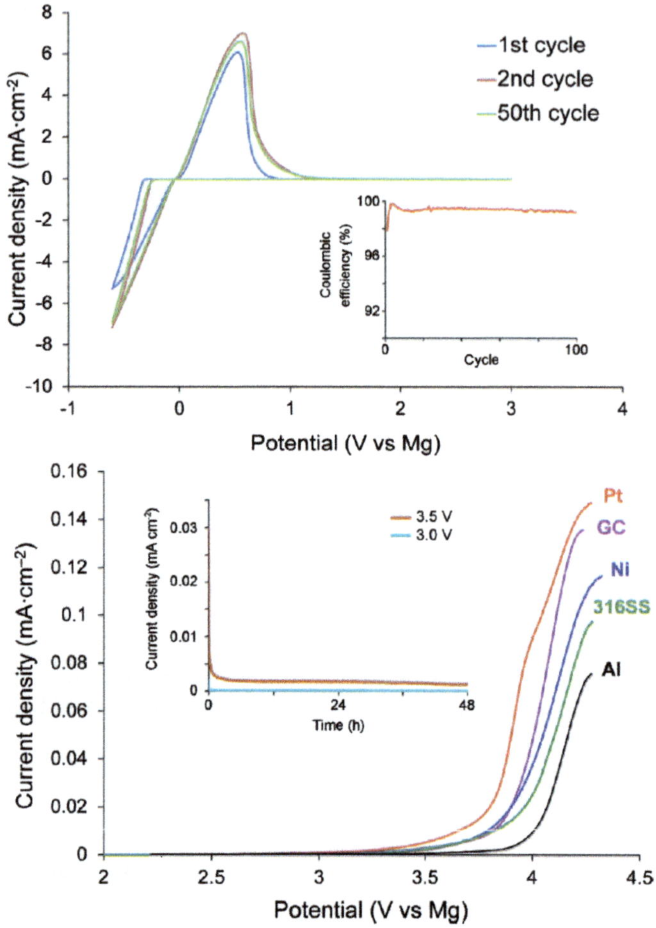

Figure 6.6 Left: selected cyclic voltammogram of a 0.75 mol L^{-1} MMC/G4 electrolyte solution on a Pt electrode with a sweep rate of 5 mV s^{-1}, right: linear sweep voltammograms of a 0.75 mol L^{-1} MMC/G4 electrolyte solution on different electrode materials with a sweep rate of 5 mV s^{-1} (inset: chronoamperometry of a 0.75 mol L^{-1} MMC/G4 electrolyte solution on 316SS disc electrodes at 3.0 and 3.5 V *vs.* Mg). Reproduced from ref. 19 with permission from John Wiley and Sons, © 2015 WILEY-VCH Verlag GmbH & Co. KGaA, Weinheim.

Major progress in halide-free electrolytes was made by Tutusaus *et al.* in 2015.[19] They developed a novel electrolyte salt magnesium monocarborane (MMC) and confirmed excellent reversibility without corrosion, as shown in Figure 6.6. They also analyzed the coordination structure of the Mg^{2+} ion using single-crystal XRD. The Mg^{2+} cation is well solvated with monoglyme and diglyme molecules without the formation of the dimer species. Herb *et al.* designed a fluorinated alkoxyaluminate electrolyte $Mg[Al(HFIP)_4]_2$, which also shows a highly reversible deposition/dissolution process over 500

Table 6.2 Three categories of the electrolyte solutions.

GEN	Electrolyte	Composition /Mg ions	Coulombic efficiency	Overpotential	Corrosion	Reference
1	$EtMgCl–AlCl_3$	1.5 mol L^{-1}	99.5%	N/A	Decompose	Gregory *et al.* 1990[2]
2	$Mg(AlCl_2BuEt)_2$ DCC	0.25 mol L^{-1}	100%	Low	Decompose	Aurbach *et al.* 2000[8]
2	$PhMgCl–AlCl_3$ (APC)	0.4 mol L^{-1}	100%	Low	Corrosive	Aurbach *et al.* 2007[11]
2	$[Mg_2(\mu\text{-}Cl)_3][HMDSAlCl_3]$	0.4 mol L^{-1}	100%	Low	Corrosive	Kim *et al.* 2011[12]
2/3	$Mg(TFSA)_2$ + $MgCl_2$	0.75 mol L^{-1}	100%	Low	Corrosive	Shterenberg *et al.* 2015[18]
3	$Mg(TFSA)_2$	0.5 mol L^{-1}	60%	High	Non corrosive	Ha *et al.* 2014[15] Fukutsuka *et al.* 2014[17]
3	MCC/G4	0.75 mol L^{-1}	94.4%	Low	Non corrosive	Tutusaus *et al.* 2015[19]
3	$Mg[B(hfip)_4]$	0.8 mol L^{-1}	>98%	Low	Non corrosive	Zao-Karger 2017[21]

cycles.[20] A fluorinated alkoxyborate electrolyte $Mg[B(HFIP)_4]_2$ developed by Zhao-Karger *et al.* also shows stable cycling performance with high oxidation stability.[21]

More detailed information is shared in the other chapters on electrolytes. We think the overall progress of the electrolyte solutions for the magnesium metal anode can be summarized into three categories, as shown in Table 6.2. The 1st GEN electrolytes are "organomagnesium compounds", just for electrodeposition. The 2nd GEN electrolytes are "controlled complex compounds", for improved reversibility and voltage window, and the 3rd GEN 3 electrolytes are "novel anions" for a halide-free system, including $Mg(TFSA)_2$–glyme solutions. One important point to remember is that all of the electrolytes are dissolved in "ether-based solvents", which is due to the reduction stability of the solvent molecules. Since the cathodic decomposition of the electrolyte solution can passivate the surface of the magnesium metal anode, all of the chemical species in the electrolyte solution need to be thermodynamically stable at the equilibrium potential of the magnesium metal. Ether-based solvents are the only choice to avoid passivation layer formation. The passivation layer is one of the most critical problems of the magnesium metal anode and will be discussed in the following section.

6.3 Deposition Mechanism

One of the most important questions in the electrodeposition process of magnesium metal is what the actual electrochemically active species in the electrolyte solution is, because the electrolyte solution containing the organomagnesium compounds has complicated equilibria. In the early work by Aurbach *et al.*, the DCC electrolyte solution containing only 0.25 mol L^{-1} of Mg^{2+} showed a higher deposition/dissolution current than a 2.0 mol L^{-1} of BuMgCl solution.[8] The result clearly suggests that the complex species is preferred in forming the electrochemically active species. In other words, the majority of the magnesium species in the BuMgCl solution are electrochemically inactive. They further performed electrochemical studies of the complex electrolyte solutions to optimize the composition of the Lewis base and Lewis acid and proposed the dimer complex structure as a possible electrochemically active species.[22] Later, Kim *et al.* successfully characterized an active dimer complex: $[Mg_2(\mu\text{-}Cl)_3\cdot 6THF][HMDSAlCl_3]$. They also proposed the formation process of the dimer complex. The source material RMgCl solution contains R_2Mg and $MgCl_2$ by Schlenk equilibrium. Also, the aluminate-based Lewis acid reacts with the R group in the RMgCl resulting in the formation of a $[MgCl]^+$ cation and a $[RAlCl_3]^-$ anion by transmetallation. Then, the $[Mg_2(\mu\text{-}Cl)_3]^+$ dimer cation is formed *via* reaction of $MgCl_2$ and $[MgCl]^+$.

The electrodeposition process also remains a fundamental issue. An electrochemical quartz crystal microbalance (EQCM) study by Aurbach *et al.* revealed that the deposition process is not based on a simple Mg^{2+}/Mg redox couple, but is instead controlled by adsorption of the complex species.[23] They proposed the adsorption of a positively charged Mg ionic species such as $[Mg_2Cl_3\cdot 6THF]^+$ or $[MgCl\cdot 5THF]^+$, as shown in Figure 6.7.[24] A similar reaction process was also

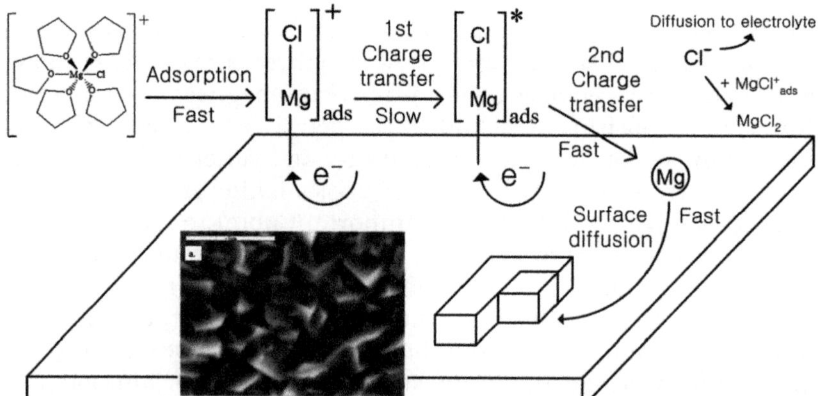

Figure 6.7 Possible Mg electrodeposition mechanism from dimer complex electrolyte solutions. Reproduced from ref. 24 with permission from the Royal Society of Chemistry.

claimed by Benmayza *et al.*[25] Once a cathodic potential was applied to the electrode, an intermediate species of magnesium with a reduced first-shell could be detected by *in situ* soft X-ray absorption spectroscopy (XAS). The spectra suggested that the dimer complex cation $[Mg_2(\mu\text{-}Cl)_3\cdot 6THF]^+$ is broken to form an intermediate species like $[MgCl\cdot 5THF]^+$ at the electrode surface.

It is easy to imagine that the electrochemically active species in the halide-free electrolyte solutions are not "dimer complexes". Recently, several research groups revealed the coordination structure of $Mg(TFSA)_2$ in triglyme (G3) or tetraglyme (G4) using pair distribution function (PDF) analysis, Raman spectroscopy and single-crystal XRD.[26–30] The Mg^{2+} cation and glyme molecule form complex cations: $[Mg(G3)_2]^{2+}$, $[Mg(G3)]^{2+}$ or $[Mg(G4)]^{2+}$ form either contact ion pairs (CIP) or solvent-separated ion pairs (SSIP) with the $TFSA^-$ anion in the electrolyte solutions. The MMC electrolyte and $Mg[B(hfip)_4]_2$ also form a similar types of complex cations such as $[Mg(G1)_3]^{2+}$, $[Mg(G2)_2]^{2+}$ and $[Mg(G3)_2]^{2+}$. Since these electrolytes do not form the dimer complexes, the electrodeposition mechanism could be different from that of the magnesium organohaloaluminate electrolytes.

6.4 Surface Morphologies of Electrodeposited Magnesium Metal

"No dendritic growth" are the most attractive words that can be used to describe the motivation to investigate "rechargeable magnesium batteries". The cycling performance of the prototype Mg/DCC electrolyte/Mo_3S_4 cell by Aurbach *et al.* already proved that the magnesium metal anode could be a practical battery anode. However, we should remember that in principle all metals have some chance of forming dendrites during the crystal growth process. Next, we review the surface morphology of the electrodeposited magnesium metal.

Although the early work of Gregory *et al.* shows a scanning electron microscope (SEM) image of the electrodeposited magnesium with hexagonal shaped crystals, as shown in Figure 6.8, they also reported the dendritic growth of magnesium metal from two electrolyte solutions: 1.5 mol L^{-1} MeMgCl with 0.1 mol L^{-1} $AlCl_3$ in THF and 2.0 mol L^{-1} EtMgCl with 0.2 mol L^{-1} $AlCl_3$ in THF, as shown in Table 6.1.[2] There is no discussion concerning the dendritic growth of magnesium metal in the paper. The results, however, indicate that the electrolyte composition has some influence on the crystal growth process. Aurbach *et al.* reported surface morphology changes during the electrodeposition process of magnesium metal using *in situ* STM.[9] The deposited magnesium metal formed "pyramidal shaped" crystals, which also reflects the hcp structure of the magnesium metal. Similar surface morphologies of the electrodeposited magnesium have been reported by the same group and other groups.[31,32]

Matsui carried out a comparative study of the surface morphology of magnesium and lithium metals,[33] where the electrodeposited magnesium

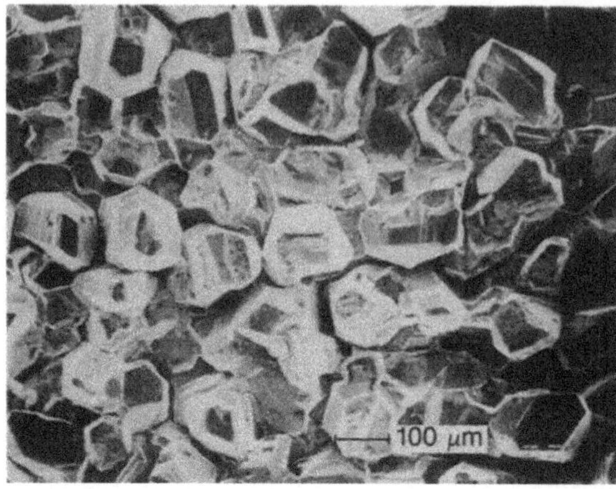

Figure 6.8 SEM image of electrodeposited magnesium from an organomagnesium solution. Reproduced from ref. 2 with permission from The Electrochemical Society, Copyright 1990.

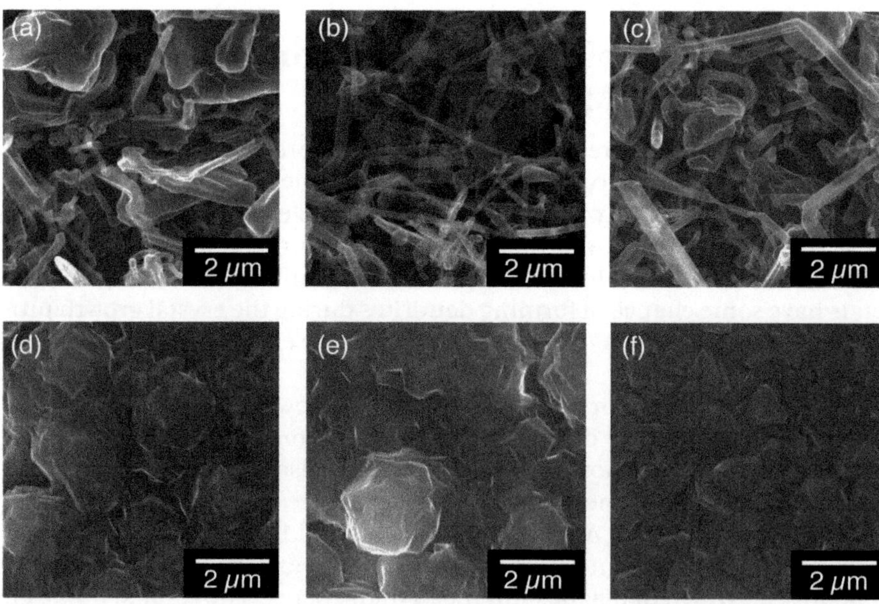

Figure 6.9 SEM images of electrodeposited (a)–(c) lithium and (d)–(f) magnesium metal at various deposition currents: (a), (d) 0.5 mA cm^{-2}, (b), (e) 1.0 mA cm^{-2} and (c), (f) 2.0 mA cm^{-2}. Reproduced from ref. 33 with permission from Elsevier, Copyright 2011.

showed smooth surface morphologies without any dendritic growth, while the lithium films deposited under the same current density had whisker-shaped surface morphologies, as shown in Figure 6.9. Further crystallographic study of the electrodeposited magnesium metal provided helpful information concerning the crystal growth process. The preferred orientation

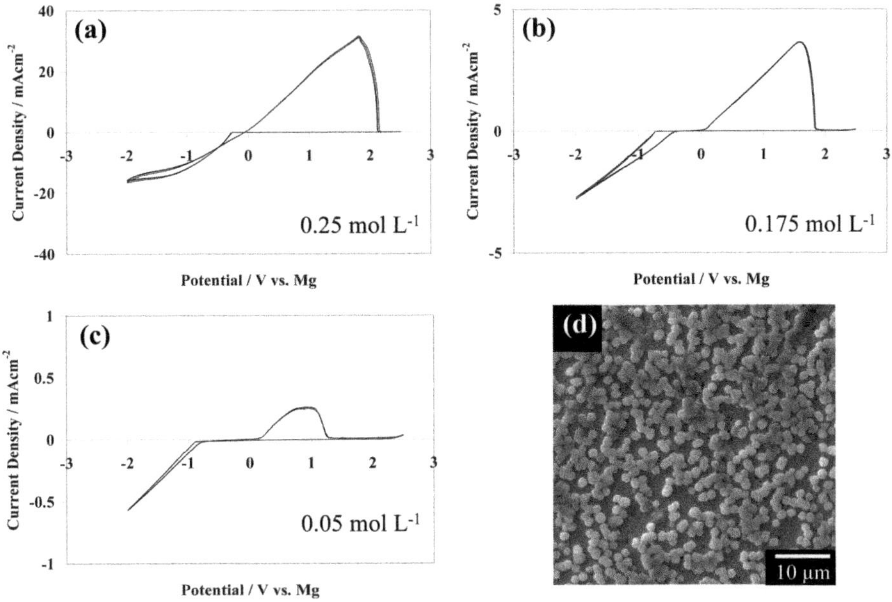

Figure 6.10 Cyclic voltammograms of a DCC type solution with different Mg^{2+} concentrations (a)–(c) and (d) SEM image of the early stages of the Mg deposition process. Reproduced from ref. 33 with permission from Elsevier, Copyright 2011.

of the electrodeposited magnesium changed from the [001] to [100] plane, when increasing the deposition current. When the crystal growth rate is slow enough, the crystals tend to minimize the surface energy. Hence, the deposited magnesium metal at a low current density (0.5 or 1.0 mA cm^{-2}) shows the preferred orientation of the [001] plane, which is a closed packed layer of the hcp structure. On the other hand, the crystal growth rate along the *c*-axis is slower than that along the *a*/*b*-axis, therefore the [001] preferred orientation cannot be maintained at a high current density (2.0 mA cm^{-2}). The fast electrodeposition of magnesium consequently initiates the preferred orientation of the [100] plane to maximize the crystal growth rate. The change in the preferred orientation indicates that faster electrodeposition may trigger the dendritic growth of magnesium metal. In this study, a possible reason for the non-dendritic growth of the magnesium metal was also discussed. It was suggested that the overpotential of the magnesium deposition/dissolution process significantly increases at a low concentration of the electrolyte solution, as shown in Figure 6.10(a)–(c). Since the locally focused current can be avoided by increasing the overpotential by consumption of Mg^{2+} ions, a random nucleation process consequently takes place during the initial stage of the electrodeposition process, as shown in Figure 6.10(d).

Another reason for the non-dendritic magnesium metal deposition was reported by Ling *et al.* using density functional theory (DFT) modeling.[34] Figure 6.11(a) shows a schematic representation of the high dimensional

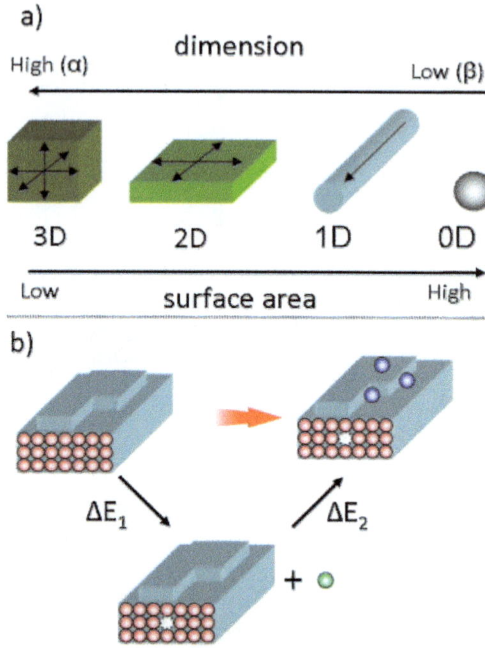

Figure 6.11 (a) Schematic of high dimensional phases (phase α) and low dimensional phases (phase β). (b) Movement of one atom from the bulk phase to the surface. The red, blue and green circles show the atom in the bulk phase. (c) Free energy change for different coordination numbers of the adsorbed atom (*i*) and surface atoms (*j*). Solid squares: Mg, open symbols: Li. Reproduced from ref. 34 with permission from Elsevier, Copyright 2012.

phases (2D plates, 3D bulk particles) and low dimensional phases (1D, whiskers/dendrites). They calculated the change in the free energies with different morphologies: the high dimensional phases and the low dimensional phases for the magnesium and lithium metals. Since the low dimensional phases have more surface area in the cluster, the energy change is represented by removing one atom from the bulk phase to the surface, as shown in Figure 6.11(b). The DFT calculation results show that magnesium metal should intrinsically prefer high dimensional growth compared with lithium metal, due to the strong Mg–Mg bond.

The surface morphology of magnesium metal is also dependent on the electrolyte. The electrolyte containing the dimer complex typically shows pyramidal shaped grains, as we already observed in the previous section, while the magnesium metal deposited in the $Mg(TFSA)_2$ in glyme solution has aggregated hemispherical particles, as shown in Figure 6.12.[15] The pyramidal shaped grain reflects the good crystalline hcp structure of the magnesium metal, while the hemispherical particles are formed by the aggregation

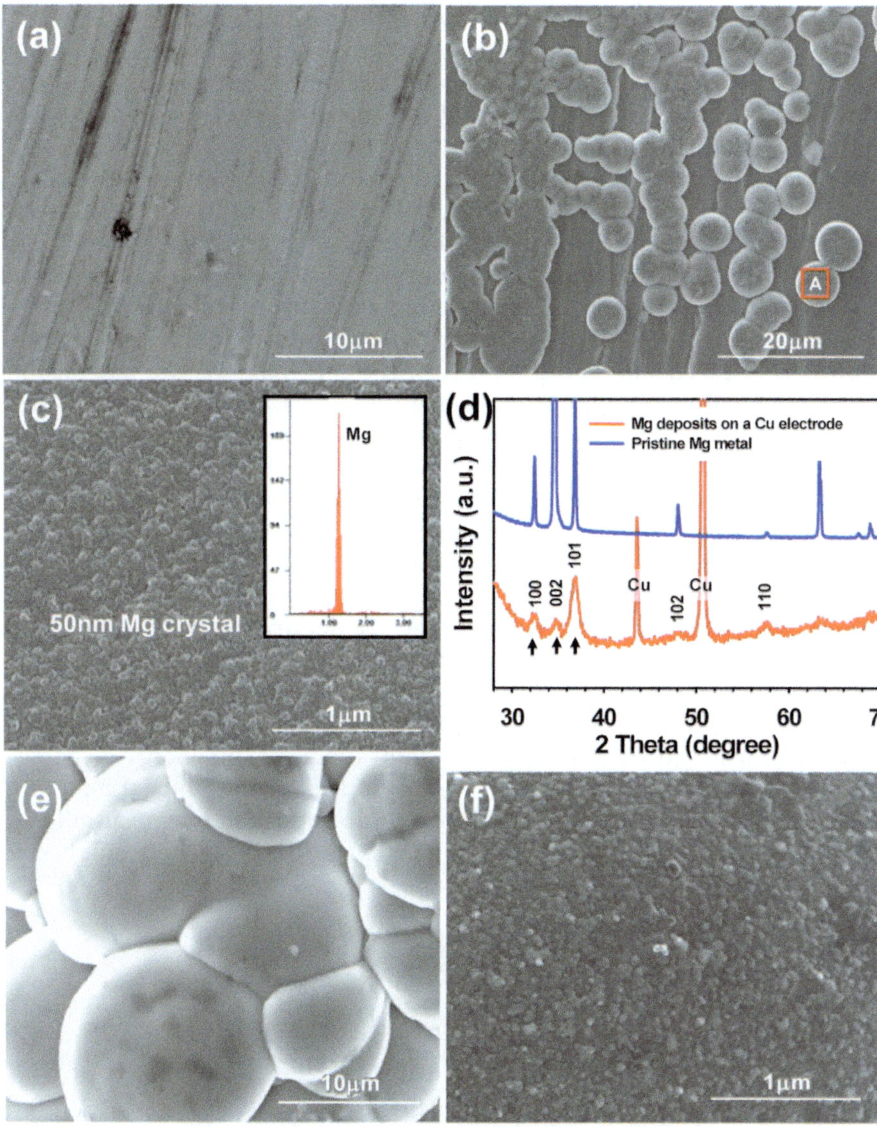

Figure 6.12 SEM images and XRD patterns of magnesium metal deposited in a $Mg(TFSA)_2$–Glyme solution. (a) Pristine Mg electrode, (b) after the 1st deposition of magnesium, (c) magnified image of section A of (b), (d) XRD pattern of electrodeposited magnesium on copper substrate, (e) SEM image of electrodeposited magnesium on copper substrate, (f) magnified image of (e). Reproduced from ref. 15 with permission from the American Chemical Society, Copyright 2014.

of nano-sized grains due to the simultaneous passivation of the surface of the grains during the electrodeposition process. The XRD pattern in Figure 6.12(d) shows the poor crystallinity of the electrodeposited magnesium. On the other hand, Shterenberg *et al.* reported highly crystalline magnesium deposition from 0.25 mol L^{-1} of $Mg(TFSA)_2$ and 0.5 mol L^{-1} of $MgCl_2$ in DME solution.[18] This suggests that high coulombic efficiencies of >90% are necessary for the smooth crystal growth process of magnesium metal. They also confirmed an edgy morphology of the magnesium grains deposited at 5.0 mA cm^{-2}. The trend in the morphology change is consistent with the studies with organohaloaluminate electrolytes. The magnesium metal deposited in a state-of-the-art electrolyte MMC solution shows an even more rough and edgy surface morphology.[19] An SEM image of the magnesium metal deposited at a high current density of 5.0 mA cm^{-2} from the MMC electrolyte solution is shown in Figure 6.13. The increased surface area of the plate-shaped magnesium grains suggests that the deposition rate had almost approached the limit of the growth rate. Once the current density exceeds the growth rate of the metal, dendritic growth is supposed to occur. Therefore, advanced electrolyte solutions may also reinitiate the dendritic growth of magnesium metal.

A locally focused current may the trigger the dendritic growth of the metal deposition. In the case of the magnesium metal, an unexpected passivation layer also results in a locally focused current. Ding *et al.* reported the short circuiting of a magnesium/$Mg(TFSA)_2$–glyme/magnesium symmetric cell *via* dendritic growth of the magnesium metal.[35] The energy-dispersive X-ray spectroscopic (EDS) mapping indicated the formation of the passivation layer limiting the active surface area of the deposited magnesium metal resulting in the initiation of a focused current in the cell. The result clearly

Figure 6.13 SEM image of electrodeposited magnesium from a 0.75 mol L^{-1}MMC/G4 solution at 5.0 mA cm^{-2}. Reproduced from ref. 19 with permission from John Wiley and Sons, © 2015 WILEY-VCH Verlag GmbH & Co. KGaA, Weinheim.

Figure 6.14 Magnesium dendrite deposited from 0.5 mol L^{-1} MeMgCl in THF solution. Reproduced from ref. 36 with permission from the American Chemical Society, Copyright 2019.

suggests that a passivation-free surface is crucial to avoid a focused current and dendritic growth during the deposition process. Even with a surface free from passivation, a focused current may initiate structured growth of the magnesium. A recent report by Davidson *et al.* proved that the strong electric field at the corner of the electrode becomes a preferred site of the focused current, resulting in the formation of a dendrite of magnesium metal, as shown in Figure 6.14.[36]

In this section, we reviewed the surface morphologies of electrodeposited magnesium metal as a potential issue for later battery applications. In summary, even though all metals have the possibility to form dendritic structures, magnesium metal shows relatively uniform deposition even at a high current density compared to other metals. Furthermore, the sluggish mobility of Mg^{2+} ions in the solid phase limits the current density of the magnesium metal anode. Therefore, dendrite formation is still not a critical problem compared with the passivation issues discussed in the following section.

6.5 Passivation Layer and Possible SEI Layer

Magnesium metal is stable in an ambient atmosphere because a stable passivation layer is formed. This is a significant advantage when using magnesium metal as a light structural material. However, the passivation layer may be a severe issue when using magnesium metal as a battery anode, also in nonaqueous systems. Although Peled claims that all alkaline metal and alkaline earth metal form SEI layers,[37] he reports that a passivation layer is formed, rather than an SEI layer, at the surface of the Mg anode in a Mg–MnO_2 dry cell. The poor Mg^{2+} mobility in the passivation layer limits the electrochemical activity of the magnesium metal.

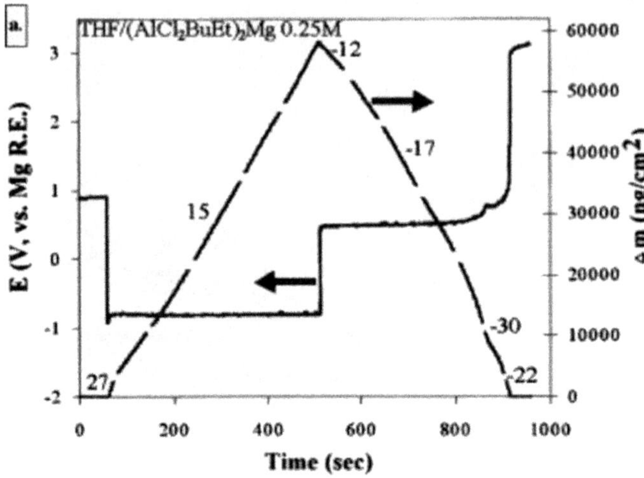

Figure 6.15 Typical chronopotentiogram and mass response of gold on a quartz crystal electrode in EQCM experiments on the Mg deposition/dissolution process using DCC electrolyte solution. Reproduced from ref. 10 with permission from The Electrochemical Society, Copyright 2001.

Lu *et al.* conducted electrochemical and spectroscopic studies of the magnesium metal in various common electrolyte solutions such as $Mg(ClO_4)_2$ in acetonitrile (ACN) or propylene carbonate (PC). The Mg dissolution can take place only *via* a breakdown of the passivation layer during the anodic polarization.[10,23] In addition, since the passivation layer may be restored during the cathodic scan, the electrodeposition of magnesium may not take place again. They also confirmed that the no surface layer is formed during the electrodeposition/dissolution process of magnesium in a DCC electrolyte solution using EQCM, as shown in Figure 6.15. Gofer *et al.* performed a series of X-ray photoelectron spectroscopic (XPS) studies to characterize the passivation layer on Mg metal immersed in pure THF, THF/Bu_2Mg, THF/BuMgCl, THF/DCC and pure PC.[38] Interestingly, all of the obtained XPS spectra (Mg2p, C1s O1s) from these electrolytes qualitatively showed similar shapes, indicating that the surface layer is formed in the post preparation process. Also, the passivation layer of the magnesium is very thin and not like the SEI layer on lithium metal. Hence, once the surface of the magnesium metal is passivated, continuous decomposition of the electrolyte solution does not take place.

Kuwata *et al.* performed a comparative study of the passivation layer formation in an organohaloaluminate-based electrolyte and a $Mg(TFSA)_2$-based electrolyte using *in situ* FTIR and XPS, as shown in Figure 6.16.[39] The *in situ* FTIR spectra clearly show the decomposition process of the $TFSA^-$ anion during the initial cathodic scan, and flat spectra in the following scans. The flat FTIR spectra indicate that the electrode surface is completely passivated. On the contrary, the spectra in the organohaloaluminate-based electrolyte

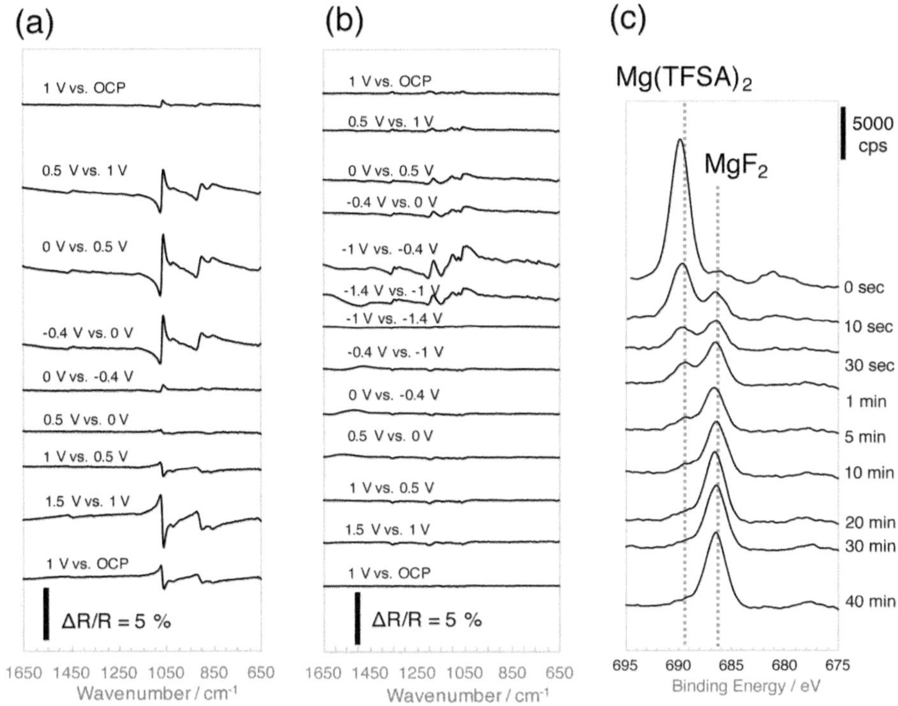

Figure 6.16 *In situ* FTIR spectra and XPS spectra for magnesium metal anodes. (a) *In situ* FTIR spectra in organohaloaluminate-based electrolyte, (b) *in situ* FTIR spectra in glyme-based electrolyte, and (c) F1s XPS spectra of magnesium metal. Reproduced from ref. 39 with permission from The Electrochemical Society, Copyright 2017.

maintain a passivation free surface during the cathodic/anodic polarization. The XPS spectra prove the formation of a thick passivation layer of MgF_2 *via* decomposition of the $TFSA^-$ anion in the $Mg(TFSA)_2$-based electrolyte, while the remaining Al or Cl species are almost negligible in the case of the organohaloaluminate-based electrolyte. Thus, the reductive stability of the anion can influence the reversibility of the electrodeposition/dissolution process.

A theoretical study by Rajput *et al.* predicted the anodic and cathodic stability of anions.[40] In this work, the $TFSA^-$ anion showed good cathodic/anodic stability compared with other anions. However, once the $TFSA^-$ anion and Mg^{2+} cation formed contact ion pairs (CIP): $[Mg^{2+}TFSA^-]^+$, the dissociation energy of the C–S bond in the $TFSA^-$ anion significantly dropped *via* the formation of a partially reduced CIP: $[Mg^+TFSA^-]$. The results show that the $TFSA^-$ anion in the CIP is easily reduced during the electrodeposition process. The work proposed that the solvation structure needs to be considered to design electrolyte solutions with high cathodic stability against magnesium metal.

The surface passivation of magnesium metal anode is also affected by the small amount of water in the electrolyte solution. Connell *et al.* conducted a series of studies concerning the passivation layer formation.[41] They reported that even trace levels of H_2O (≤3 ppm) can affect the kinetics of the magnesium deposition/dissolution process in a $Mg(TFSA)_2$–glyme system. Moreover, the passivation layer is simultaneously formed on the electrodeposited magnesium metal. They also found that the formation of the $Mg–Cl^+$(ad.) or $MgCl_2$ layer suppresses the passivation of the magnesium metal *via* H_2O molecules.

Cathodic decomposition of solvent molecules is another trigger for the passivation process. Lowe and Siegel investigated possible reaction pathways of the reduction process of DME molecules at the surface of magnesium metal using first-principles calculations.[42] They proved highly exothermic and rapid decomposition of DME molecules at the surface of the magnesium metal with an evolution of ethylene gas. On the other hand, the surfaces of the ionic MgO and $MgCl_2$ have a limited impact on solvent decomposition. It was claimed that there is some possibility that the $MgCl_2$ is integrated into the SEI layer.

Arthur *et al.* reported another possible SEI component formed in a $Mg(BH_4)_2$ solution using soft X-ray XAS. The *in situ* soft X-ray XAS data reveal the formation of borohydride cluster compounds, at the initial deposition/dissolution process.[43] The ionic conductive properties of the borohydride cluster compounds are unknown at this moment, however, some reports concerning borohydride-based solid-state magnesium ion conductors have indicated that borohydride cluster compounds may also work as an SEI layer of the magnesium metal anode.

Again, the formation of the passivation layer at the surface of the magnesium metal is a critical issue in the overall direction for the development of an electrode/electrolyte interphase focused on a "passivation-free" system. The naked surface of the magnesium metal is, however, highly reactive against electrolyte solutions. Therefore, further materials design for controlled formation of the interphase may be crucial for the development of rechargeable magnesium batteries.

6.6 Intermetallic Anodes

The problems of the passivation layer of magnesium metal anodes have initiated another research direction toward "Mg-ion" batteries using alternate anode active materials. As already discussed in the previous sections, since magnesium metal easily forms a passivation layer in most conventional electrolyte solutions, intermetallic compounds have been proposed as possible alternative anodes. Arthur *et al.* first reported that electrodeposited $Bi_{1-x}Sb_x$ alloys show a reversible magnesiation/demagnesiation process in half-cell tests.[44] The alloys, except pure Sb, showed excellent cyclability at a 1C rate. Furthermore, the Mg_3Bi_2 intermetallic anode showed reversible magnesiation/demagnesiation in a conventional electrolyte solution, 1.0 mol L^{-1} $Mg(TFSA)_2$ in acetonitrile solution, which normally does not show reversible deposition/dissolution of the magnesium metal. It should be mentioned that the Mg_3Bi_2 anode compromises the energy density of the battery due to its equilibrium potential of 0.28 V *vs.* Mg and its low specific capacity of 385 mA h g^{-1}/Bi. Mg_3Bi_2, however, still exhibits a high volumetric capacity of 1906 mA h cm^{-3}/Mg_3Bi_2 due to the high atomic weight of bismuth. The volumetric capacity is comparable to that of lithium metal, whose volumetric capacity is approximately 2000 mA h cm^{-3}, thus the Mg_3Bi_2 intermetallic anode remains a high-capacity anode active material.

It is well known that alloy/intermetallic anodes typically show volume expansion/shrinkage during the charging/discharging process, resulting in a pulverization of the anode particles. The Mg_3Bi_2 shows a volume expansion of 196% accompanied by the magnesiation process, thus particle size control improves the capacity retention during the charge/discharge cycling. Shao *et al.* synthesized Bi nanotubes and confirmed the improvement in the rate capability and capacity retention.[45] The TEM image after the discharging process shows that nanosized domains of ≤10 nm are formed during the initial magnesiation process, without losing electrical contact. The reduced diffusion length contributes towards the better rate capability of the electrode. The reaction mechanism of the Mg_3Bi_2 intermetallic anode during the electrochemical magnesiation/demagnesiation process was studied by Murgia *et al.*[46] A typical two-phase reaction, which is normally predicted from the binary phase diagram of the Bi–Mg system, was confirmed by galvanostatic intermittent titration technique (GITT) analyses and *in operando* XRD, as shown in Figure 6.17. Recent DFT calculation studies indicate a low migration barrier of the Mg^{2+} ions in the Mg_3Bi_2 solid phase.[47] Between two neighboring tetrahedral sites the migration barrier of the Mg^{2+} ion is 0.32 eV, while it is 0.80 eV between two octahedral sites. Lee *et al.* reported that the migration barrier of the Mg^{2+} ions between the 1a octahedral site and the 2d tetrahedral site is also comparably low: 0.34 eV.[48] This shows that the anisotropic diffusion pathway along the (100) plane through tetrahedral 2d sites of the Mg_3Bi_2 is preferably formed.

The biggest advantage of the intermetallic anode is compatibility against conventional electrolyte solutions such as $Mg(TFSA)_2$ dissolved in acetonitrile. Matsui *et al.* performed a comparative study of Mg_3Bi_2 and Mg_3Sb_2 synthesized by a solid-state process.[49] Despite Mg_3Sb_2 and Mg_3Bi_2 having the same crystal structure, the electrochemical activity of Mg_3Sb_2 was almost negligible in an acetonitrile-based electrolyte solution. The electrochemical inactivity of Mg_3Sb_2 is not owing to the passivation layer, because both surfaces of the Mg_3Bi_2 and Mg_3Sb_2 are not essentially passivated even in an ambient atmosphere. Rather, the bond valence sum (BVS) mapping of the Mg_3Bi_2 and Mg_3Sb_2 indicates that relatively fast Mg^{2+} diffusion in Mg_3Bi_2 stimulates the electrochemical activity, as shown in Figure 6.18. In

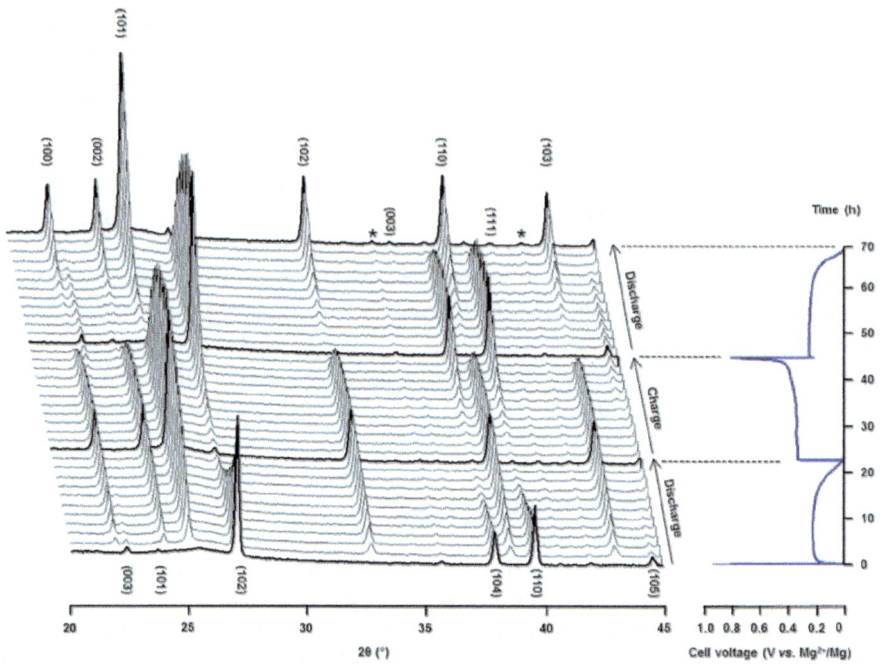

Figure 6.17 *In operando* XRD patterns recorded during the first magnesiation/demagnesiation of a Bi/Mg half-cell. A clear biphasic reaction is directly confirmed. Reproduced from ref. 46 with permission from the Royal Society of Chemistry.

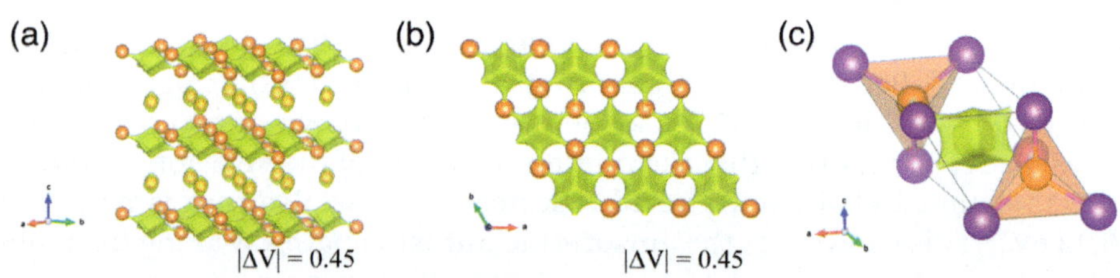

Figure 6.18 BVS mapping of a Mg_3Bi_2 intermetallic anode visualizing the migration pathways of the Mg^{2+} ions. Reproduced from ref. 49, https://doi.org/10.3389/fchem.2019.00007, under the terms of the CC BY 4.0 licence, https://creativecommons.org/licenses/by/4.0/.

addition, the migration pathways of the Mg^{2+} ions in Mg_3Bi_2 are in good agreement with the DFT calculation results.

A series of other intermetallic anodes have been investigated by different groups. Mg_2Sn reported by Singh *et al.* has a theoretical capacity of 900 mA h g^{-1}/Sn during the magnesiation process. The problem with the Mg_2Sn intermetallic anode is the reported sluggish rate capability and reversibility. The theoretical value of the magnesiation capacity was obtained only at 0.002C and only 35% of the magnesium could be extracted. This is probably due to the slow kinetics of the Mg^{2+} in the inverse-fluorite structure of Mg_2Sn. On the other hand, Mg_2Pb, which has the same inverse-fluorite structure as that of Mg_2Sn, shows excellent reversibility, delivering ≈ 90% of the theoretical value. The Pb/Mg half-cell exhibited a reversible capacity of approx. 450 mA h g^{-1}/Pb at 60 °C at a current density of C/50. Since Pb is also a heavy element similar to Bi, the volumetric capacity of the Mg_2Pb is high, theoretically 2300 mA h L^{-1}. Its low equilibrium potential of 121 mV *vs.* Mg metal is also an advantage in terms of energy density. MgIn shows a reversible capacity of 450 mA h g^{-1} at a charge/discharge rate of C/100, reported by Murgia *et al.* They confirmed the expected two-phase reaction of the Mg–In binary system using *in operando* XRD. The MgIn anode showed severe capacity fading at a high charge/discharge rate.

Mg_3Bi_2, as the first investigated alternative intermetallic anode, still shows the most preferable electrochemical properties among all of the intermetallic compounds. A series of studies have suggested that the mobility of the Mg^{2+} in the solid phase is the key property of the intermetallic anodes, rather than the passivation layer discussed for the magnesium metal anode.

6.7 Summary

In this chapter, we discussed the electrochemical properties of magnesium metal and intermetallic anodes. In the early stages of this research, since few organomagnesium compounds are available for electrolyte solutions, the deposition mechanism of magnesium metal is only based upon speculation. Also, it is not easy to claim the advantage of rechargeable magnesium batteries as "beyond lithium-ion" systems. During the last decade, several new classes of electrolytes have improved the electrochemical properties of magnesium anodes. A series of contributions from advanced spectroscopy and theoretical studies in the last decade have given deep insights into the reaction mechanism of magnesium-based anodes. Electrolyte solutions containing dimer complexes: $[Mg_2(\mu\text{-}Cl)_3]^+$ show excellent reversibility, while the practical electrochemical windows are limited by corrosion problems. Despite the wide electrochemical window of the glyme coordinates: $[Mg\text{–}Glymes]^{2+}$ in a $Mg(TFSA)_2$-based system, CIP formation sacrifices the coulombic efficiency of the deposition/dissolution process. Even though state-of-the-art electrolyte solutions solve these problems, still, ether-based molecules are the choice solvents. Intermetallic anodes expand the choice of electrolyte solutions, with a sacrifice in the energy density. We think those contributions enable us to propose a possible rechargeable magnesium battery system in the near future.

References

1. G. J. Kipouros and D. R. Sadoway, in *Advances in Molten Salt Chemistry*, ed. G. Mamantov, C. B. Mamantov and J. Braunstein, Elsevier, 1987, vol. 6, p. 127.
2. T. D. Gregory, R. J. Hoffman and R. C. Winterton, *J. Electrochem. Soc.*, 1990, **137**, 775.
3. D. Aurbach, I. Weissman, Y. Gofer and E. Levi, *Chem. Rec.*, 2003, **3**, 61.
4. M. P. Jolibois, *Comptes Rendus*, 1912, **155**, 353.
5. L. W. Gaddum and H. E. French, *J. Am. Chem. Soc.*, 1927, **49**, 1295.
6. J. H. Connor, W. E. Reid and G. B. Wood, *J. Electrochem. Soc.*, 1957, **104**, 38.
7. J. D. Genders and D. Pletcher, *J. Electroanal. Chem. Interfacial Electrochem.*, 1986, **199**, 93.
8. D. Aurbach, Z. Lu, A. Schechter, Y. Gofer, H. Gizbar, R. Turgeman, Y. Cohen, M. Moshkovich and E. Levi, *Nature*, 2000, **407**, 724.
9. D. Aurbach, Y. Cohen and M. Moshkovich, *Electrochem. Solid-State Lett.*, 2001, **4**, A113.
10. D. Aurbach, A. Schechter, M. Moshkovich and Y. Cohen, *J. Electrochem. Soc.*, 2001, **148**, A1004.
11. D. Aurbach, G. S. Suresh, E. Levi, A. Mitelman, O. Mizrahi, O. Chusid and M. Brunelli, *Adv. Mater.*, 2007, **19**, 4260.
12. H. S. Kim, T. S. Arthur, G. D. Allred, J. Zajicek, J. G. Newman, A. E. Rodnyansky, A. G. Oliver, W. C. Boggess and J. Muldoon, *Nat. Commun.*, 2011, **2**, 427.
13. J. Muldoon, C. B. Bucur, A. G. Oliver, T. Sugimoto, M. Matsui, H. S. Kim, G. D. Allred, J. Zajicek and Y. Kotani, *Energy Environ. Sci.*, 2012, **5**, 5941.
14. S. Yagi, A. Tanaka, Y. Ichikawa, T. Ichitsubo and E. Matsubara, *J. Electrochem. Soc.*, 2013, **160**, C83.
15. S.-Y. Ha, Y.-W. Lee, S. W. Woo, B. Koo, J.-S. Kim, J. Cho, K. T. Lee and N.-S. Choi, *ACS Appl. Mater. Interfaces*, 2014, **6**, 4063.
16. Y. Orikasa, T. Masese, Y. Koyama, T. Mori, M. Hattori, K. Yamamoto, T. Okado, Z. D. Huang, T. Minato, C. Tassel, J. Kim, Y. Kobayashi, T. Abe, H. Kageyama and Y. Uchimoto, *Sci. Rep.*, 2014, **4**, 5622.
17. T. Fukutsuka, K. Asaka, A. Inoo, R. Yasui, K. Miyazaki, T. Abe, K. Nishio and Y. Uchimoto, *Chem. Lett.*, 2014, **43**, 1788.
18. I. Shterenberg, M. Salama, H. D. Yoo, Y. Gofer, J.-B. Park, Y.-K. Sun and D. Aurbach, *J. Electrochem. Soc.*, 2015, **162**, A7118.
19. O. Tutusaus, R. Mohtadi, T. S. Arthur, F. Mizuno, E. G. Nelson and Y. V. Sevryugina, *Angew. Chem., Int. Ed.*, 2015, **54**, 7900.
20. J. T. Herb, C. A. Nist-Lund and C. B. Arnold, *ACS Energy Lett.*, 2016, **1**, 1227.
21. Z. Zhao-Karger, M. E. Gil Bardaji, O. Fuhr and M. Fichtner, *J. Mater. Chem. A*, 2017, **5**, 10815.

22. D. Aurbach, H. Gizbar, A. Schechter, O. Chusid, H. E. Gottlieb, Y. Gofer and I. Goldberg, *J. Electrochem. Soc.*, 2002, **149**, A115.
23. Z. Lu, A. Schechter, M. Moshkovich and D. Aurbach, *J. Electroanal. Chem.*, 1999, **466**, 203.
24. H. D. Yoo, I. Shterenberg, Y. Gofer, G. Gershinsky, N. Pour and D. Aurbach, *Energy Environ. Sci.*, 2013, **6**, 2265.
25. A. Benmayza, M. Ramanathan, T. S. Arthur, M. Matsui, F. Mizuno, J. Guo, P.-A. Glans and J. Prakash, *J. Phys. Chem. C*, 2013, **117**, 26881.
26. S. H. Lapidus, N. N. Rajput, X. Qu, K. W. Chapman, K. A. Persson and P. J. Chupas, *Phys. Chem. Chem. Phys.*, 2014, **16**, 21941.
27. A. Kitada, Y. Kang, K. Matsumoto, K. Fukami, R. Hagiwara and K. Murase, *J. Electrochem. Soc.*, 2015, **162**, D389.
28. T. Kimura, K. Fujii, Y. Sato, M. Morita and N. Yoshimoto, *J. Phys. Chem. C*, 2015, **119**, 18911.
29. S. Terada, T. Mandai, S. Suzuki, S. Tsuzuki, K. Watanabe, Y. Kamei, K. Ueno, K. Dokko and M. Watanabe, *J. Phys. Chem. C*, 2016, **120**, 1353.
30. K. Hashimoto, S. Suzuki, M. L. Thomas, T. Mandai, S. Tsuzuki, K. Dokko and M. Watanabe, *Phys. Chem. Chem. Phys.*, 2018, **20**, 7998.
31. N. Amir, Y. Vestfrid, O. Chusid, Y. Gofer and D. Aurbach, *J. Power Sources*, 2007, **174**, 1234.
32. Y. Guo, J. Yang, Y. NuLi and J. Wang, *Electrochem. Commun.*, 2010, **12**, 1671.
33. M. Matsui, *J. Power Sources*, 2011, **196**, 7048.
34. C. Ling, D. Banerjee and M. Matsui, *Electrochim. Acta*, 2012, **76**, 270.
35. M. S. Ding, T. Diemant, R. J. Behm, S. Passerini and G. A. Giffin, *J. Electrochem. Soc.*, 2018, **165**, A1983.
36. R. Davidson, A. Verma, D. Santos, F. Hao, C. Fincher, S. Xiang, J. V. Buskirk, K. Xie, M. Pharr, P. P. Mukherjee and S. Banerjee, *ACS Energy Lett.*, 2019, **4**, 375.
37. E. Peled, *J. Electrochem. Soc.*, 1979, **126**, 2047.
38. Y. Gofer, R. Turgeman, H. Cohen and D. Aurbach, *Langmuir*, 2003, **19**, 2344.
39. H. Kuwata, M. Matsui and N. Imanishi, *J. Electrochem. Soc.*, 2017, **164**, A3229.
40. N. N. Rajput, X. Qu, N. Sa, A. K. Burrell and K. A. Persson, *J. Am. Chem. Soc.*, 2015, **137**, 3411.
41. J. G. Connell, B. Genorio, P. P. Lopes, D. Strmcnik, V. R. Stamenkovic and N. M. Markovic, *Chem. Mater.*, 2016, **28**, 8268.
42. J. S. Lowe and D. J. Siegel, *J. Phys. Chem. C*, 2018, **122**, 10714.
43. T. S. Arthur, P.-A. Glans, N. Singh, O. Tutusaus, K. Nie, Y.-S. Liu, F. Mizuno, J. Guo, D. H. Alsem, N. J. Salmon and R. Mohtadi, *Chem. Mater.*, 2017, **29**, 7183.
44. T. S. Arthur, N. Singh and M. Matsui, *Electrochem. Commun.*, 2012, **16**, 103.
45. Y. Shao, M. Gu, X. Li, Z. Nie, P. Zuo, G. Li, T. Liu, J. Xiao, Y. Cheng, C. Wang, J. G. Zhang and J. Liu, *Nano Lett.*, 2014, **14**, 255.

46. F. Murgia, L. Stievano, L. Monconduit and R. Berthelot, *J. Mater. Chem. A*, 2015, **3**, 16478.
47. S. C. Jung and Y.-K. Han, *J. Phys. Chem. C*, 2018, **122**, 17643.
48. J. Lee, B. Monserrat, I. D. Seymour, Z. Liu, S. E. Dutton and C. P. Grey, *J. Mater. Chem. A*, 2018, **6**, 16983.
49. M. Matsui, H. Kuwata, D. Mori, N. Imanishi and M. Mizuhata, *Front. Chem.*, 2019, **7**, 7.

CHAPTER 7

Insertion Electrodes for Magnesium Batteries: Intercalation and Conversion

H. D. YOO*[a] AND S. H. OH*[b]

[a]Pusan National University, Department of Chemistry, Busan 46241, Republic of Korea; [b]Korea Institute of Science and Technology, Centre for Energy Storage Research, Seoul 02792, Republic of Korea
*E-mail: hyundeog.yoo@pusan.ac.kr, sho74@kist.re.kr

7.1 Introduction

Layered materials are composed of two-dimensional (2D) layers, which are weakly bound by van der Waals forces. In contrast to 1D and 3D networks with rigid structures, weakly bound 2D networks provide structural flexibility to intercalate various kinds of ions or molecules, such as protons, alkali metal ions, amines, and organic cations. Intercalation of organic species expands the interlayer distance in a controllable manner without exfoliating the structure into single layers, signifying structural flexibility.[1–3] Likewise, intercalation of ions into layered materials undergoes topotactic reactions, which maintain the orientations of the structure upon intercalation. This aspect signifies that the intercalation into layered materials is prone to be highly reversible, making the interlayers versatile hosts for the insertion of various ions and molecules. This is a characteristic feature that distinguishes layered materials from other types of hosts with 1D or 3D channels for guest

Energy and Environment Series No. 23
Magnesium Batteries: Research and Applications
Edited by Maximilian Fichtner

Published by the Royal Society of Chemistry, www.rsc.org

ions. As a result, layered materials have been key players in many successful devices for electrochemical energy storage, such as graphite and lithium transition metal oxides in lithium-ion batteries, just to name a few.

Considering these aspects, it was a natural consequence to consider layered materials as promising cathodes for rechargeable magnesium batteries. Since magnesium rechargeable batteries utilize a magnesium metal anode, it was preferable to consider bare transition metal chalcogenides without Mg in their pristine structure. Such transition metal chalcogenides include layered materials such as titanium disulfide (TiS_2), molybdenum disulfide (MoS_2), and vanadium pentoxide (V_2O_5), which have been efficient intercalation hosts for Li^+. However, non-layered, Chevrel-phase molybdenum chalcogenides ($Mo_6S_{8-x}Se_x$) were the only known cathodes that intercalate Mg^{2+} efficiently,[4,5] until the recent discoveries of other selenides and sulfides that intercalate Mg^{2+} at 25 and 60 °C, respectively.[6–8]

Despite the highest redox potentials and theoretical capacities of oxides among the chalcogenides, the highly electrophilic nature of the oxides results in severe side reactions with magnesium ion electrolytes that are often nucleophilic. Moreover, the smallest ionic diameter and number of core electrons in oxygen among the group VI elements brings about strong electrostatic interactions of divalent Mg^{2+} ions with oxide ions in the solid-state, as well as with counterions in electrolytes. As a result, the intercalation of divalent Mg^{2+} is greatly hindered in layered oxides, despite the diagonal relationship between Li^+ and Mg^{2+} that predicts similar chemical properties (Figure 7.1a). In more detail, Mg^{2+} must be desolvated or dissociated from counterions to

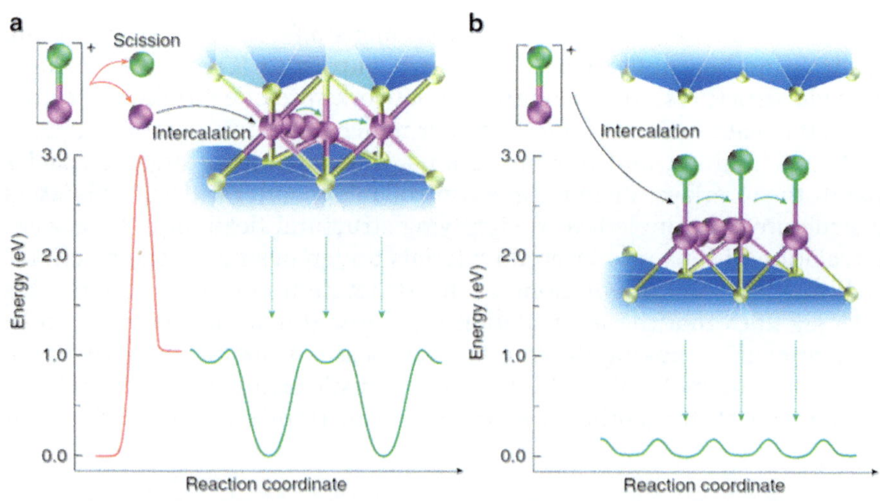

Figure 7.1 Energy diagrams for the intercalation and diffusion of (a) Mg^{2+} and (b) $MgCl^+$ in a sulfide host. Purple, green, and yellow balls represent Mg, Cl, and S atoms, respectively. Reproduced from ref. 9, https://doi.org/10.1038/s41467-017-00431-9, under the terms of the CC BY 4.0 licence, http://creativecommons.org/licenses/by/4.0/.

initiate the intercalation, which often requires more than 3 eV per Mg atom. Moreover, the desolvated Mg^{2+} must overcome a large diffusion barrier of *ca.* 1 eV per Mg atom, which is far beyond what can be readily achieved at room temperature.[9]

Due to the sluggish kinetics of the intercalation and diffusion of divalent Mg^{2+} into layered oxides, other more facile side reactions often prevail. This aspect provides a significant possibility for the co-intercalation of undesirable species, which interferes with the genuine intercalation of Mg^{2+}. Recent rigorous compositional and structural characterization found that a substantial portion of the electrochemical activity in hydrated Mg^{2+} electrolytes is actually due to the intercalation of protons,[10–12] although previous literature studies have assumed that the electrochemical activity was evidence for the intercalation of Mg^{2+}. In this respect, it will be meaningful to look back on the history of research on layered hosts for the intercalation of Mg^{2+}. In addition to the retrospective information on layered cathodes for Mg^{2+}, this chapter also summarizes and discusses conversion-type cathodes for Mg^{2+} as a newer avenue.

7.2 Materials for Intercalation

Because a separate chapter will be dedicated to cathodes with spinel structures, this part will mainly focus on layered intercalation hosts.

7.2.1 Layered Sulfides and Selenides

Sulfides and selenides are compatible in Mg^{2+} electrolytes wherein a Mg metal anode can be operated, because they are less electrophilic compared with oxides. This aspect enables full cell operation using cathodes based on sulfides or selenides, despite their relatively lower theoretical voltage and capacity than oxides. Moreover, the larger ionic diameter leads to weaker electrostatic interactions with Mg^{2+}, enabling more facile kinetics to intercalate Mg^{2+}. As a result, these materials are relatively free from the debates on genuine Mg^{2+} intercalation. There is solid proof for genuine intercalation of Mg^{2+} in sulfides and selenides, such as in Chevrel-phase Mo_6S_8, TiS_2, and MoS_2. On the other hand, the operational voltage is limited to *ca.* 1 V *vs.* Mg/Mg^{2+}, making them impractical for commercialization. Therefore, the main usage of sulfides and selenides has been the testing of general strategies and ideas to enhance the electrochemical performances to store Mg^{2+}.

Layered TiS_2 is a representative intercalation host that was reported by Whittingham in 1976.[13] However, intercalation of Mg^{2+} into layered TiS_2 has been substantially difficult because of the large migration energy barrier.[9] As a result, the reversible capacity is limited to *ca.* 20 mA h g^{-1} at room temperature. On the other hand, more facile intercalation has been proven at 60 °C, giving rise to *ca.* 160 mA h g^{-1} by overcoming the migration barrier at a higher operational temperature. Expansion of the interlayer by larger organic cations, namely 1-butyl-1-methylpyrrolidinium ($PY14^+$) has enabled

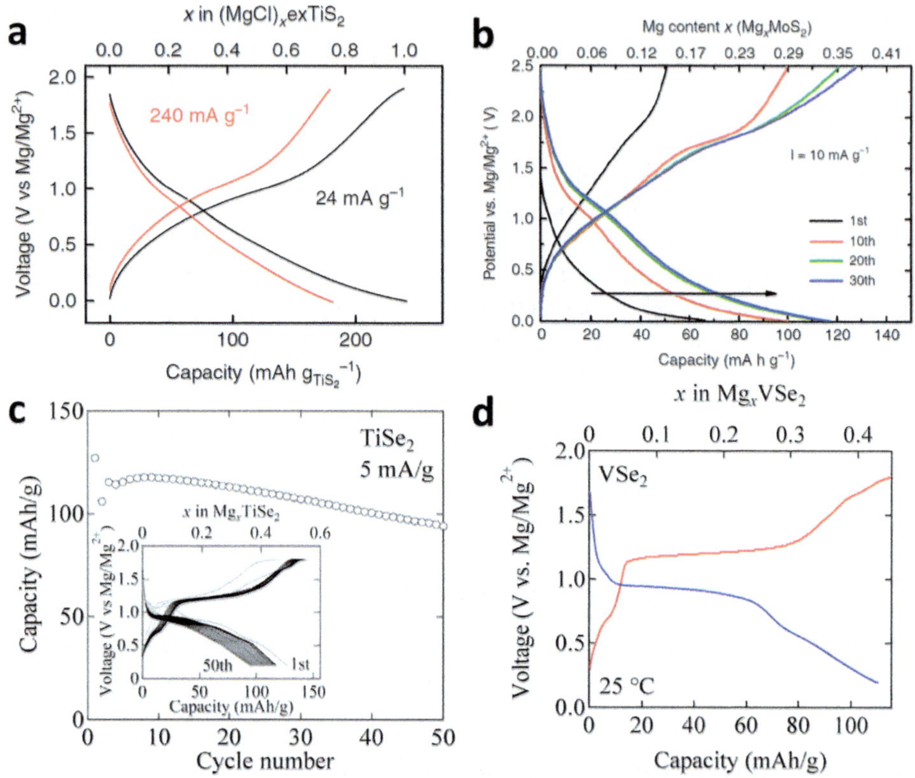

Figure 7.2 Voltage profiles of layered sulfides and selenides in a full cell coupled with a Mg metal anode at room temperature. (a) Expanded TiS_2, (b) MoS_2/C composite, (c) $TiSe_2$ and (d) VSe_2. Reproduced from ref. 6, https://doi.org/10.1038/s41467-018-07484-4, ref. 9, https://doi.org/10.1038/s41467-017-00431-9, and ref. 16, https://doi.org/10.1038/srep12486, under the terms of the CC BY 4.0 licence, http://creativecommons.org/licenses/by/4.0/.

an enhanced capacity that is close to the theoretical value of *ca.* 240 mA h g^{-1} and excellent rate capability at room temperature by the intercalation of $MgCl^+$ (Figures 7.1b and 7.2a).[9]

MoS_2 is another representative layered sulfide that is known to intercalate lithium ions reversibly. However, Mg^{2+} intercalation is sluggish due to the strong electrostatic forces that exist between the divalent Mg^{2+} ions and sulphur anions. The interlayer expansion through pillaring with polymer was shown to be effective to facilitate the diffusion of Mg^{2+} in the host materials,[14] although the strategy of expanding the interlayer has drawbacks in the form of a lower cell voltage.[15] The utilization of solvated Mg^{2+} into nanosized 1T-phase MoS_2 also enabled a much larger capacity at room temperature, by coupling of three factors: screening of divalent ions, high electronic conductivity, and decrease in the diffusion length (Figure 7.2b).[16] It is possible that further engineering of the size of organic cations or solvent molecules may enhance the electrochemical performance. On the other hand, the selenides

$TiSe_2$ and VSe_2 deliver capacity of *ca.* 110 mA h g^{-1} at room temperature without any special treatment, in accordance with general tendencies of chalcogenides (Figure 7.2c and d).[6] Also, doping selenium into other chalcogenides may further facilitate the kinetics of Mg^{2+} intercalation, as shown in the previous example with Chevrel phase $Mo_6S_{8-x}Se_x$.[5]

7.2.2 Layered Oxides

7.2.2.1 *Vanadium Oxide (V_2O_5)*

Since the pioneering electrochemical and chemical screening of various cathodes by T.D. Gregory *et al.*,[17] a group of European researchers took initiative to find potential cathodes for Mg batteries. P.G. Bruce *et al.* conducted more rigorous chemical magnesiation using di-*n*-butylmagnesium ($(C_4H_9)_2Mg$) as a reducing agent, with which he found several candidates, including V_2O_5, which can possibly intercalate Mg^{2+}.[18] P. Novák *et al.* focused on electrochemical studies on oxides based on vanadium and molybdenum, using organic electrolytes based on magnesium perchlorate ($Mg(ClO_4)_2$) salt and propylene carbonate (PC) or acetonitrile (AN) solvents.[19–21] They concluded that a crystal of V_2O_5 hardly intercalates Mg^{2+}, but porous electrodes based on powder deliver a high capacity of *ca.* 180 mA h g^{-1} if water is added in the electrolyte to a concentration of *ca.* 1 M. However, it was unclear whether the capacity originates from the intercalation of Mg^{2+} or not. During the initial stage of research, only a few researchers were aware of the possible side reactions other than the intercalation of Mg^{2+} in non-aqueous electrolytes containing water. The electrochemical tests with $Mg(ClO_4)_2$ salt had severe pitfalls as the salt is highly hygroscopic, so the electrolytes are prone to contain water moieties. In this respect, less hygroscopic magnesium trifluoromethylsulfonylimide ($MgTFSI_2$) salt was more useful in the electrochemical tests.

Amatucci *et al.* claimed that nanosized V_2O_5 is capable of intercalating Mg^{2+} based on electrochemical tests only.[22] Similarly, Gershinsky *et al.* utilized a thin film electrode with facilitated electrode kinetics.[23] However, there was a pitfall in that the thin film was much more susceptible to trace amounts of water because the total amount of water in the electrolyte was sufficient enough to interfere with such small amount of electrode material. Based on the initial results, a couple of companies targeted the development of full battery cells with a V_2O_5 cathode and Mg anode. Pellion technologies invented a method that is claimed to facilitate the solid-state diffusion of Mg^{2+} by expanding the interlayer distance with the intercalation of larger organic cations in the interlayer.[24] They claimed that partial expansion of the interlayer distance resulted in an enhancement of the reversible capacity based on the electrochemical and spectroscopic characterizations; however, more rigorous studies are needed regarding the potentially profitable roles of organic cations in V_2O_5 for the intercalation of Mg^{2+}. T. S. Arthur *et al.* at the Toyota Research Institute North America claimed enhanced electrochemical storage of Mg^{2+} by amorphization and manipulation of the interlayer distance of V_2O_5 by compositing with phosphorus pentoxide (P_2O_5).[25]

There have been continuous reports that water in the electrolyte or in the crystal structure plays significant roles in enhancing the capacity of V_2O_5 in Mg^{2+} electrolytes.[26–28] In particular, a V_2O_5 aerogel containing structural water has shown a large capacity of up to 542 mA h g^{-1} in AN-based Mg^{2+} electrolytes.[29–32] In some cases, the Mg content was shown to increase in accordance with the electrochemical reduction, although some perchlorate ions were detected along with Mg^{2+}, leaving the possibility for co-intercalation of $MgClO_4^+$ ions.[30]

Since 2013, the Joint Center for Energy Storage Research (JCESR) team initiated rigorous and critical studies on α-V_2O_5. However, the electrochemical and spectroscopic tests revealed that the actual Mg content was small and the intercalation of protons prevailed, especially as observed from solid-state nuclear magnetic resonance (NMR) spectroscopic measurements.[10] Two research groups conducted rigorous structural analysis of the electrochemically reduced product in hydrated Mg^{2+} electrolytes with XRD, finding that the intercalated species was dominated by protons rather than Mg^{2+} ions.[11,12] These findings even led to the serious question of whether α-V_2O_5 is able to intercalate Mg^{2+} or not. Recently, S.-T. Hong *et al.* claimed that Mg content was linearly proportional to the state of charge (SoC) for layered $V_3O_7{\cdot}H_2O$ that contains structural water (Figure 7.3).[28] However, it has not been possible to cycle such hydrated materials with structural water in Mg^{2+} electrolytes that can operate a Mg metal anode, probably due to the chemical reactivity of such Mg^{2+} electrolytes towards oxides or due to the leakage of structural water into the electrolyte that may lead to severe degradation of the Mg metal anode.

In summary, layered V_2O_5 has been regarded as a potentially ideal candidate for a Mg cathode; however, it is under a serious test to determine if it can be a Mg battery cathode or not. The added water may have been the source of protons, against the previous explanation of enhanced capacity in

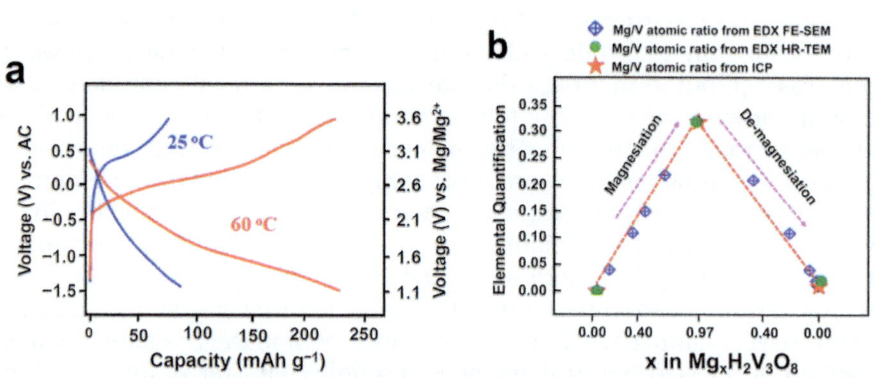

Figure 7.3 Voltage profiles of the layered $V_3O_7{\cdot}H_2O$ at 25 and 60 °C, and (b) the linearly proportional Mg content to the state of charge. Reproduced from ref. 28 with permission from American Chemical Society, Copyright 2018.

wet electrolytes by electrostatic screening of divalent Mg^{2+} ions with water dipoles.[26] These debates await direct proof or disproof of its inherent ability to intercalate Mg^{2+} in the absence of water moieties.

7.2.2.2 Molybdenum Oxide (MoO_3)

Generally, the strategies to enhance V_2O_5 have also been applied to layered MoO_3, such as a thin film approach and interlayer expansion.[23,24] Theoretical and experimental studies have shown that doping with fluorine could enhance the reversible capacity.[33,34] However, protons may have played a significant role in the reversible capacity as in the case of V_2O_5.

7.2.3 Graphite

To overcome the limitations in the available electrolytes that are compatible with magnesium metals, insertion electrodes of Mg^{2+} ions with a low redox potential have emerged as alternative negative electrode materials. These insertion electrodes are capable of being applied in conventional electrolytes such as $Mg(TFSI)_2$, $Mg(PF_6)_2$ or $Mg(CF_3SO_3)_2$, free from corrosive chloride ions (Cl^-).[35–37] Moreover, the use of intercalation electrodes is beneficial in terms of structural and chemical integrity compared to alloying compounds such as Sn, and Bi, which suffer from drastic pulverization caused by large volumetric changes upon repeated alloying and de-alloying processes. Pontiroli *et al.*[38] first discovered that Mg^{2+} ions can reversibly intercalate into fullerene (C_{60}), forming Mg_2C_{60}, which has the same bonding architecture as the lithiated fulleride Li_4C_{60}.[38,39] This report initiated other research studies on the possibility of the formation of a Mg^{2+}-intercalated graphite intercalation compound (GIC) in a similar manner to how Li^+ forms LiC_6 with graphite.[40,41] K. T. Lee *et al.*[40] have found that Mg^{2+} ions are reversibly intercalated into graphite together with linear ether solvents such as diethylene glycol dimethyl ether (DEGDME) and 1,2-dimethoxyethane (DME). A similar phenomenon has also been reported in the incorporation of Li^+ and Na^+ ions into graphite layers.[42–46] For example, solvated Li^+ ions in propylene carbonate (PC) are known to intercalate into graphite to form a ternary compound, $Li_x(PC)_yC_n$.[47] Density functional theory (DFT) calculations shows that Mg^{2+} ions strongly bind to linear ethers, particularly DEGDME and DME (binding energies are estimated to be 10.05 and 7.60 eV, respectively), enabling co-intercalation of these ion–solvent pairs into the graphene layers. On the other hand, the binding energies for Mg^{2+}–ethylene carbonate (EC) and Mg^{2+}–diethyl carbonate (DEC) complexes are substantially lower, alluding that de-solvation of Mg^{2+} ions should occur at the electrolyte/graphite interface in this case. The DFT calculations also revealed that the intercalation of Mg^{2+}–DEGDME into graphite is a thermodynamically favorable process and furthermore, the formation of a double-layer structure between graphene layers is favored over that of a single-layer (Figure 7.4), similar to the case

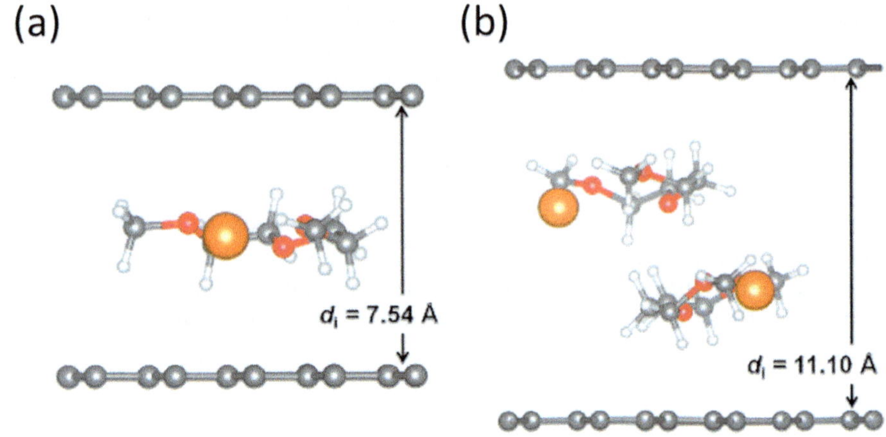

Figure 7.4 Schematic of the Mg^{2+}–DEGDME co-intercalated graphite: (a) single- and (b) double-layer structures. Orange, white, gray, and red balls represent Mg, H, C, and O atoms, respectively. d_i represents the intercalant gallery height. Reproduced from ref. 40 with permission from American Chemical Society, Copyright 2018.

of Na^+–DEGDME co-intercalation into graphite.[48] It is noteworthy that the diffusivity of the Mg^{2+}–DEGDME complex into graphite layers is estimated to be 1.5×10^{-8} cm^2 s^{-1} at room temperature, comparable to that of Li^+ in graphite (1.8×10^{-9} cm^2 s^{-1}).[40] This indicates that the kinetics involved in the co-intercalation of the Mg^{2+}–DEGDME complex are as fast as Li^+ ion intercalation into graphite, suggesting that graphite may be utilized as a high-rate anode material. The galvanostatic cycling of a Mg/graphite cell with 0.3 M of $Mg(TFSI)_2$ in DME/DEGDME solvent exhibited reversible voltage profiles with a constant capacity of ~180 mA h g^{-1}. A rather high polarization (>2 V) even at a low current rate during cycling is speculated to originate from the slow kinetics for Mg stripping/plating on the Mg electrode in this electrolyte, not from the intercalation of Mg^{2+}–DEGDME complexes into graphite layers. A galvanostatic intermittent titration technique (GITT) measurement indicated that the approximate redox potential for co-intercalation of Mg^{2+}–DEGDME into graphite is surprisingly 1 V *vs.* Mg/Mg^{2+}.[40] Therefore, a drastic improvement in the performance actually enables graphite to be considered as a positive electrode material. The mechanistic study analyzed by *ex situ* X-ray diffraction patterns of the electrodes at various charge–discharge states reveals that co-intercalation of Mg^{2+}–DEGDME exhibits a well-known staging behavior of graphite,[46,49,50] signifying that Mg^{2+} ions intercalate every *n*th space between the graphene layers. The theoretical intercalant gallery height from the double-layer structure (11.10 Å) is in good agreement with the value experimentally estimated (11.45 Å) of the fully discharged graphite at stage number 5.[46] Fourier-transform infrared (FT-IR) studies also indicated the presence of DEGDME molecules in the fully discharged electrode. Transmission electron microscope studies showed that a Mg-rich disordered region is observed after the intercalation of Mg^{2+} ions. These amorphous phases account for a rather high discharge capacity (180 mA h g^{-1}) compared with the calculated value from the graphite intercalation compound at stage number 5. Schmuck *et al.*[41] investigated Mg^{2+} intercalation into natural graphite in 0.5 M $Mg(TFSI)_2$/*N*,*N*-dimethylformamide (DMF) electrolyte, delivering a low but steady reversible discharge capacity of ~30 mA h g^{-1}. From *ex situ* XRD pattern analysis, they observed that Mg^{2+} intercalation produces a staging compound, but it was not clearly identified whether co-intercalation of DMF solvent occurs. It is noteworthy that in many cases, co-intercalation of solvent molecules with metal cations is prone to triggering irreversible exfoliation of graphite that will gradually deteriorate the electrochemical performance.[51,52] This aspect of the co-intercalation of Mg^{2+}–solvent pairs should be further investigated.

7.2.4 $VOPO_4$

A report by Yao *et al.*[53] proposed that the expansion of the interlayer spacing of layered TiS_2 by pre-intercalation of large $PY14^+$ cations enables the intercalation of a $MgCl^+$ species rather than Mg^{2+} ions. This lowers the activation barrier for ion migration and saves the dissociation energy (E_a) related to the dissociation of $MgCl^+$ into its component species, which could amount for at least 3 eV. This report opened up a new opportunity in that other layered compounds (metal chalcogenides or metal oxides) could also be potentially utilized as positive electrode materials by applying a similar strategic approach. $VOPO_4$ has a two-dimensional layered structure consisting of a polyanion framework, where VO_6 octahedra are linked to PO_4 tetrahedra by sharing corners to form two-dimensional layered structures (Figure 7.5).[54–56] This material has been proven to be an excellent cathode material for Li- or Na-ion batteries.[57–59] The hydrated phase with interlayer water molecules, $VOPO_4 \cdot nH_2O$ (termed as OH–$VOPO_4$, n is typically 2) can be readily prepared *via* a hydrothermal method.[60] The typical interlayer distance in OH–$VOPO_4$ is 7.41 Å.[56,58] These $VOPO_4 \cdot 2H_2O$ particles were subjected to ultrasonic agitation to be exfoliated in water, followed by incorporation of phenylamines, allowing a self-assembly process to occur in phenylamine, producing $VOPO_4$

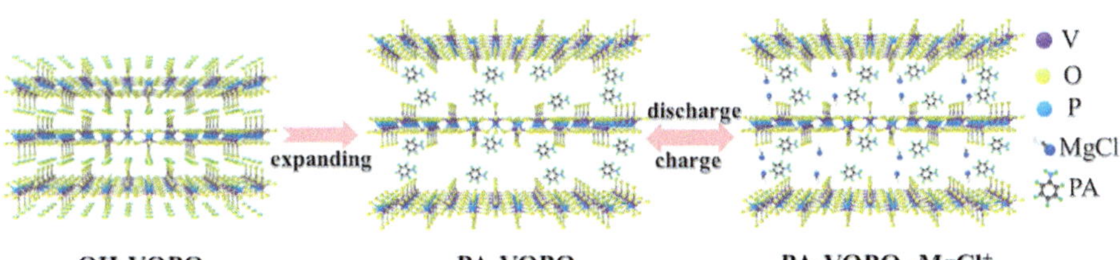

Figure 7.5 Schematic illustration of the layer expanded $VOPO_4$ by phenylamine, and proposed reaction mechanism of PA–$VOPO_4$ nanosheets as Mg^{2+} ion-storage materials. Reproduced from ref. 60 with permission from John Wiley and Sons, © 2018 WILEY-VCH Verlag GmbH & Co. KGaA, Weinheim.

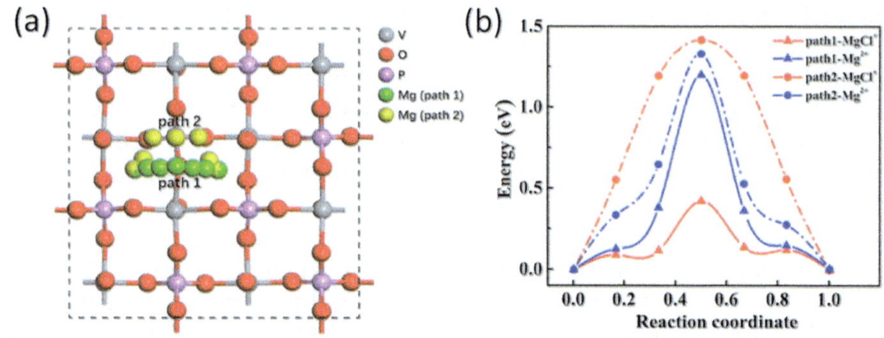

Figure 7.6 (a) Possible diffusion paths and (b) corresponding activation energy barrier profiles for Mg^{2+} or $MgCl^+$ migration in PA–$VOPO_4$ nanosheets. Reproduced from ref. 60 with permission from John Wiley and Sons, © 2018 WILEY-VCH Verlag GmbH & Co. KGaA, Weinheim.

nanosheets with a vastly expanded interlayer spacing of 14.2 Å (termed as PA–$VOPO_4$). The expansion of layers was confirmed by a significant shift in the (001) reflection in the XRD diffraction pattern to a lower angle and measurement of the interlayer distance directly from high-resolution TEM images. The existence of phenylamine in the interlayer was clearly evidenced by comparative analyses of EDS, FT–IR and TGA results for OH– and PA–$VOPO_4$. This structural reconfiguration of $VOPO_4$ led to remarkable enhancement in the electrochemical performance, especially in the rate performance, where the reversible discharge capacity was observed to be 310 mA h g^{-1} at 50 mA g^{-1} even after 500 cycles. The discharge capacity at 2000 mA g^{-1} was still found to be 100 mA h g^{-1}. The change in the electrode mass containing PA–$VOPO_4$ after various stages of discharge reactions indicates that it is $MgCl^+$ that intercalates into layers for PA–$VOPO_4$, while un-solvated Mg^{2+} ions are inserted into the layers of OH–$VOPO_4$. Furthermore, X-ray photoelectron spectroscopic (XPS) spectra and energy dispersive X-ray spectroscopic (EDS) analysis for the discharged electrode concurrently showed the existence of both Mg and Cl species in a roughly 1 : 1 ratio in the discharged electrode. First-principles calculations based on DFT showed a substantial difference in the activation energy, E_a, for ion migration in PA–$VOPO_4$, *i.e.* 0.42 eV for $MgCl^+$ diffusion and 1.20 eV for Mg^{2+} along the P1 pathway shown in Figure 7.6,[60] indicating that much higher diffusion kinetics are expected for $MgCl^+$ than for Mg^{2+}. This study hints that other layered compounds could be expanded to improve their performance by taking a similar approach.

7.2.5 VS_4

Z. Jin *et al.*[61] reported that vanadium tetrasulfide (VS_4) is a favorable cathode material that acts as a facile intercalation host for Mg^{2+} ions. VS_4 has a peculiar one-dimensional chain-like crystal structure similar to the

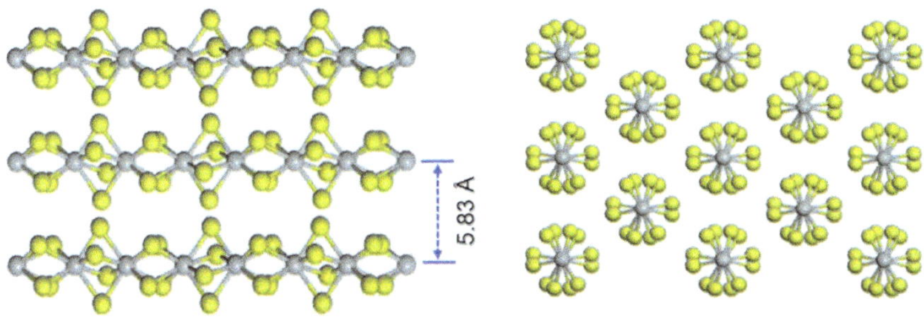

Figure 7.7 Schematic crystal structure of VS_4. (mineral patronite). Lateral-view (left) and vertical-view (right) of the one-dimensional chain-like crystalline structure of VS_4, exhibiting an interval of 5.83 Å between the atomic chains. Reproduced from ref. 61 with permission from John Wiley and Sons, © 2018 WILEY-VCH Verlag GmbH & Co. KGaA, Weinheim.

mineral patronite, where V^{4+} is coordinated to four disulfide anions (S_2^{2-}) along the c-axis (Figure 7.7).[62–64] These atomic chains are 5.83 Å apart from each other, which is much larger than the ionic diameter of Mg^{2+} ion (~1.44 Å), and are only weakly bound by van der Waals forces. They have shown that Mg^{2+} ions can readily pass through the open channels formed by these atomic chains.[61] Electrochemical insertion of Li^+ and Na^+ ions into these 1D channels in VS_4 was investigated previously.[65–68] Nanodendrites of VS_4 prepared by a solvothermal method exhibited an initial discharge capacity of 251 mA h g^{-1} at a current rate of 100 mA g^{-1} and stable cycling performance up to 800 cycles with a current density of 500 mA g^{-1}. XPS studies on the electrode at various stages of discharge and charge states indicated that the insertion of Mg^{2+} ions into VS_4 led to the partial oxidation of V^{4+} to V^{5+}, and a partial reduction of S_2^{2-} to S^{2-}, while the extraction of Mg^{2+} ions from the channels caused V^{5+} and S^{2-} to revert back to V^{4+} and S_2^{2-}. Insertion of Mg^{2+} ions into channels in VS_4 was verified analytically by *ex situ* TEM, XRD and Raman spectroscopy for the electrode at various stages of discharge and charge states. From the XRD and HR–TEM studies on both the discharged and charged electrodes, little variation in the lattice parameters was observed after the insertion of Mg^{2+} ions into VS_4, in an accordance with the results from DFT calculations. The fact that no significant morphological changes were observed during the insertion process, and that MgS and elemental vanadium were not observed after full discharge strongly indicates that the intercalation of Mg^{2+} ions is mainly responsible for the reaction mechanism. This is distinctly different from the behavior observed for a Li–VS_4 battery,[64] where Li_2S and elemental V were observed after full reduction by means of a conversion process. In a Mg–VS_4 battery, the original topotactic structure is well maintained, which is beneficial to the reversibility and cycling life of this battery system.

7.2.6 Prussian Blue Analogues

Prussian Blue Analogues (PBAs) feature a wide open framework with the general formula, $A_xMM'(CN)_6{\cdot}yH_2O$ (A = metal ions, M = Fe, Ni, Mn, V, Mo, Cu, Co; M′ = Fe, Co, Cr, Ru).[69,70] The crystal structure of PBAs are analogous to those of ReO_3 or ABX_3 perovskites, where the A sites are filled with A cations as well as zeolitic water molecules, and the B sites are systematically occupied by M^{m+} and M'^{n+} ions bridged by cyano (C≡N) ligands along the edges (Figure 7.8). It is known that small molecules or ions, 0 to 2 per formula unit, can be incorporated into the A sites of the large open cages, leading to corresponding changes in the oxidation states of the M and M′ ions.[71–74] Y. Cui *et al.*[75–77] reported various types of PBAs as possible cathode materials for monovalent (Li^+, Na^+, K^+) or multivalent (Mg^{2+}, Ca^{2+}, Sr^{2+}, Ba^{2+}) cathode materials with a high rate capability and a long cycling life. In particular, they showed that nickel hexacyanoferrate ($KNiFe^{3+}(CN)_6$, NiHCF) can function as an efficient cathode material with a long cycling life and high energy efficiency in aqueous electrolyte.[75] In NiHCF, half of the A sites are occupied by K^+ ions, and Ni^{2+} and Fe^{3+} fill the N-coordinated M sites, and C-coordinated M′ sites, respectively. The discharge–charge curves for NiHCF with an aqueous electrolyte show a sloping potential, indicating that a single phase reaction occurs. They postulated that the partial shielding of electrostatic interactions by the water molecules located at the A sites and ferricyanide vacancies with a diameter of 5 Å are responsible for the facile intercalation kinetics of divalent cations into NiHCF. Hong *et al.* carried out a related study using NiHCF with a composition of $K_{0.86}Ni[Fe(CN)_6]_{0.954}(H_2O)_{0.766}$, but in an organic electrolyte, 0.5 M $Mg(ClO_4)_2$ in acetonitrile with a thick carbon black anode.[70] Xia employed a NiHCF cathode and polyimide anode to construct a

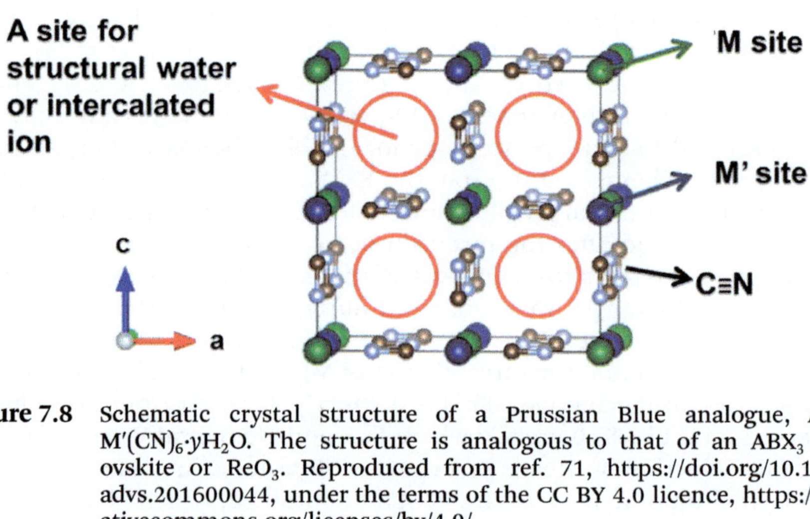

Figure 7.8 Schematic crystal structure of a Prussian Blue analogue, $A_xMM'(CN)_6{\cdot}yH_2O$. The structure is analogous to that of an ABX_3 perovskite or ReO_3. Reproduced from ref. 71, https://doi.org/10.1002/advs.201600044, under the terms of the CC BY 4.0 licence, https://creativecommons.org/licenses/by/4.0/.

supercapacitor-like high-power cell with a 1 M $MgSO_4$ aqueous electrolyte.[78] They obtained a remarkable cycling performance of up to 5000 cycles with an excellent rate performance.

7.3 Materials Based on Conversion and Displacement Reactions

7.3.1 Advantages of Conversion/Displacement Reactions for Mg^{2+} Storage

The adoption of intercalation compounds (graphite and layered transition metal oxides) as electrode materials plays a crucial role in securing the high performance and safety of battery cells required for the commercialization of Li-ion batteries. However, that is not the case with the intercalation of divalent Mg^{2+} ions into host materials. One of the main roadblocks is that it induces a strong coulombic interaction with the framework of the host materials, leading to a high activation barrier for ion migration and consequently, poor reaction kinetics.[79,80] Expanded layered compounds with a large interlayer spacing, such as $PY14^+$-intercalated TiS_2, showed one possible direction to pursue in order to overcome this issue through facilitating the incorporation of $MgCl^+$ instead of Mg^{2+} ions, markedly lowering the electrostatic interactions between the host and guest.[53] However, from a practical point of view, this approach requires concentrated electrolytes, that are abundant in chloride ions, otherwise chloride ions may be depleted from the electrolyte causing changes in the insertion mode or a sudden drop in the ionic conductivity of the electrolyte as the discharge process proceeds. Furthermore, a wide expansion of the interlayer spacing may result in low volumetric capacity, even lower than the conventional Mo_6S_8 Chevrel phase (519 Ah L^{-1}). Conversion-type reactions could provide an alternative reaction pathway to circumvent this situation.[81–83] In this case, a new phase, which bears little or no topotactic relationship to the original material, arises as a result of the reaction with Mg^{2+} ions. Conversion reactions usually proceed slowly as the reaction occurs only at the phase boundaries among the involved phases. However, the reaction kinetics of a conversion-type reaction can be significantly enhanced by reducing the particle size to the nanoscale as they provide much larger reaction sites and a shorter diffusion length for the full utilization of the original particles.[84,85] For some transition metal chalcogenides bearing a close structural kinship with MgS or MgSe with a rock-salt structure, a displacement reaction has been proposed to account for the high reversible capacity and fast diffusion kinetics at room temperature. In this reaction, insertion of Mg^{2+} ions result in the replacement of transition metal ions in the cathode material by Mg^{2+} ions, and the extrusion of transition metal in the form of nanocrystals. The high mobility of some transition metal ions and weak interaction of hard acid Mg^{2+} ions with soft base anions such as S^{2-} or Se^{2-} (based on hard–soft acid–base (HSAB) theory) can promote the displacement reaction.[85–87] Since the overall anion structure is preserved during the process, a displacement reaction typically involves less irreversible capacity and a smaller overpotential compared to conventional conversion reactions. In both cases, the preparation of nanosized or nanostructured materials is the most important factor to successfully accomplish fast reaction kinetics. Recently, copper chalcogenides with a variety of crystal structures have been extensively studied for the possibility of using these compounds as conversion-type cathode materials for rechargeable Mg batteries.

7.3.2 Copper Chalcogenides

A series of copper chalcogenides, $Cu_\alpha X$ ($1< \alpha < 2$, X = S, Se), such as CuS, Cu_2S, and Cu_2Se have been proposed as promising cathode materials for rechargeable Mg batteries.[88–91] When applied as cathode materials in rechargeable Mg batteries, these are known to undergo conversion or displacement reactions, producing Cu metal and MgS or MgSe as discharge products. These materials have gained attention since they show a relatively small overpotential at room temperature, comparable to that of other efficient cathode materials based on intercalation chemistry. The practical voltage from these conversion reactions typically ranges from 1.0 to 1.5 V.

CuS has a hexagonal crystal structure analogous to the mineral covellite, and its theoretical capacity from full conversion into elemental Cu and MgS amounts to 560 mA h g^{-1}.[92] Nazar *et al.*[88] was the first to report that CuS can store Mg ions up to 200 mA h g^{-1} in an all-phenyl-complex electrolyte at 150 °C. Mai *et al.*[89] demonstrated that CuS nanospheres can deliver a high discharge capacity of over 360 mA h g^{-1} at room temperature by applying $Mg(ClO_4)_2$ in acetonitrile as an electrolyte system, although they had to use a thick carbon black anode due to the incompatibility of their electrolyte with Mg metals. XRD and XPS studies on the electrodes at various stages verified that this electrode works based on a conversion reaction mechanism. Two-step reactions as suggested below have been proposed as Cu_2S appears as an intermediate species during the electrochemical process. Cu_2S is speculated to undergo a conversion reaction with Mg^{2+} ions.[89]

$$2CuS + Mg^{2+} + 2e^- \leftrightarrow Cu_2S + MgS$$

$$Cu_2S + Mg^{2+} + 2e^- \leftrightarrow 2Cu + MgS$$

$Cu_{2-\delta}S$ has a wide variety of crystal structures with significant non-stoichiometry in its composition. The theoretical capacity from its full conversion into Cu and MgS amounts to 337 mA h g^{-1}.[93] Miyasaka *et al.*[90]

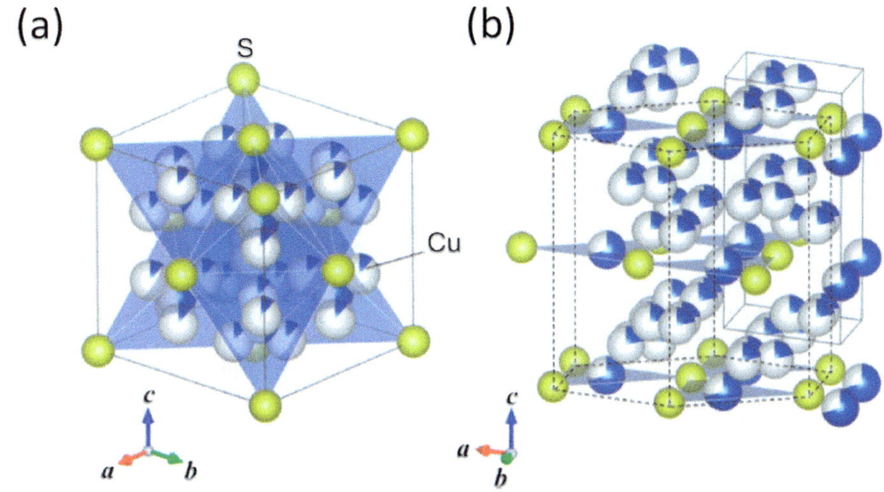

Figure 7.9 A comparison of the crystal structure of cubic and hexagonal $Cu_{2-\delta}S$. (a) In cubic digenite (h-$Cu_{2-\delta}S$), anions are arranged in a face-centered cubic lattice, where Cu ions are distributed at tetrahedral or trigonal sites. The blue polyhedra represent tetrahedral sites. (b) In hexagonal chalcocite (h-$Cu_{2-\delta}S$), the anions make up a hexagonal close-packed lattice, where Cu ions are distributed at the interstitial sites. Reproduced from ref. 90 with permission from The Chemical Society of Japan, Copyright 2017.

evaluated the effect of these crystal structures of $Cu_{2-\delta}S$ on its performance as a cathode material for a rechargeable Mg battery. They proposed that the cubic phase (mineral digenite), which have an analogous anion lattice to that of MgS (Figure 7.9), undergoes a facile displacement reaction to account for its low overpotential, even at room temperature. On the other hand, a conversion mechanism also works for the hexagonal phase (mineral chalcocite), whose structure is unrelated to that of MgS. A nanocomposite formation with conductive carbon improved the reversible capacity up to ~200 mA h g^{-1}.

Miyasaka *et al.*[91] investigated β-Cu_2Se as a displacement-type cathode material for Mg batteries. In β-Cu_2Se, Se atoms are arranged in a face centered cubic structure, which is the same arrangement as for MgSe (Figure 7.10). Cu atoms are located randomly at tetrahedral and trigonal sites. In this structure, Cu^+ ions are known to be highly mobile, so that β-Cu_2Se is considered as a superionic conductor. The displacement reaction between Li^+ and Na^+ ions was reported previously.[94,95] Nanocrystallites (~100 nm) of β-Cu_2Se were prepared *via* a solution-based method, delivering a reversible discharge capacity of around 230 mA h g^{-1}. From the XRD observations, MgSe and Cu were mainly found in the fully discharged electrode, although several other phases such as α-Cu_2Se, and Cu_3Se_2, which is structurally related to β-Cu_2Se, were also found.

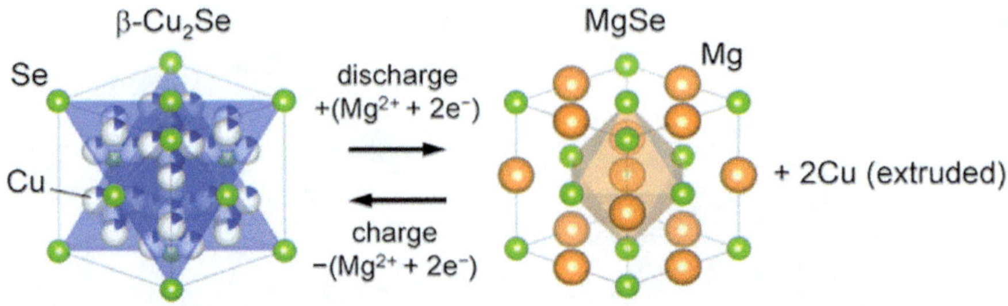

Figure 7.10 Crystal structure of β-Cu_2Se and a schematic for the displacement reaction with Mg^{2+} ions. In β-Cu_2Se, the Se anions take a face-centered cubic arrangement, where Cu ions are distributed randomly at the trigonal or tetrahedral sites. MgSe has the same anion arrangement as β-Cu_2Se. During the displacement reaction, Mg^{2+} ions are incorporated into the octahedral sites, while Cu ions are extruded from the interstitial sites to form Cu metal. Reproduced from ref. 91 with permission from Elsevier, Copyright 2016.

7.4 Conclusion

Insertion cathodes for magnesium batteries are under continuous investigation by researchers worldwide who are aiming at understanding and developing multivalent battery chemistry as a more sustainable technology for energy storage and electromobility. Recent clarification of the previous ambiguity about the origin of the capacity for V_2O_5 in hydrated electrolytes represents the indispensable necessity of distinguishing between the genuine insertion of Mg^{2+} and various side reactions. Currently, new ideas and strategies are tested in sulfides or selenides as model compounds that are chemically stable in Mg^{2+} electrolytes, while oxides are awaiting proof or disproof of their inherent ability to intercalate Mg^{2+} in non-nucleophilic electrolytes. Alternatively, insertion cathodes based on conversion or displacement mechanisms are under initial development. Appreciating the history of the research on insertion cathodes for magnesium batteries, the present is a time for serious retrospect after a surge in preliminary and unverified results in the past. Such rigorous verification is essential to build a solid foundation for a genuine breakthrough in Mg^{2+} insertion cathodes for high energy, low cost rechargeable batteries.

Acknowledgement

H.D.Y. was supported by the National Research Foundation (NRF-2018R1C1B6004808 and NRF-2018R1A5A1025594) of the Korean Ministry of Science and ICT. S.H.O. acknowledges financial support from the KIST institutional program (Project No. 2E29641).

References

1. S. Jeong, D. Yoo, M. Ahn, P. Miró, T. Heine and J. Cheon, *Nat. Commun.*, 2015, **6**, 5763.
2. C. Wan, X. Gu, F. Dang, T. Itoh, Y. Wang, H. Sasaki, M. Kondo, K. Koga, K. Yabuki, G. J. Snyder, R. Yang and K. Koumoto, *Nat. Mater.*, 2015, **14**, 622.
3. A. V. Powell, *Annu. Rep. Prog. Chem., Sect. C: Phys. Chem.*, 1993, **90**, 177.
4. D. Aurbach, Z. Lu, A. Schechter, Y. Gofer, H. Gizbar, R. Turgeman, Y. Cohen, M. Moshkovich and E. Levi, *Nature*, 2000, **407**, 724.
5. D. Aurbach, G. S. Suresh, E. Levi, A. Mitelman, O. Mizrahi, O. Chusid and M. Brunelli, *Adv. Mater.*, 2007, **19**, 42607.
6. Y. Gu, Y. Katsura, T. Yoshino, H. Takagi and K. Taniguchi, *Sci. Rep.*, 2015, **5**, 12486.
7. X. Sun, P. Bonnick, V. Duffort, M. Liu, Z. Rong, K. A. Persson, G. Ceder and L. F. Nazar, *Energy Environ. Sci.*, 2016, **9**, 2273.
8. X. Sun, P. Bonnick and L. F. Nazar, *ACS Energy Lett.*, 2016, **1**, 297.
9. H. D. Yoo, Y. Liang, H. Dong, J. Lin, H. Wang, Y. Liu, L. Ma, T. Wu, Y. Li, Q. Ru, Y. Jing, Q. An, W. Zhou, J. Guo, J. Lu, S. T. Pantelides, X. Qian and Y. Yao, *Nat. Commun.*, 2017, **8**, 339.
10. N. Sa, H. Wang, D. L. Proffit, A. L. Lipson, B. Key, M. Liu, Z. Feng, T. T. Fister, Y. Ren, C.-J. Sun, J. T. Vaughey, P. A. Fenter, K. A. Persson and A. K. Burrell, *J. Power Sources*, 2016, **323**, 44.
11. S.-C. Lim, J. Lee, H. H. Kwak, J. W. Heo, M. S. Chae, D. Ahn, Y. H. Jang, H. Lee and S.-T. Hong, *Inorg. Chem.*, 2017, **56**, 7668.
12. R. Verrelli, A. P. Black, C. Pattanathummasid, D. S. Tchitchekova, A. Ponrouch, J. Oró-Solé, C. Frontera, F. Bardé, P. Rozier and M. R. Palacín, *J. Power Sources*, 2018, **407**, 162–172.
13. M. S. Whittingham, *Science*, 1976, **192**, 1126.
14. Y. Liang, H. D. Yoo, Y. Li, J. Shuai, H. A. Calderon, F. C. Robles Hernandez, L. C. Grabow and Y. Yao, *Nano Lett.*, 2015, **15**, 2194.
15. S. Jing, Y. Hyun Deog, L. Yanliang, L. Yifei, Y. Yan and C. G. Lars, *Mater. Res. Express*, 2016, **3**, 064001.
16. Z. Li, X. Mu, Z. Zhao-Karger, T. Diemant, R. J. Behm, C. Kübel and M. Fichtner, *Nat. Commun.*, 2018, **9**, 5115.
17. T. D. Gregory, R. J. Hoffman and R. C. Winterton, *J. Electrochem. Soc.*, 1990, **137**, 775.
18. P. G. Bruce, F. Krok, J. Nowinski, V. C. Gibson and K. Tavakkoli, *J. Mater. Chem.*, 1991, **1**, 705.
19. P. Novák and J. Desilvestro, *J. Electrochem. Soc.*, 1993, **140**, 140.
20. V. Shklover, T. Haibach, F. Ried, R. Nesper and P. Novák, *J. Solid State Chem.*, 1996, **123**, 317.
21. P. Novák, R. Imhof and O. Haas, *Electrochim. Acta*, 1999, **45**, 351.
22. G. G. Amatucci, F. Badway, A. Singhal, B. Beaudoin, G. Skandan, T. Bowmer, I. Plitz, N. Pereira, T. Chapman and R. Jaworski, *J. Electrochem. Soc.*, 2001, **148**, A940.

23. G. Gershinsky, H. D. Yoo, Y. Gofer and D. Aurbach, *Langmuir*, 2013, **29**, 10964.
24. R. E. Doe, C. M. Downie, C. Fischer, G. H. Lane, D. Morgan, J. Nevin, G. Ceder, K. A. Persson and D. Eaglesham, Pellion Technologies, Inc., *US Pat.*, US9240612B2, 2015.
25. T. S. Arthur, K. Kato, J. Germain, J. Guo, P.-A. Glans, Y.-S. Liu, D. Holmes, X. Fan and F. Mizuno, *Chem. Commun.*, 2015, **51**, 15657.
26. E. Levi, Y. Gofer and D. Aurbach, *Chem. Mater.*, 2010, **22**, 860.
27. S. Tepavcevic, Y. Liu, D. Zhou, B. Lai, J. Maser, X. Zuo, H. Chan, P. Král, C. S. Johnson, V. Stamenkovic, N. M. Markovic and T. Rajh, *ACS Nano*, 2015, **9**, 8194.
28. M. Rastgoo-Deylami, M. S. Chae and S.-T. Hong, *Chem. Mater.*, 2018, **30**, 7464.
29. D. B. Le, S. Passerini, F. Coustier, J. Guo, T. Soderstrom, B. B. Owens and W. H. Smyrl, *Chem. Mater.*, 1998, **10**, 682.
30. D. Imamura and M. Miyayama, *Solid State Ionics*, 2003, **161**, 173.
31. Q. Y. An, Y. F. Li, H. D. Yoo, S. Chen, Q. Ru, L. Q. Mai and Y. Yao, *Nano Energy*, 2015, **18**, 265.
32. A. Moretti and S. Passerini, *Adv. Energy Mater.*, 2016, **6**, 1600868.
33. J. T. Incorvati, L. F. Wan, B. Key, D. Zhou, C. Liao, L. Fuoco, M. Holland, H. Wang, D. Prendergast, K. R. Poeppelmeier and J. T. Vaughey, *Chem. Mater.*, 2016, **28**, 17.
34. L. F. Wan, J. T. Incorvati, K. R. Poeppelmeier and D. Prendergast, *Chem. Mater.*, 2016, **28**, 6900.
35. T. S. Arthur, N. Singh and M. Matsui, *Electrochem. Commun.*, 2012, **16**, 103.
36. N. Singh, T. S. Arthur, C. Ling, M. Matsui and F. Mizuno, *Chem. Commun.*, 2013, **49**, 149.
37. Y.-H. Tan, *et al.*, *ACS Nano*, 2018, **12**, 5856.
38. D. Pontiroli, *et al.*, *Carbon*, 2013, **51**, 143.
39. S. Margadonna, *et al.*, *J. Am. Chem. Soc.*, 2004, **126**, 15032.
40. D.-M. Kim, *et al.*, *Chem. Mater.*, 2018, **30**, 3199.
41. C. God, *et al.*, *RSC Adv.*, 2017, **7**, 14168.
42. P. Schoderböck and H. P. Boehm, *Synth. Met.*, 1991, **44**, 239.
43. T. Abe, N. Kawabata, Y. Mizutani, M. Inaba and Z. Ogumi, *J. Electrochem. Soc.*, 2003, **150**, A257.
44. H. Moon, *et al.*, *J. Phys. Chem. C*, 2014, **118**, 20246.
45. B. Jache and P. Adelhelm, *Angew. Chem., Int. Ed.*, 2014, **53**, 10169.
46. H. Kim, *et al.*, *Energy Environ. Sci.*, 2015, **8**, 2963.
47. M. R. Wagner, J. H. Albering, K.-C. Moeller, J. O. Besenhard and M. Winter, *Electrochem. Commun.*, 2005, 7, 947.
48. H. Kuwata, M. Matsui and N. Imanishi, *J. Electrochem. Soc.*, 2017, **164**, A3229.
49. A. P. Cohn, *et al.*, *J. Mater. Chem. A*, 2016, **4**, 14954.
50. G. A. Elia, *et al.*, *J. Mater. Chem. A*, 2017, **5**, 9682.

51. M. Kawaguchi and A. Kurasaki, *Chem. Commun.*, 2012, **48**, 6897.
52. Y. Maeda and P. Touzain, *Electrochim. Acta*, 1988, **33**, 1493.
53. H. D. Yoo, *et al.*, *Nat. Commun.*, 2017, **8**, 339.
54. K. Goubitz, P. Capkova, K. Melanova, W. Molleman and H. Schenk, *Acta Crystallogr.*, 2001, **B57**, 178.
55. C. Wu, *et al.*, *Nat. Commun.*, 2013, **4**, 2431.
56. T. Nakato, Y. Furumi, N. Terao and T. Okuhara, *J. Mater. Chem.*, 2000, **10**, 737.
57. Y. Zhu, L. Peng, D. Chen and G. Yu, *Nano Lett.*, 2017, **16**, 742.
58. L. Peng, *et al.*, *Nano Lett.*, 2017, **17**, 6273.
59. G. He, W. H. Kan and A. Manthiram, *Chem. Mater.*, 2016, **28**, 682.
60. L. Zhou, *et al.*, *Adv. Mater.*, 2018, **30**, 1801984.
61. Y. Wang, *et al.*, *Adv. Mater.*, 2018, **30**, 1802563.
62. W. F. Hillebrand, *J. Am. Chem. Soc.*, 1907, **29**, 1019.
63. R. Allmann, I. Baumann, A. Kutoglu, H. Rösch and E. Hellner, *Die Naturwissenschaften*, 1964, **51**, 263.
64. S. Britto, *et al.*, *J. Am. Chem. Soc.*, 2015, **137**, 8499.
65. C. S. Rout, *et al.*, *J. Am. Chem. Soc.*, 2013, **135**, 8720.
66. X. Xu, *et al.*, *J. Mater. Chem. A*, 2014, **2**, 10847.
67. Y. L. Zhou, *et al.*, *ACS Appl. Mater. Interfaces*, 2016, **8**, 18797.
68. R. M. Sun, *et al.*, *ACS Appl. Mater. Interfaces*, 2015, **7**, 20902.
69. A. Kumar and S. M. Yusuf, *Phys. Rev. B*, 2005, **71**, 054414.
70. M. S. Chae, J. Hyoung, M. Jang, H. Lee and S.-T. Hong, *J. Power Sources*, 2017, **363**, 269.
71. X. Sun, V. Duffort and L. F. Nazar, *Adv. Sci.*, 2016, **3**, 1600044.
72. D. Asakura, *et al.*, *J. Am. Chem. Soc.*, 2013, **135**, 2793.
73. J. Song, *et al.*, *J. Am. Chem. Soc.*, 2015, **137**, 2658.
74. A. L. Lipson, *et al.*, *Chem. Mater.*, 2015, **27**, 8442.
75. R. Y. Wang, C. D. Wessells, R. A. Huggins and Y. Cui, *Nano Lett.*, 2013, **13**, 5748.
76. M. Pasta, C. D. Wessells, R. A. Huggins and Y. Cui, *Nat. Commun.*, 2012, **3**, 1149.
77. H.-W. Lee, *et al.*, *Nat. Commun.*, 2014, **5**, 5280.
78. L. Chen, *et al.*, *ACS Energy Lett.*, 2017, **2**, 1115.
79. H. D. Yoo, *et al.*, *Energy Environ. Sci.*, 2013, **6**, 2265.
80. M. Mao, T. Gao, S. Hou and C. Wang, *Chem. Soc. Rev.*, 2018, **47**, 8804.
81. P. Poizot, S. Laruelle, S. Grugeon, L. Dupont and J.-M. Tarascon, *Nature*, 2000, **407**, 496.
82. J. Cabana, L. Monconduit, D. Larcher and M. R. Palacin, *Adv. Mater.*, 2010, **22**, E170.
83. F. Klein, B. Jache, A. Bhide and P. Adelhelm, *Phys. Chem. Chem. Phys.*, 2013, **15**, 15876.
84. M. Armand and J.-M. Tarascon, *Nature*, 2008, **451**, 652.
85. P. G. Bruce, B. Scrosati and J.-M. Tarascon, *Angew. Chem., Int. Ed.*, 2008, **47**, 2930.

86. M. N. Obrovac, R. A. Dunlap, R. J. Sanderson and J. R. Dahn, *J. Electrochem. Soc.*, 2001, **148**, A576.
87. O. Crosnier and L. F. Nazar, *J. Electrochem. Soc.*, 2004, **7**, A187.
88. V. Duffort, X. Sun and L. F. Nazar, *Chem. Commun.*, 2016, **52**, 12458.
89. F. Xiong, *et al.*, *Nano Energy*, 2018, **47**, 210.
90. Y. Tashiro, K. Taniguchi and H. Miyasaka, *Chem. Lett.*, 2017, **46**, 1240.
91. Y. Tashiro, K. Taniguchi and H. Miyasaka, *Electrochim. Acta*, 2016, **210**, 655.
92. J. S. Chung and H. J. Sohn, *J. Power Sources*, 2002, **108**, 226.
93. C.-H. Lai, *et al.*, *J. Mater. Chem.*, 2010, **20**, 6638.
94. M.-Z. Xue, *et al.*, *J. Electrochem. Soc.*, 2006, **153**, A2262.
95. J.-L. Yue, Q. Sun and Z.-W. Fu, *Chem. Commun.*, 2013, **49**, 5868.

CHAPTER 8

High Energy Density Insertion Cathode Materials

BRIAN J. INGRAM*

Argonne National Laboratory, Chemical Sciences and Engineering Division, Lemont, IL 60439, USA
*E-mail: ingram@anl.gov

8.1 Introduction

Electrochemical energy storage technologies, exemplified in our daily lives by the ubiquitous use of batteries for an ever greater variety of consumer devices, have been a critical component of enabling the widespread mobility of human activities. From the earliest adaptation of energy storage in primary "wet cells," dating back to the 19th century, and the original secondary (*i.e.* rechargeable) battery based on lead–acid chemistry, storing energy in chemical bonds required a chemical reaction to physically change the state of materials from a high energy (*i.e.* charged) state to a lower energy (*i.e.* discharged) state. The advancement in intercalation chemistry and batteries thus inspired in the latter half of the 20th century – made famous by the now omnipresent Li-ion battery – was a significant breakthrough that has led to widespread adoption of portable electronics, and, indirectly, social media platforms. Continued advancements in Li-ion batteries are now supporting electrification of transportation, renewable energy generation, and implantable medical devices that may lead to the next unforeseen advancement in social engagement and technology.

Energy and Environment Series No. 23
Magnesium Batteries: Research and Applications
Edited by Maximilian Fichtner

Published by the Royal Society of Chemistry, www.rsc.org

Intercalation batteries, based on the shuttling of a working mobile cation, offer the opportunity for significant scientific breakthroughs since they convert chemical energy into electricity while allowing for minimal structural, chemical, and thermal changes. This enables systems with large storage capacities to cycle for hundreds to thousands of charge/discharge cycles. When coupled with the lightweight and highly electropositive element lithium, intercalation batteries support the highest energy densities in the marketplace. In other words, the ability to store and transport energy electrochemically in light and small packages.

As designed in the simplest intercalation battery system, the battery consists of two electrodes, known as the anode and cathode, which are separated by an electron-blocking electrolyte. In a typical discharged lithium-ion interaction system, lithium starts within the crystal structure of the cathode, *e.g.* $LiCoO_2$. In order to store energy, known as charging, an external electrical current is applied to extract lithium from the cathode structure, transport it through the electrolyte, and finally insert it between atomic layers of the anode (*e.g.* graphite). During a discharge cycle, lithium is shuttled in the reverse direction from the anode back to the cathode material. During this process, an electrical current in an external circuit is generated.

Attributed to the economies of scale associated with increased production volumes and manufacturing advances in battery cells and packs, the cost of Li-ion batteries has decreased dramatically over the last decade. Significant breakthroughs at the fundamental or materials level, have trailed off over the same period of time, and it is thought that the theoretical limits of materials are being approached. In light of this, alternative chemistries and energy storage concepts have been under consideration and are collectively known as "beyond Li-ion." This terminology may represent an unfair grouping of many diverse scientific directions with rich and nuanced details. With that said, it is clear that Li-ion batteries face challenges in supporting the next generation of electrical vehicles, large scale and inexpensive grid-storage, or unforeseen applications which will require even higher energy densities, costs restrictions, or operation under extreme environmental conditions. Just in considering the raw material requirements for elemental lithium and cobalt, fulfilling the worldwide penetration of 50% electric vehicles (EV) into the market will be challenging.

Alternative intercalation working cations, such as magnesium, are an attractive alternative to lithium within the scope of beyond Li-ion options. As is well documented here and elsewhere, magnesium faces numerous scientific and technical challenges; however, it supports increased energy densities relative to lithium, and implementation of magnesium-ion technologies can in principle smoothly integrate into Li-ion battery cell architecture and design by leveraging the significant depth of knowledge of lithium-ion manufacturing.

As will be discussed in this chapter, the benefit of high energy densities from magnesium systems relative to today's state-of-the-art Li-ion batteries will be realized by utilization of metallic anodes and high voltage, high capacity, high mobility cathodes. A series of well-resourced review articles, all published in the last half-decade, highlight the many challenges and breakthroughs in magnesium-ion batteries; the absence of a feasible high energy dense magnesium insertion cathode and corresponding stable electrolyte is evident in these reviews.[1–10] As such, development of materials with high voltage magnesium insertion reactions is vital for achieving truly transformative energy storage systems.

8.2 Techno-economic Modelling

8.2.1 Adapting Li-ion Models

Techno-economic (TE) models are important tools for predicting and evaluating the potential of new materials against established engineered or mature technologies. TE models predict the cost of a complete system coupled to defined performance requirements. Assumptions can be included to incorporate a fully refined and engineered manufacturing process at high volume production and competitive market economics.

Of course, the nascent nature of magnesium batteries requires significant assumptions and estimations of input parameters and battery structures. For instance, it is very reasonable to expect that practical magnesium batteries will utilize vastly different current collector metals, separators, and electrolytes relative to today's Li-ion systems. Not to mention, it is unclear what the future market or production costs will be; however, similarities do exist between lithium- and magnesium-ion systems that can be exploited for an effective comparison.

The battery performance and cost (BatPaC) model[11] is one such model that can be used to evaluate the cost-effectiveness of magnesium battery technologies. The BatPaC model has been peer-reviewed and validated material-to-system for lithium-ion batteries. Additionally, it supports maximum flexibility in input parameters through a bottom-up lithium-ion battery design and cost calculation. The advantage of this approach is the ability to probe the complete power and energy space in order to understand the complicated performance and cost relationship, while factoring impracticable designs consisting of physical limitations or electrochemical reaction constraints. These details can be adapted to provide flexibility in evaluating magnesium-ion battery systems, and indeed the BatPac model was used by Canepa *et al.* as an effective tool in predicting metrics to assess magnesium systems and their potential commercial usefulness in energy storage chemistry.[1] Descriptions and examples of the use of the BatPaC model are available;[12–16] it can be downloaded from http://www.cse.anl.gov/batpac as a public domain model that relates the design and cost of Li-ion batteries for transportation and vehicle applications.

As a final word, techno-economic models are well established and refined for Li-ion systems to precisely identify unique variations in cost and performance. In the case of magnesium, a level of suspicion is warranted for any finite value predicted by the models. As such, any TE model used should be flexible enough to incorporate such parameters as costs predictions, volume and density considerations, and electrical conductivity of components in order to evaluate general trends and ranges in performance. Values discussed in this chapter are intended to provide a general trend and potential merits of magnesium. All assumptions utilized are conservative, yet untested and certainly not demonstrated at a commercial scaled-up level.

8.2.2 Establish the Materials Requirements for Transformative Batteries

Using the BatPac model, Canepa *et al.* analyzed metal–oxide cathodes *vs.* a metallic magnesium anode to consider the upper-bound of energy densities in magnesium-ion batteries. The balance of cell structural components were reported to be: 10 and 15 μm thick Cu and Al current collectors for the anode and cathode, respectively, as well as a 15 μm thick separator. They reported and compared results from various design assumptions fulfilling technical performance requirements. By leveraging knowledge of impedance and electrolyte transport properties in state-of-the-art lithium-ion batteries, optimum electrode thicknesses were calculated that fulfil the pulse power requirements at 80% open circuit voltage (OCV) and sustained discharge at a C/2 rate.[12,15,17] The maximum theoretical cell-level energy density (defined as the current collector to current collector, inclusive, and excluding packaging, balance of plant components, *etc.*) was reported as a function of the open cell voltage for a given Mg metal/cathode couple of materials-only specific energy densities. The TE analysis provides a description of the baseline materials parameters for multiple couples that satisfy arbitrary cell-level energy density requirements.

Figure 8.1 compiles the TE modelling results from Canepa *et al.*[1] and relates the materials-level cathode capacitance (mAh g^{-1}) required to achieve a particular cell-level energy density at different open cell voltages, effectively determined by the choice of cathode material (assuming a metallic Mg anode). This is an instructive plot when considering cathode materials for research and development efforts. Estimations of average (or more detailed) voltage profiles[18] and materials-level charge capacity are relatively straight forward and can direct targeted efforts to technological critical systems. As is readily observable, despite their promising kinetics and cycle lifetime capabilities,[19–23] for typical sulphide materials that are active <2.0 V *vs.* Mg^0/Mg^{2+}, Herculean efforts are required to identify and develop charge capacities of >350 mA h g^{-1} to obtain a cell level energy density in excess of 600 W h L^{-1}. For instance, for two promising sulphide insertion Mg-ion cathodes, $Mg_{2-2x}Mo_6S_8$ and $Mg_{1-x}Ti_2S_4$,[22,23] the maximum theoretical capacities (cycling $x = 0 \rightarrow 1$) are 130 mA h g^{-1} per Mo_6S_8 formula unit and 240 mA h g^{-1} per Ti_2S_4 formula unit.

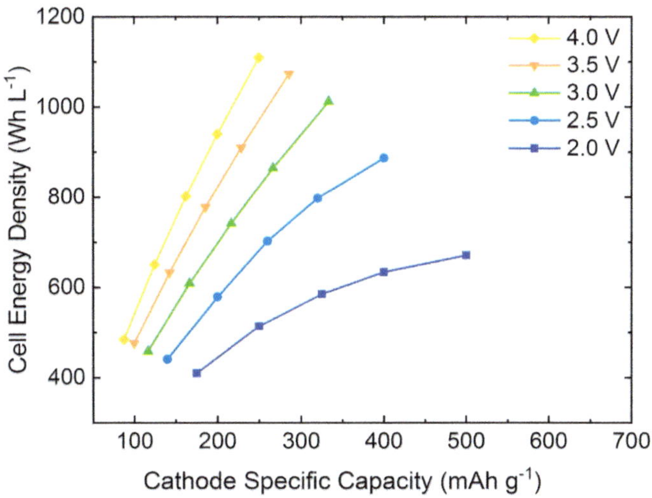

Figure 8.1 A summary of TE modelling results from Canepa *et al.* showing the relationship between the cathode specific capacity and the cell energy density (consisting of current collector to current collector, inclusive).[1] The voltages are average values across the cycle *vs.* Mg^0/Mg^{2+}.

Developing Mg-insertion cathodes with high average insertion/extraction voltages is advantageous for many reasons. It can be inferred from the free energy of Mg insertion *vs.* $Mg^{2+} + 2e^- \leftrightarrow Mg_{(s)}$, that the electrochemical potential of the cell is a primary driving factor, more so than the cathode specific charge capacity, to increasing the cell energy density. This is observed in the TE derived data in Figure 8.1. An additional advantage of higher voltage cells is the mitigation of impedance effects, which are more predominant at lower voltages. Mg insertion metal–oxides, which inherently should result in higher cell OCVs than sulphides,[24] can achieve theoretical charge capacities of ~300 mA h g^{-1} and practical capacities (assuming 70% Mg cycling) of 210 mA h g^{-1}. It is clear from Figure 8.1 that at these cathode specific capacity values a cell energy density representing a transformative technology, *e.g.* 750 W h L^{-1}, requires an average voltage of ≥3.0 V *vs.* Mg^0/Mg^{2+}.

8.2.3 Predicting and Comparing Technology Performances

TE modelling is also important when evaluating how newly proposed/discovered materials will perform in full-scale battery packs in order to provide realistic energy densities and price estimates. Broad categories of cathode materials, therefore, can be compared for critical performance at the pack-level, which includes multiple cell stacking, packaging, and current wiring. Figure 8.2 presents cathode (*vs.* Mg-metal anode) development pathways for transformative energy storage systems. A comparison of TE-model derived energy densities and specific energies for sulphides (*e.g.* $MgTi_2S_4$ and

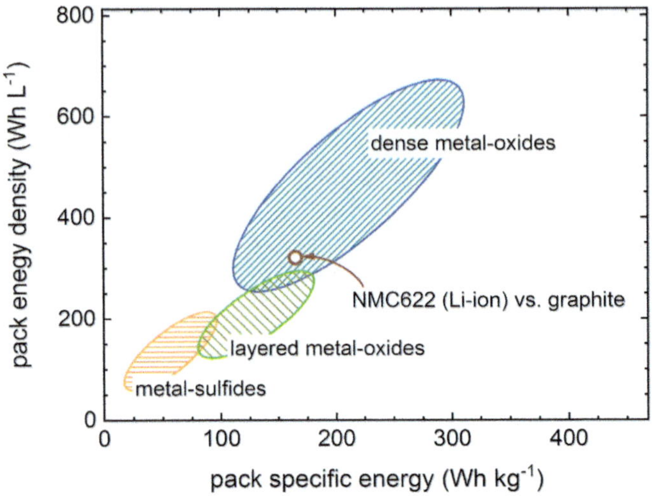

Figure 8.2 Techno-economic modelling indicating magnesium intercalation cathode design pathways for a commercially competitive system (*vs.* Mg-metal anode). The projections are shown at the battery pack-level. Calculated values for a representative lithium-ion system. NMC622 *vs.* graphite cells arranged in a similar pack design, is indicated for reference.

$Mg_2Mo_6S_8$), layered metal–oxide (*e.g.* $MoO_{2.8}F_{0.2}$ and V_2O_5), and dense metal–oxide (*e.g.* $MgMn_2O_4$) magnesium intercalation cathodes are compared to a BaTPaC modelled result of a typical lithium metal system (NMC622: $LiNi_{0.6}Mn_{0.2}Co_{0.2}O_2$ *vs.* graphite) in a similar battery pack configuration. Increased uncertainty is noted as a larger projected performance window for dense metal oxides, since they prove to be challenging to assess in terms of actual performance due to a large variation in the reported literature results.

Taken together, Figures 8.1 and 8.2 establish a picture of the need for developing a high-density metal–oxide cathode for realizing practical magnesium-ion battery systems that can compete with current state-of-the-art lithium-ion technologies. A comparison with layered metal–oxides and metal–sulphides is illustrative of the qualitative challenges faced in achieving commercially viable magnesium-ion battery cathodes. Figure 8.3 compares five features critical for practical implementation of battery technologies: TE-modelled battery pack-level energy density, specific energy, and cost, as well as experimentally determined Coulombic efficiency and cycle life. These are not comprehensive, as temperature, vibration, and other mechanical limitations are not considered; however, the parameters depicted in each of the five spokes of Figure 8.3 do present a good overall picture of progress in magnesium-ion cathode development. The TE calculated values for each metric are normalized to a typical lithium-ion system performance value (represented by the central pentagon) with the absolute Li-ion performance values quantified in the label of each individual spoke. The outer pentagon ring corresponds to a 2 × "improvement" in a particular metric.

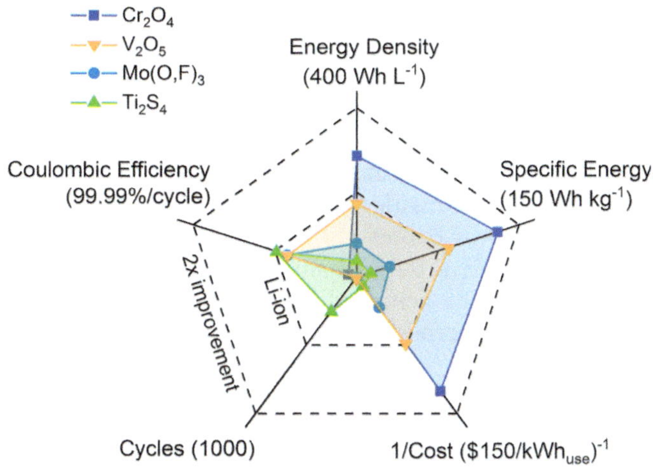

Figure 8.3 A representation of the trade-offs in performance metrics: TE-modeled battery pack-level energy density, specific energy, and cost, as well as experimentally determined coulombic efficiency and cycle life. The performance of several Mg-ion intercalation cathodes (*vs.* Mg metal anodes) are normalized by a representative current state-of-the-art lithium ion system with absolute performance values shown in parentheses. The outer pentagon represents a two-fold "improvement" along a given axes relative to the Li-ion value.

A trade-off between the reversibility and lifetime of known magnesium-battery cathodes (coupled with a Mg-metal anode) *vs.* energy density, specific energy, and cost of a battery pack is readily apparent in Figure 8.3. Established magnesium-ion battery cathodes that have shown reasonably fast kinetics and good lifetimes (*i.e.* $MgTi_2S_4$) cannot compete with lithium-ion batteries in terms of cost and energy density. As reversibility and lifetime are sacrificed, potentially to the point of impracticality, energy density and predicted cost values become more attractive, as can be seen by the areas mapped in the "spider web" diagram in Figure 8.3, which stretches from left to right, moving from the thio- to oxo-spinels, $MgTi_2S_4$ to $MgCr_2O_4$.

8.3 High Energy Density Materials for Magnesium Insertion Cathodes

In the previous chapter, several existing Mg intercalation and conversion cathode materials are discussed that have been widely studied in the literature. Developing a sound understanding of Mg intercalation processes and transfer kinetics is critical to enable a breakthrough in Mg intercalation cathodes for high energy, low cost rechargeable batteries. As established in the previous section, significant commercially relevant gains will be achieved by developing a high energy dense metal–oxide magnesium intercalation cathode active above 3.0 V *vs.* Mg^0/Mg^{2+} when coupled with a Mg metal anode. In general, the relative ionic character of metal–oxides provides a higher cell voltage than their chalcogenide cousins,[24] which is a key factor controlling the cell-level energy density. This benefit, however, typically comes at the detriment of bulk magnesium diffusion rates, which manifest as large overpotentials and low Mg^{2+} insertion (*i.e.* low charge capacity) levels.[19,25] The electrostatics of Mg^{2+} are considered to be the primary culprit behind the slow migration kinetics in oxides with highly localized electronic charge. Mitigation strategies, such as delocalizing charge[1,20] as in sulphides or other chalcogenides,[21–23,26] can be achieved by materials selection and engineering materials complexity across length scales (*e.g.* composition, defects, or morphology).

To date, there are limited reports on Mg intercalation host structures, and fewer still that have demonstrated successful reversible Mg intercalation (see Figure 8.3). Additionally, unlike lithium systems which after nearly 50 years of effort are fully characterized and their behavior well validated, magnesium systems suffer from the interrelated challenges of electrolyte stability, Mg plating and stripping, current collector corrosion, and Mg intercalation. As such, electrochemical performance, especially at voltages >3.0 V *vs.* Mg^0/Mg^{2+} can be particularly difficult to analyze with certainty. Emphasis must be placed on the importance of coupling electrochemistry experiments with both computational as well as varied and complete characterization techniques to establish, without a doubt, that Mg intercalation actually occurs.

The effort to realize commercially important Mg-ion intercalation cathodes is two-fold: first, materials development (both selection and engineering) and second, in-depth characterization. The remainder of this chapter provides perspectives on high energy density materials, their history, recent progress, and their limitations. There is significant progress that remains before commercial realization can be achieved; however, there is also a lot of unexplored design space which provides an opportunity for discovery.

The absence of functional high voltage Mg-ion based energy storage devices is attributable to poor Mg^{2+} mobility within close-packed oxides, in addition to the lack of an oxidative stable electrolyte and stable metallic anode. Ultimately low mobility rates limit intercalation reactions at the cathode and manifest as a non-functioning system, in the worst case, or a limited charge/discharge rate capability. Indeed, direct experimental verification of Mg^{2+} cation mobility in these compounds is incomplete at this time. Nudged elastic band (NEB) simulations coupled with density functional theory (DFT), however, have been widely used to calculate the activation barriers for migration for Mg^{2+} (and other multivalent) cations in oxide host lattices, *e.g.* including spinels.[27–32] The approach is critical for identifying materials crystal structures[27,33] and compositions[28,34] that support feasible Mg^{2+} mobility.

This can be illustrated in a practical sense to make sense of the relationship between the Mg^{2+} mobility activation energy, cathode particle size, and achievable C-rate (*i.e.* the rate at which the material is charged/discharged relative to its maximum capacity in 1 h). By modelling the cathode particles as spheres with radius *r*, an effective diffusion rate can be determined using:

$$D = D_o \exp(-E_m/k_B T)$$

where D_o is the pre-exponential factor proportional to the atomic jump frequency and the square of the atomic jump distance, ν and a^2, respectively. Figure 8.4 provides guidance for the relationship of required particle size and Mg^{2+} cation mobility, given the values of $\nu = 10^{12}\ s^{-1}$ and $a = 3$ Å, for various C-rates. It should be noted that by using C-rates, the analysis by Canepa *et al.*[1] in Figure 8.4 is independent of the material's volumetric capacity; therefore, coupling this analysis of fundamental materials properties to battery performance metrics is critical in the "first-order" design of practical materials. Additional challenges, such as kinetically controlled interfacial reactions (*e.g.* desolvation and charge/mass transfer) or charge trapping mechanisms (*e.g.* polaron transport) must also be resolved before realizing a high energy density for Mg-ion batteries.

Typical particles for Li-ion batteries are in the range of 1 to 10 μm requiring ≤500 meV to achieve a 1C discharge rate. This can readily be realized in several Li-ion oxide cathodes;[30,35] however, magnesium is expected to have higher energy barriers for migration.[30] According to Figure 8.4, smaller particle sizes of approaching 10 nm may be necessary to achieve E_m values of as high as 700 meV.

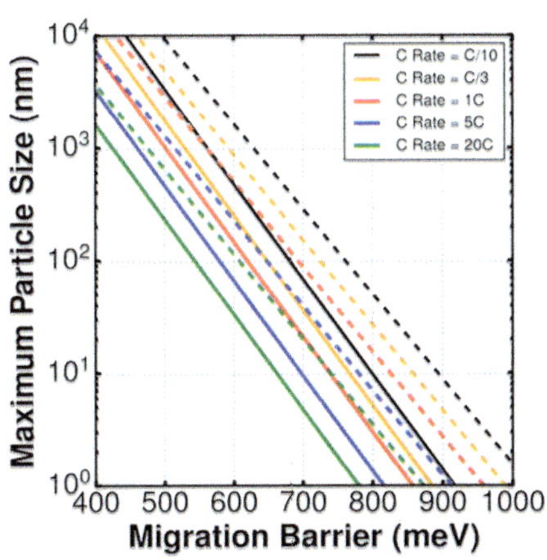

Figure 8.4 Relationship between the MV [*i.e.* Mg^{2+}] ion migration barrier E_m and the maximum particle size permitting reasonable diffusivity in the context of battery performance. Various charging rates are displayed and color-coded as indicated in the figure legend. Solid lines indicate the relationship between the migration barrier and particle size at 298 K, while the dashed lines indicate 333 K. Reproduced from ref. 1, https://pubs.acs.org/doi/abs/10.1021%2Facs.chemrev.6b00614, with permission from the American Chemical Society, Copyright 2017.

8.3.1 Oxo-Spinel Structures

The spinel structure, with the general formula AB_2X_4 (X = chalcogenide, oxygen), has proven to be a promising host lattice for magnesium insertion cathodes. The structure gains its name from the archetypical mineral magnesium aluminate, $MgAl_2O_4$, and ideally possesses cubic symmetry (space group $Fd\bar{3}m$) with oxygen ions in a face-centered cubic close packing arrangement, as shown in Figure 8.5. The A and B cations fill an eighth of the tetrahedral and half of the octahedral sites in the oxygen sublattice interstices in various configurations, depending on their crystal field stabilization energies, ranging from normal to inverse. The "normal" spinel structure has A-cations located in the tetrahedral sites (Wyckoff position: 8a) and the B-cations are in the 16d octahedral positions; whereas, in the "inverse" spinel the A and half of the B cations are on the 16d octahedral sites and the other half of the B cations are on the 8a tetrahedral sites. A continuum can exist between normal ($x = 0$) and inverse ($x = 1$) distributions in the general formula $(A_{1-x}B_x)[A_{x/2}B_{1-x/2}]_2O_4$, where the parentheses and brackets represent the tetragonal and orthorhombic sites, respectively.

A feature of the spinel structure is the BO_6 octahedra, centered on the 16c sites, to create a three-dimensional network of 8a tetrahedron (visualized along the <110> direction in Figure 8.5a), which enables cation mobility. Since the spinel is closely related to the rocksalt structure, which is achieved by filling the 16c sites in addition to the 16d sites, blockage of these percolation pathways can occur.[36–40] Although, filling these interstitial sites has been suggested as a strategy for achieving reversible Mg insertion through a "push-out" process resulting in $Mg_{1+x}M_2O_4$ ($x < 1$ and M = Co, Cr, Fe, or Mn).[36]

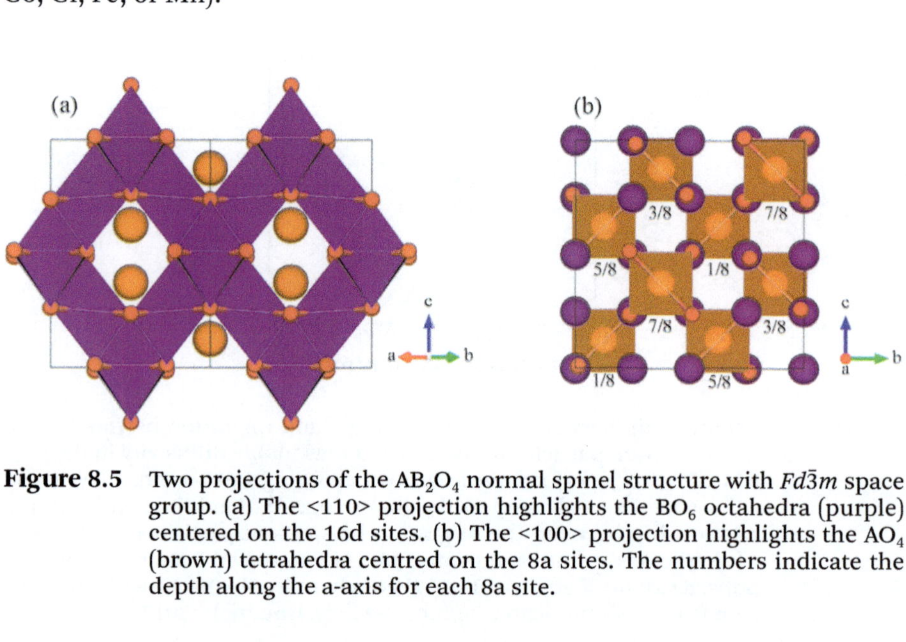

Figure 8.5 Two projections of the AB_2O_4 normal spinel structure with $Fd\bar{3}m$ space group. (a) The <110> projection highlights the BO_6 octahedra (purple) centered on the 16d sites. (b) The <100> projection highlights the AO_4 (brown) tetrahedra centred on the 8a sites. The numbers indicate the depth along the a-axis for each 8a site.

Table 8.1 Predicted materials level values for Mg-based first-row transition metal spinel materials ($Mg_{1-x}M_2O_4$) in normal distribution.[18,49]

	Average voltage (V)	Sp. Energy[a] (W h kg^{-1})	Energy Den.[a] (W h L^{-1})	Stability[b] (meV)		Structure (space group)	
				Charged (x = 1)	Discharged (x = 0)	Charged (x = 1)	Discharged (x = 0)
$Mg_{1-x}Ti_2O_4$	1.31	381	1492	180	10	[227]	[227]
$Mg_{1-x}V_2O_4$	2.36	665	2664	100	0	[74]	[227]
$Mg_{1-x}Cr_2O_4$	3.53	983	4102	170	0	[74]	[227]
$Mg_{1-x}Mn_2O_4$	2.86	775	3189	30	0	[227]	[141]
$Mg_{1-x}Co_2O_4$	3.35	872	4389	60	0	[227]	[227]
$Mg_{1-x}Ni_2O_4$	3.39	885	4201	100	0	[227]	[141]

[a]Discharged (x = 0) basis.
[b]Energy above the hull.

The high density of oxygen close packing coupled with interconnected disfavored 4-fold coordinated sites for Mg (*i.e.* 8a tetrahedral), is an advantage for high energy density cathodes with potentially high Mg-cation mobility. It should be mentioned that the discussion above assumed a perfect cubic lattice symmetry; however, it is common for local distortions or cation ordering, particularly in the case of Jahn–Teller active transition metals, *e.g.* Mn^{3+}, to result in lower symmetries, *e.g.* space groups $F\bar{4}3m$, $I4_1/amd$, or $P4_2/nnm$ are known (and highlighted in Table 8.1 for $MgMn_2O_4$).[34,41,42] This variation in symmetry may result in controlling cation migration properties and could play a critical role in materials design.[34,43,44] Additionally, the spinel structure supports a rich chemistry of elements and complex compositions to tailor the materials properties, making it an ideal structure for consideration. For example, in Li-ion systems, substituting the B-cation with various transition metals results in altered Li insertion properties.[45–48]

The first analysis of defining an oxo-spinel material for a Mg-ion insertion cathode, as has been discussed above, is confirming its average voltage, charge capacity (or energy density), and stability for use in a high energy electrochemical storage system. For instance, Figure 8.1 suggests that 3.0 V and 300 mA h g^{-1} (theoretical materials-level) are required for obtaining a cell-level (current collector to current collector, inclusive) energy density of 750 W h L^{-1}. Table 8.1 compares average Mg insertion voltages, materials energy densities, and stability for a series of first-row transition metal oxo-spinels, which identifies that several are close to meeting these metrics.[18,49,50] The stability, as defined by the energy above the convex hull is compared for the charged and discharged compounds. The convex hull is defined by the lower boundary of compounds formed between the component elements that cannot be decomposed to lower energy; thereby, the energy above the hull represents the energy saved by decomposing to the hull boundary. This is a guideline for stability, and values up to 100 meV per atom have been shown to be metastable during electrochemical cycling.[18,49,51–54]

The spinel compounds compared in Table 8.1 present a series of potential oxygen close-packed materials for high energy density Mg-ion battery cathodes. As a second level analysis of spinel structures, NEB simulations coupled with DFT were used to identify the spinel structure-types that support practical Mg^{2+} mobility at feasible particle sizes for energy storage needs.[28–30,55] Neglecting any likely kinetic limiting reactions, a migration barrier of <600–750 meV, as shown in Figure 8.4, is needed for a particle size of 100 nm to achieve charge/discharge rates of C/3. A review of a series of cations (A = Li, Mg, Ca, Zn, or Al) in $A_{1-x}M_2O_4$ (where M = Mn, Co, Cr, or Ni) is shown in Figure 8.6.[30] The mobilities calculated for Mg^{2+} are low (*i.e.* high activation energy), which is a result of transitioning between the 8a tetrahedral and interstitial 16c octahedral sites (at path distances of 0 and 50%, respectively, as shown in Figure 8.6). The mobility barriers for particle sizes of ~100 nm, however, indeed warrant further experimental and computational investigations (*i.e.* <750 meV).

Indeed, building on these analyses, spinel-type compounds based on single first-row transition metals are promising as electrochemically Mg-ion

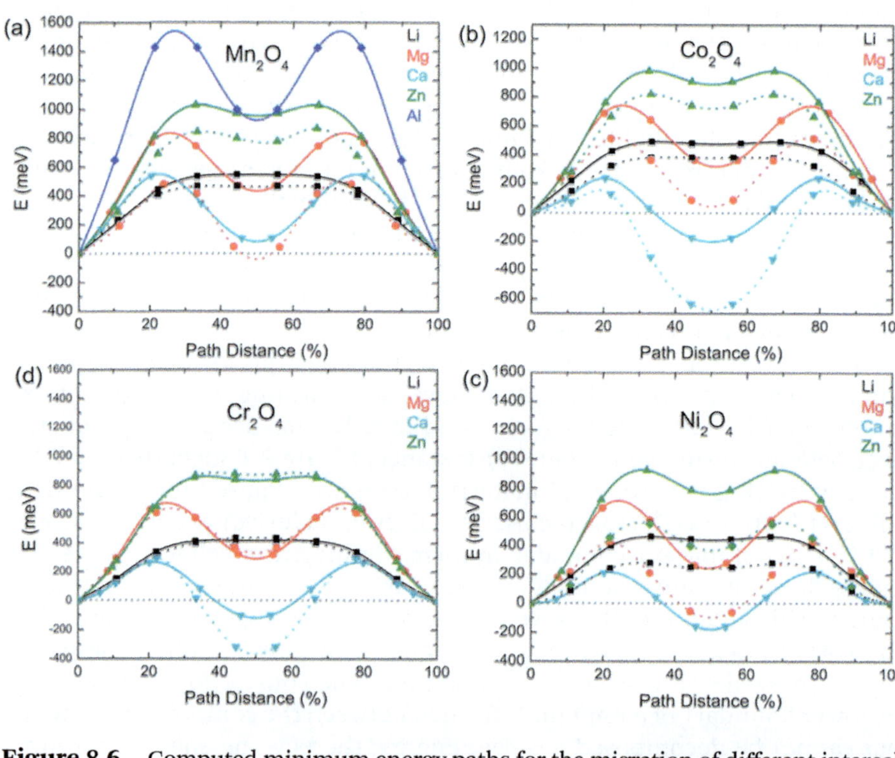

Figure 8.6 Computed minimum energy paths for the migration of different intercalants between the tetrahedral sites in the spinel B_2O_4 (B ¼ Mn, Co, Ni, Cr) at the high vacancy limit (solid line) and dilute vacancy limit (dotted line), *i.e.* one mobile specie per supercell (2 × 2 × 1 of the primitive cell). Reprinted from ref. 30 with permission from the Royal Society of Chemistry.

insertion cathodes. As mentioned previously, high voltage electrolytes stable to the charged cathode have proven to be elusive, frustrating these investigations. Progress has been made, nonetheless, in electrochemically inserting Mg into oxo–spinel host structures. Careful and precise characterization must be completed in these studies to verify, without a doubt, the reversible electrochemical insertion of Mg in the host lattice.

Mn_2O_4: A wide range of Mn–O polymorphs are known; λ-MnO_2 is the designation for the fully charged $A_{1-x}Mn_2O_4$ (x = 1) spinel structure. Indeed, $LiMn_2O_4$ was identified decades ago as a viable Li-ion cathode material[56,57] and is regarded as a high rate capable cathode for reversible Li energy storage. DFT calculations identified λ-Mn_2O_4 as an important polymorph for high energy density Mg insertion cathode *vs.* conversion,[49,58,59] and it has gained a lot of recent attention in this area.[28,33,34,36–39,44,55,60–65]

Early reports on the electrochemical activity of Mg in oxo–spinel materials focused on delithiating[60,66] the known Li-ion cathode $LiMn_2O_4$[56,57] followed by electrochemical cycling in aqueous $Mg(NO_3)_2$ or $MgSO_4$ electrolytes. Reversible electrochemical results were claimed based on the conditions of the studies; however, the diffraction data, in combination with the presented electrochemical CV and cycle data, left an ambiguous conclusion as to the level and certainty of Mg insertion. For example, the electrochemical cycling data for λ-Mn_2O_4 suggested discharge (*i.e.* Mg insertion) >400 mA h g^{-1} at currents of up to 27.2 mA g^{-1},[60] which is grossly in excess of the theoretical capacity of 308 mA h g^{-1} (1 mole Mg per mole of Mn_2O_4). Electrolyte decomposition is most likely contributing to the observed capacity, due to voltages utilized above the electrolysis of water (*i.e.* ≈1.0 V *vs.* SCE), and potentially results in proton cycling. Additional materials characterization beyond the presented structural investigations are required to develop a definitive conclusion on the electrochemical mechanisms.

A complete and systematic characterization study combining scanning transmission electron microscopy (STEM), X-ray absorption near edge structure (XANES), X-ray absorption, and X-ray diffraction, on similarly prepared and cycled $LiMn_2O_4$ in aqueous $Mg(NO_3)_2$ electrolyte, verified the inclusion of Mg^{2+} on the 8a tetrahedral sites, Mn reduction from Mn^{4+} to Mn^{3+}, and subsequent symmetry change from cubic to tetragonal upon electrochemical reduction (discharge).[55] Figure 8.7 shows the microscopy characterization providing direct evidence of Mg^{2+} insertion. Interestingly, ^{25}Mg magic-angle spinning (MAS) quadrupolar Carr–Purcell–Meiboom–Gill (qCPMG) NMR, which is elemental species and site dependent, was used to definitely and quantitatively indicate the presence of Mg ions in the paramagnetic spinel lattice (x = 0.11 in $Mg_xMn_2O_4$). The multipronged characterization approach detailed by Kim *et al.*[55] is required to substantiate conclusions of Mg insertion in λ-Mn_2O_4.

Kim *et al.* utilized aqueous electrolytes and suggested that the spinel structure is maintained during magnesium insertion.[55] However, other studies on λ-Mn_2O_4, also in aqueous electrolytes, provide evidence of Mg insertion coupled with spinel-to-layered transition.[40,67] Water from the aqueous environment was hypothesized to support this structural transition.[67] A

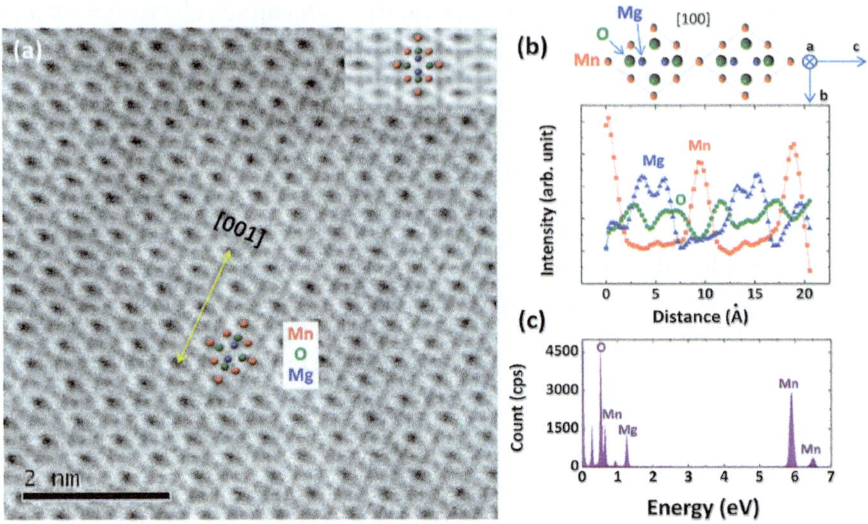

Figure 8.7 a) ABF STEM image of the acid-treated spinel reduced to −0.2 V *vs.* SCE (sample e′) aligned along a [100] zone axis. Atomic positions of Mn (red), O (green), and Mg (blue) are clearly identified in the raw image, while the inset, a filtered and averaged/cross-correlated sub image, further highlights the atomic occupancies; b) an EDX line scan along the [001] direction (yellow line in a), confirming the presence of Mg, as well as the consecutive atomic arrangements of Mn–O–Mg–Mg–O–Mn, in accordance with the $MgMn_2\ O_4$ structure; c) an integrated EDX spectrum obtained from a large field of view confirms the existence of ≈11 at% Mg in the discharged spinel host. Reproduced from ref 55 with permission from John Wiley and Sons, © 2015 WILEY-VCH Verlag GmbH & Co. KGaA, Weinheim.

spinel-to-rocksalt structural transition has also been reported, and in some cases actively utilized for Mg insertion.[36–38] Interestingly, water was reported to not participate directly in the magnesium insertion reaction, reported by Kim *et al.*, thereby implying that non-aqueous electrolytes can support similar reactions;[55] however, the to-date results for spinel $Mg_{1-x}Mn_2O_4$ have been less convincing in non-aqueous environments.

A spinel-to-rocksalt transition results when the 16c octahedral sites at the center of the channels observed in Figure 8.5a become occupied. This will clearly negatively impact upon the magnesium mobility, and indeed, was shown to be detrimental in sulphide spinel analogs.[21,22] Rocksalt conversion was observed at the surface of $Mg_{1-x}Mn_2O_4$ particles during cycling (in 0.5 M $Mg(ClO_4)_2$ in acetonitrile) as Mg diffused to the surface and relocated to the 16c octahedral sites forming $(Mg_xMn_{1-x})O$.[37] Sustained magnesium insertion into the bulk of the particles was thereby cutoff and the cell impedance dramatically increased. Currently, the complex relationship of spinel to related phases under reversible electrochemical (or chemical) magnesium cycling, especially in the Mn–O composition phase space, is not yet fully understood.

In addition to structural conversions, as mentioned above, $Mg_{1-x}Mn_2O_4$ readily supports site-mixing in the form of cation inversion. Aqueous chemical methods were used to demagnesiate $Mg_{1-x}Mn_2O_4$ with various degrees of inversion (from 34 to 1%) with clear evidence that removal of Mg *via* acid leaching was severely retarded under a high degree of inversion. Inversion is expected to significantly influence the magnesium mobility by blocking migration pathways. Gautum *et al.* investigated the roll inversion has on mobility by calculating migration energy barriers with DFT.[28] They concluded that $MgMn_2O_4$ spinel inversion hinders the electrochemical performance related to magnesium migration rates, affects the average magnesium insertion voltage (as described in Figure 8.8), and limits the extent of reversible magnesium extraction/insertion capability, estimating a 15% decrease in the available capacity with every 10% increase in inversion.[28] Finally, they suggested stabilization strategies must be made to promote the normal spinel structure.[28]

Engineering $Mg_{1-x}Mn_2O_4$ is not limited to including lattice structure and composition complexity to stabilize favorable qualities, but also to tailoring morphological complexity. For instance, delithiated $LiMn_2O_4$ cycled in non-aqueous electrolytes, *i.e.* $Mg(TFSI)_2$ in diglycol methyl ether (diglyme), resulted in a much lower magnesiation rate as characterized by a suite of techniques, than in an aqueous electrolyte.[61] In order to rule out the role of slow diffusion, nanostructured Mn-based oxo-spinel compositions have been investigated.[34,58,61]

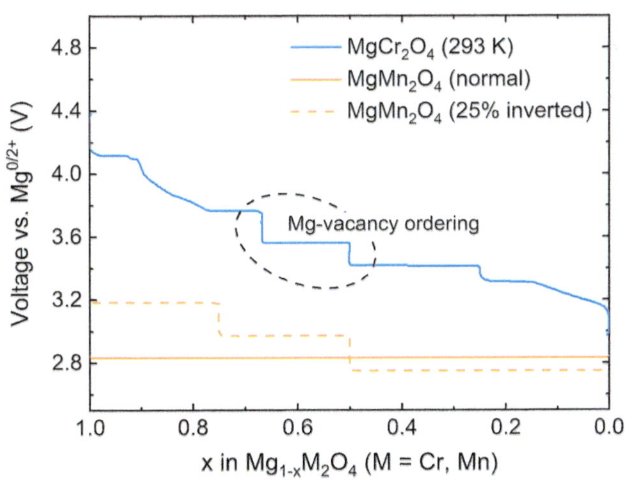

Figure 8.8 DFT calculated DFT (0 K) voltage curves for Mg insertion/extraction in $Mg_{1-x}Cr_2O_4$ and $Mg_{1-x}Mn_2O_4$ are compared.[28,71] The room temperature (293 K) Monte Carlo-calculated voltage curve indicates that the voltage profile $x > 0.8$ of the chromite phase is dramatically smoothed, whereas the steps at $x = 0.67$ and 0.5 are retained and correspond to magnesium vacancy ordering. Comparison of the two Mn-phase profiles indicates the effect that spinel inversion and Mn oxidation state dissociation has on the voltage profile.

The experimental results for Mg^{2+} insertion in λ-Mn_2O_4 have been highly variable with long term performance challenges (*i.e.* capacity fade).[1,9] Evidence for reversible magnesium insertion in aqueous environments appears to be promising; however, in non-aqueous electrolytes, progress remains elusive. In the case of aqueous environments, further investigations should exclude the possibility of protons or smaller diffusion species. The structural distortions, rocksalt formation from interstitial octahedral 16c site filling, and inversion all affect the magnesium insertion mechanisms and kinetics. This highlights the needs for precise and extensive characterization approaches as well as informed materials engineering of λ-Mn_2O_4 to stabilize the optimal performance.

Cr2O4: The majority of exploratory research on oxo–spinels has focused on the manganese system; however, spinels provide a rich chemistry and space for materials engineering. The comparison of $Mg_{1-x}M_2O_4$ spinels in Table 8.1 indicates that the chromite-based spinel is another composition of interest. The average voltage and energy density of this material is highly attractive towards the technological goals, and the calculated migration barrier of $Mg_{1-x}Cr_2O_4$, as shown in Figure 8.6c, is feasible for a working magnesium-ion cathode.[30]

The chromite spinel has several advantages beyond energy density, despite the calculated instability of the charged (demagnesiated) Cr_2O_4 structure indicated in Table 8.1. Hannah *et al.* calculated the thermodynamic relationships for conversion and insertion energetics of cathode hosts across chemical and structural motifs.[59] They concluded that Cr-based cathodes were the optimal systems to resist conversion reactions. Furthermore, both oxidation states of Cr expected in cycling $Mg_{1-x}Cr_2O_4$ (*i.e.* Cr^{3+} and Cr^{4+}) prefer 6-fold coordination based on large crystal field stabilization energies;[68–70] therefore, $Mg_{1-x}Cr_2O_4$ is resistant to inversion, unlike its manganese counterpart during magnesium extraction/insertion.

In light of the high operation voltages and the complete absence of suitable electrolytes, computational approaches are crucial for evaluating the chromite spinel system. DFT calculated voltage profiles at 0 K are shown in Figure 8.8 and indicate a sudden drop in voltage at $Mg_{0.33}Cr_2O_4$ and $Mg_{0.5}Cr_2O_4$ ($x = 0.67$ and 0.5, respectively) corresponding to a proposed stable ordering configuration of magnesium vacancies, which subsequently lead to energy barriers to magnesium extraction/insertion. Room temperature discharge voltage profiles were also calculated based on the 0 K results coupled with grand-canonical Monte Carlo (GMC) simulations on a cluster expansion Hamiltonian to include configurational entropy in the finite temperature free energy.[71] The room temperature profile maintains the edge steps corresponding to the vacancy ordering, while smoothing the profile above $x = 0.8$. The authors, therefore, proposed cation substitution strategies (*i.e.* doping or solid-solution) in order to mitigate the vacancy ordering.[71] As mentioned above, the chromite spinel is resistant to inversion, unlike its manganese counterpart, yet due to vacancy site ordering it still presents a complex voltage profile and impediments to magnesiation. Figure 8.8, therefore, highlights the multiple challenges predicted by computational modelling to magnesium insertion processes by comparing the structure and composition in spinel materials.[28,71]

Recent experimental work has looked past the deficiencies of the available electrolytes, and begun to investigate the electrochemical properties of the chromite spinel material. Particles of <10 nm were synthesised by a continuous hydrothermal flow approach and evaluated in non-aqueous ionic liquid electrolytes comprising 0.5 M $Mg[N(SO_2)_2(CF_3)_2]_2$–$(C_9H_{20}N)$–$N(SO_2)_2(CF_3)_2$ ($MgTFSI_2$–$PY_{14}TFSI$) at 110 °C.[72] This work shows that Mg can be extracted/inserted into the bulk structure of nanosized, highly-defective $Mg_{1-x}Cr_2O_4$. In a separate experiment using the stationary probe rotating disk electrode coupled with inductively coupled plasma mass spectrometry (SPRDE-ICP-MS) method, which decouples any electrolyte decomposition reaction currents from the desired electrochemical electrode processes,[73] magnesium extraction from $Mg_{1-x}Cr_2O_4$ was investigated using an Li-based electrolyte (0.2 M LiTFSI in diglycol methyl ether) in order to directly monitor the Mg^{2+} ion current from extraction from the chromite spinel. The results[73] clearly indicated Mg^{2+} extraction commences at 3.8 V *vs.* Mg^0/Mg^{2+} (approximately equivalent to the value detailed in Table 8.1) with no current attributed to chromium corrosion. The Mg current is only 3–4% of the total current. This implies that significant electrolyte oxidation occurs. These two early results provide evidence of the importance of the chromite spinel as a magnesium insertion cathode. Just as in the manganese analogue, improvements through materials engineering and inclusion of tailored structural, composition, or morphological complexity may open the door for this material.

8.4 Conclusion

Research and development benefits greatly from investigating materials that do not meet technological goals in order to develop fundamental principles, mechanisms, or understanding of phenomena not readily understood under more challenging conditions (*i.e.* high voltage Mg insertion processes). As such, this chapter specifically explores technologically useful Mg-ion insertion cathodes, rather than dismissing their scientifically valuable brethren with less transformative attributes.

As described in this chapter, the electrostatic interactions between magnesium and metal–oxide host lattices result in significant challenges for effective reversible extraction/insertion – much less long term cyclability. Designing and introducing materials complexity (*e.g.* defects, ordering, and changes in symmetry) in periodic lattices of bulk inorganic materials will instil more precise control of materials behaviour. This will require fundamental understanding of the role and relative hierarchy of the numerous varieties of crystalline complexities as they contribute towards and control magnesium cation intercalation, transport mechanisms, and the stability of a wide array of materials classes. Additionally, there is a need to codify a mechanistic framework to guide the formation and stabilization of specific inorganic phases with tailored structures with desired qualities.

Material characterization is critical in understanding the nature of complexities in materials and their interactions with magnesium insertion and transport mechanisms. More practically, due to an array of undesired parasitic reactions in these high energy dense magnesium systems, isolated electrochemical measurements of magnesium cells will most certainly lead to erroneous conclusions. In order to establish, beyond doubt, reversible magnesium extraction/insertion, as well as a fundamental understanding of the mechanisms guiding these reactions, a wide range of *in situ*, *operando*, and *ex situ* characterization approaches is necessary to probe the evolution of, and interactions with, the host material.

In light of this, a breakthrough in the elusive quest for a transformative high energy density magnesium insertion cathode would permanently alter the face of battery and electrochemical science and lead to a blossoming of new materials with unique electroactive functions.

Acknowledgement

This work was supported as part of the Joint Center for Energy Storage Research, an Energy Innovation Hub funded by the US Department of Energy, Office of Science, Basic Energy Sciences. The submitted chapter has been created by UChicago Argonne, LLC, Operator of Argonne National Laboratory ("Argonne"). Argonne, a US Department of Energy Office of Science laboratory, is operated under Contract No. DE-AC02-06CH11357.

References

1. P. Canepa, G. Sai Gautam, D. C. Hannah, R. Malik, M. Liu, K. G. Gallagher, K. A. Persson and G. Ceder, *Chem. Rev.*, 2017, **117**, 4287.
2. J. Muldoon, C. B. Bucur and T. Gregory, *Angew. Chem., Int. Ed.*, 2017, **56**, 12064.
3. J. Song, E. Sahadeo, M. Noked and S. B. Lee, *J. Phys. Chem. Lett.*, 2016, **7**, 1736.
4. M. M. Huie, D. C. Bock, E. S. Takeuchi, A. C. Marschilok and K. J. Takeuchi, *Coord. Chem. Rev.*, 2015, **287**, 15.
5. P. Saha, M. K. Datta, O. I. Velikokhatnyi, A. Manivannan, D. Alman and P. N. Kumta, *Prog. Mater. Sci.*, 2014, **66**, 1.
6. Z. Zhao-Karger and M. Fichtner, *Front. Chem.*, 2019, **6**, 656.
7. R. C. Massé, E. Uchaker and G. Cao, *Sci. China Mater.*, 2015, **58**, 715.
8. A. J. Crowe and B. M. Bartlett, *J. Solid State Chem.*, 2016, **242**, 102.
9. C. B. Bucur, T. Gregory, A. G. Oliver and J. Muldoon, *J. Phys. Chem. Lett.*, 2015, **6**, 3578.
10. M. Mao, T. Gao, S. Hou and C. Wang, *Chem. Soc. Rev.*, 2018, **47**, 8804.

11. P. A. Nelson, S. Ahmed, K. A. Gallagher and D. W. Dees, *Modeling the Performance and Cost of Lithium-Ion Batteries for Electric-Drive Vehicles*, 3rd edn, United States, https://www.osti.gov/servlets/purl/1503280.
12. K. G. Gallagher, P. A. Nelson and D. W. Dees, *J. Power Sources*, 2011, **196**, 2289.
13. R. M. Darling, K. G. Gallagher, J. A. Kowalski, S. Ha and F. R. Brushett, *Energy Environ. Sci.*, 2014, **7**, 3459.
14. D. Eroglu, S. Ha and K. G. Gallagher, *J. Power Sources*, 2014, **267**, 14.
15. K. G. Gallagher, S. Goebel, T. Greszler, M. Mathias, W. Oelerich, D. Eroglu and V. Srinivasan, *Energy Environ. Sci.*, 2014, **7**, 1555.
16. P. A. Nelson, S. Ahmed, K. G. Gallagher and D. W. Dees, *J. Power Sources*, 2015, **283**, 506.
17. K. G. Gallagher, S. E. Trask, C. Bauer, T. Woehrle, S. F. Lux, M. Tschech, P. Lamp, B. J. Polzin, S. Ha, B. Long, Q. Wu, W. Lu, D. W. Dees and A. N. Jansen, *J. Electrochem. Soc.*, 2016, **163**, A138.
18. F. Zhou, M. Cococcioni, C. A. Marianetti, D. Morgan and G. Ceder, *Phys. Rev. B*, 2004, **70**, 235121.
19. P. Novák and J. Desilvestro, *J. Electrochem. Soc.*, 1993, **140**, 140.
20. E. Levi, G. Gershinsky, D. Aurbach, O. Isnard and G. Ceder, *Chem. Mater.*, 2009, **21**, 1390.
21. P. Bonnick, X. Sun, K.-C. Lau, C. Liao and L. F. Nazar, *J. Phys. Chem. Lett.*, 2017, **8**, 2253.
22. X. Sun, P. Bonnick, V. Duffort, M. Liu, Z. Rong, K. A. Persson, G. Ceder and L. F. Nazar, *Energy Environ. Sci.*, 2016, **9**, 2273.
23. D. Aurbach, Z. Lu, A. Schechter, Y. Gofer, H. Gizbar, R. Turgeman, Y. Cohen, M. Moshkovich and E. Levi, *Nature*, 2000, **407**, 724.
24. Y. Wang, W. D. Richards, S. P. Ong, L. J. Miara, J. C. Kim, Y. Mo and G. Ceder, *Nat. Mater.*, 2015, **14**, 1026.
25. E. Levi, Y. Gofer and D. Aurbach, *Chem. Mater.*, 2010, **22**, 860.
26. P. Canepa, S.-H. Bo, G. S. Gautam, B. Key, W. D. Richards, T. Shi, Y. Tian, Y. Wang, J. Li and G. Ceder, *Nat. Commun.*, 2017, **8**, 1759.
27. Z. Rong, R. Malik, P. Canepa, G. Sai Gautam, M. Liu, A. Jain, K. Persson and G. Ceder, *Chem. Mater.*, 2015, **27**, 6016.
28. S. G. Gautam, P. Canepa, A. Urban, S.-H. Bo and G. Ceder, *Chem. Mater.*, 2017, **29**, 7918.
29. A. Emly and A. Van der Ven, *Inorg. Chem.*, 2015, **54**, 4394.
30. M. Liu, Z. Rong, R. Malik, P. Canepa, A. Jain, G. Ceder and K. A. Persson, *Energy Environ. Sci.*, 2015, **8**, 964.
31. D. C. Hannah, G. Sai Gautam, P. Canepa, Z. Rong and G. Ceder, *Chem. Commun.*, 2017, **53**, 5171.
32. Z. Rong, P. Xiao, M. Liu, W. Huang, D. C. Hannah, W. Scullin, K. A. Persson and G. Ceder, *Chem. Commun.*, 2017, **53**, 7998.
33. S. K. Kolli and A. Van der Ven, *ACS Appl. Energy Mater.*, 2018, **1**, 6833.
34. J. C. Knight, S. Therese and A. Manthiram, *ACS Appl. Mater. Interfaces*, 2015, 7, 22953.

35. S. Kim, V. I. Hegde, Z. Yao, Z. Lu, M. Amsler, J. He, S. Hao, J. R. Croy, E. Lee, M. M. Thackeray and C. Wolverton, *ACS Appl. Mater. Interfaces*, 2018, **10**, 13479.
36. S. Okamoto, T. Ichitsubo, T. Kawaguchi, Y. Kumagai, F. Oba, S. Yagi, K. Shimokawa, N. Goto, T. Doi and E. Matsubara, *Adv. Sci.*, 2015, **2**, 1500072.
37. Q. D. Truong, M. Kempaiah Devaraju, P. D. Tran, Y. Gambe, K. Nayuki, Y. Sasaki and I. Honma, *Chem. Mater.*, 6245–6251.
38. K. W. Nam, S. Kim, S. Lee, M. Salama, I. Shterenberg, Y. Gofer, J.-S. Kim, E. Yang, C. S. Park, J.-S. Kim, S.-S. Lee, W.-S. Chang, S.-G. Doo, Y. N. Jo, Y. Jung, D. Aurbach and J. W. Choi, *Nano Lett.*, 2015, **15**, 4071.
39. L. Wang, K. Asheim, P. E. Vullum, A. M. Svensson and F. Vullum-Bruer, *Chem. Mater.*, 2016, **28**, 6459.
40. X. Sun, V. Duffort, B. L. Mehdi, N. D. Browning and L. F. Nazar, *Chem. Mater.*, 2016, **28**, 534.
41. S. Senz, W. Blum and D. Hesse, *Philos. Mag. A*, 2001, **81**, 109.
42. A. M. Balagurov, I. A. Bobrikov, M. S. Maschenko, D. Sangaa and V. G. Simkin, *Crystallogr. Rep.*, 2013, **58**, 710.
43. K. Kim and D. J. Siegel, *J. Mater. Chem. A*, 2019, 3216–3227.
44. Z. Feng, X. Chen, L. Qiao, A. L. Lipson, T. T. Fister, L. Zeng, C. Kim, T. Yi, N. Sa, D. L. Proffit, A. K. Burrell, J. Cabana, B. J. Ingram, M. D. Biegalski, M. J. Bedzyk and P. Fenter, *ACS Appl. Mater. Interfaces*, 2015, **7**, 28438.
45. D. Liu, Y. Lu and J. B. Goodenough, *J. Electrochem. Soc.*, 2010, **157**, A1269.
46. K. Amine, H. Tukamoto, H. Yasuda and Y. Fujita, *J. Electrochem. Soc.*, 1996, **143**, 1607.
47. Y. Terada, K. Yasaka, F. Nishikawa, T. Konishi, M. Yoshio and I. Nakai, *J. Solid State Chem.*, 2001, **156**, 286.
48. J. C. Knight, S. Therese and A. Manthiram, *J. Electrochem. Soc.*, 2015, **162**, A426.
49. A. Jain, S. P. Ong, G. Hautier, W. Chen, W. D. Richards, S. Dacek, S. Cholia, D. Gunter, D. Skinner, G. Ceder and K. A. Persson, *APL Mater.*, 2013, **1**, 011002.
50. L. Cheng, R. S. Assary, X. Qu, A. Jain, S. P. Ong, N. N. Rajput, K. Persson and L. A. Curtiss, *J. Phys. Chem. Lett.*, 2015, **6**, 283.
51. C. Masquelier, A. K. Padhi, K. S. Nanjundaswamy and J. B. Goodenough, *J. Solid State Chem.*, 1998, **135**, 228.
52. A. K. Padhi, K. S. Nanjundaswamy and J. B. Goodenough, *J. Electrochem. Soc.*, 1997, **144**, 1188.
53. B. L. Ellis, W. R. M. Makahnouk, Y. Makimura, K. Toghill and L. F. Nazar, *Nat. Mater.*, 2007, **6**, 749.
54. S.-Y. Chung and Y.-M. Chiang, *Electrochem. Solid-State Lett.*, 2003, **6**, A278.
55. C. Kim, P. J. Phillips, B. Key, T. Yi, D. Nordlund, Y.-S. Yu, R. D. Bayliss, S.-D. Han, M. He, Z. Zhang, A. K. Burrell, R. F. Klie and J. Cabana, *Adv. Mater.*, 2015, **27**, 3377.
56. M. M. Thackeray, A. de Kock, M. H. Rossouw, D. Liles, R. Bittihn and D. Hoge, *J. Electrochem. Soc.*, 1992, **139**, 363.
57. M. M. Thackeray and M. H. Rossouw, *J. Solid State Chem.*, 1994, **113**, 441.

58. C. Ling, R. Zhang and F. Mizuno, *ACS Appl. Mater. Interfaces*, 2016, **8**, 4508.
59. D. C. Hannah, G. S. Gautam, P. Canepa and G. Ceder, *Adv. Energy Mater.*, 2018, **8**, 1800379.
60. C. Yuan, Y. Zhang, Y. Pan, X. Liu, G. Wang and D. Cao, *Electrochim. Acta*, 2014, **116**, 404.
61. C. Kim, A. A. Adil, R. D. Bayliss, T. L. Kinnibrugh, S. H. Lapidus, G. M. Nolis, J. W. Freeland, P. J. Phillips, T. Yi, H. D. Yoo, B. J. Kwon, Y.-S. Yu, R. Klie, P. J. Chupas, K. W. Chapman and J. Cabana, *Chem. Mater.*, 2018, **30**, 1496.
62. W. Jin, G. Yin, Z. Wang and Y. Q. Fu, *Appl. Surf. Sci.*, 2016, **385**, 72.
63. M. Cabello, R. Alcántara, F. Nacimiento, G. Ortiz, P. Lavela and J. L. Tirado, *CrystEngComm*, 2015, **17**, 8728.
64. J.-S. Kim, W.-S. Chang, R.-H. Kim, D.-Y. Kim, D.-W. Han, K.-H. Lee, S.-S. Lee and S.-G. Doo, *J. Power Sources*, 2015, **273**, 210.
65. N. H. Zainol, D. Hambali, Z. Osman, N. Kamarulzaman and R. Rusdi, *Ionics*, 2019, **25**, 133.
66. N. N. Sinha and N. Munichandraiah, *Electrochem. Solid-State Lett.*, 2008, **11**, F23.
67. S. Kim, K. W. Nam, S. Lee, W. Cho, J.-S. Kim, B. G. Kim, Y. Oshima, J.-S. Kim, S.-G. Doo, H. Chang, D. Aurbach and J. W. Choi, *Angew. Chem., Int. Ed.*, 2015, **54**, 15094.
68. I. D. Brown, *Acta Crystallogr., Sect. B: Struct. Sci.*, 1988, **44**, 545.
69. C. Glidewell, *Inorg. Chim. Acta*, 1976, **19**, L45.
70. H. K. Mao and P. M. Bell, *Geochim. Cosmochim. Acta*, 1975, **39**, 865.
71. T. Chen, G. Sai Gautam, W. Huang and G. Ceder, *Chem. Mater.*, 2018, **30**, 153.
72. L. Hu, I. D. Johnson, S. Kim, G. M. Nolis, J. W. Freeland, H. D. Yoo, T. T. Fister, L. McCafferty, T. E. Ashton, J. A. Darr and J. Cabana, *Nanoscale*, 2019, **11**, 639.
73. P. P. Lopes, M. Zorko, K. L. Hawthorne, J. G. Connell, B. J. Ingram, D. Strmcnik, V. R. Stamenkovic and N. M. Markovic, *J. Phys. Chem. Lett.*, 2018, **9**, 4935.

CHAPTER 9

Organic Compounds as Electrodes for Rechargeable Mg Batteries

J. BITENC*[a] AND R. DOMINKO*[a,b,c]

[a]National Institute of Chemistry, Department of Materials Chemistry, Hajdrihova 19, 1000 Ljubljana, Slovenia; [b]University of Ljubljana, Faculty of Chemistry and Chemical Technology, Večna pot 113, 1001 Ljubljana, Slovenia; [c]ALISTORE - European Research Institute, 33 rue Saint-Leu, Amiens 80039 Cedex, France
*E-mail: jan.bitenc@ki.si, robert.dominko@ki.si

9.1 Introduction

Redox active organic materials are a wide and extremely variable group of materials prepared from sustainable resources typically with low carbon footprint. From the standpoint of electrochemical reactivity they can be divided according to the redox reactions that they undergo into n-type, p-type or bipolar organics. N-type redox active organics require a cation to balance their charged state, while p-type organics require an anion to balance their charged state. In the case of bipolar materials, the counterion can be either a cation or anion, depending on the charge state of the bipolar redox active organic material. When considering electrodes for Mg batteries, mostly n-type materials will be discussed since only negatively charged organics can accommodate cations, *i.e.* Mg^{2+} ions. Depending on their functional

Energy and Environment Series No. 23
Magnesium Batteries: Research and Applications
Edited by Maximilian Fichtner

Published by the Royal Society of Chemistry, www.rsc.org

groups, redox active organic materials can be further differentiated into one of the following groups: conjugated carbonyls, conjugated hydrocarbons, conjugated amines, conjugated thioethers, organodisulfides, thioethers and nitroxyl radicals. Organic electrode materials can in theory be produced from low-cost and abundant resources given the fact that they contain only ubiquitous elements such as C, O, H, S and N. This fact is becoming especially important with increasing market demand for various battery devices, which is raising concerns about price and the future availability of certain materials (Co, Ni, Cu, Li, and graphite) used in the production of contemporary Li-ion batteries.

The fact that redox active organic materials can be used with various counterions makes them interesting for application in post Li-ion batteries, where most of the potential inorganic cathode analogues suffer from poor cation insertion, especially when considering multivalent ion insertion, such as in the case of Mg^{2+}, Ca^{2+} and Al^{3+}.[1] Use of redox active organic materials opens up the possibility to effectively bypass the limitations connected with difficult insertion of highly polarised multivalent cations into the solid-state matrix and with their slow solid-state diffusion, which are typically encountered in inorganic hosts. Contrary to this, redox active organic materials with their weak intermolecular forces enable the reversible electrochemical interaction of Mg cations coupled with fast diffusion. One of the main downsides of redox active organic materials is their low volumetric capacity, but this can be effectively compensated by the high volumetric density of the Mg anode. The voltage of a Mg metal–organic battery should be approximately 0.7 V lower than its Li metal counterpart due to the difference in the redox potentials of Li and Mg metal. The biggest problem with the application of redox active organic materials in different types of batteries comprising organic solvents for electrolytes is the dissolution process, which can significantly reduce the cycle life of such materials. To mitigate this problem, different research groups have suggested either grafting of redox active organics onto a solid support,[2] preparation of insoluble polymers[3] or the use of selective separators.[4] However, all approaches come at a certain cost. Grafting introduces additional electrochemically inactive mass into the composite electrode, preparation of polymers decreases the specific capacity of the electrochemically active group, while selective separators are often bulky and available only for certain ions. In our opinion, and from the standpoint of practical applications, the use of polymers with low molecular weight linkers might be the most optimal solution. Organic materials often contain electrophilic centres, which makes them incompatible with Mg electrolytes that contain nucleophilic species, typically those synthesized from Grignard reagents. Thus, the development of non-nucleophilic electrolytes has been of primary importance to enable electrochemical testing of all organic materials in Mg metal battery setups, especially those containing conjugated carbonyl groups and different sulfides. With the recent development of different types of non-nucleophilic electrolytes, the doors have opened towards the application of different organic materials and there is room for the optimization of their electrochemical performance.

Redox-active organic materials were for the first time applied in a Mg battery as a positive electrode by NuLi *et al.* in 2007.[5] They explored three different materials containing sulfide bonds (2,5-dimercapto-1,3,4-thiadiazole, poly-2,2′-dithiodianiline and a conductive sulfur-containing material). Electrochemical tests showed some reversible activity with discharge voltages of around 1.4 V *vs.* Mg/Mg^{2+}. However, it has to be noted that their tests were carried out in Mg electrolytes prepared from Grignard compounds, which means that the presence of nucleophilic species in the final electrolyte solutions cannot be excluded. Additionally, the use of copper current collectors could be problematic since it has been demonstrated that copper is prone to corrosion in chloride-containing electrolytes.[6] This work was followed up by another report from the same group, where a redox active organic material from the class of nitroxyl radicals, poly(2,2,6,6-tetrametylpiperdinyloky-4-yl methacrylate) (PTMA), with a graphene support was utilized.[7] However, its electrochemical performance featured mostly a pseudocapacitance response with almost no voltage plateau. The reason for such behaviour can be explained by the fact that the redox potential of PTMA is close to 3 V *vs.* Mg/Mg^{2+}, well beyond the oxidative stability window of the electrolytes used in this study.

A different approach was taken by Sano *et al.*, who tested 2,5-dimethoxybenzoquinone in a different electrolyte, $Mg(ClO_4)_2$ in γ-butyrolactone (γ-GBL).[8] Since reversible Mg stripping/deposition is not possible within this electrolyte, a reversible cycling of the organic cathode was demonstrated through the use of a three electrode cell. DMBQ showed a maximum discharge capacity of 260 mA h g^{-1} (Figure 9.1a). The obtained capacity was more than 80% of the theoretical capacity with limited cycling due to the highly passivating nature of the electrolyte used in this study. DMBQ displayed

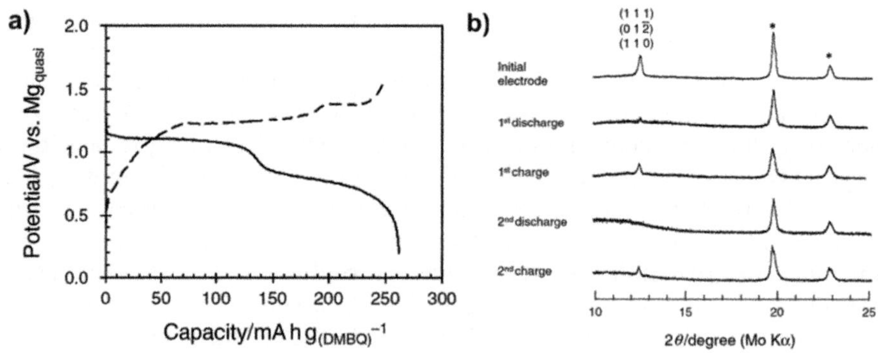

Figure 9.1 (a) Galvanostatic charge and discharge of a 2,5-dimethoxybenzene cathode in a three-electrode cell setup with 0.5 M $Mg(ClO_4)_2$ in γ-GBL electrolyte. (b) *Ex situ* X-ray diffraction of the cycled electrodes. The reflections at 20° and 23° are assigned to the stainless steel current collector and do not change during cycling, while the signal at 12° disappeared and reappeared upon discharge and charge, respectively. Adapted from ref. 8 with permission from the Chemical Society of Japan, Copyright 2012.

two voltage plateaus consistent with its behaviour in Li electrolytes. However, the voltage plateaus were located at 1.1 and 0.8 V *vs.* a Mg reference electrode, respectively. This was significantly lower than the expected value from analogy of the Li electrolyte, where plateaus are located at 2.8 and 2.5 *vs.* Li/Li^+, respectively. The reason behind this was passivation of the Mg metal, which increased the potential of the Mg reference electrode and upshifted it by approximately 0.7 V. Reversible electrochemical activity of DMBQ was confirmed through *ex situ* X-ray diffraction (XRD) measurements, which demonstrated the disappearance of the DMBQ peak at 12° after discharge and its reappearance after charge (Figure 9.1b). This work can be considered as a pioneering study on Mg organic batteries, although the battery configuration cannot be commercialized. Later, the same group applied the DMBQ cathode to other Mg electrolytes. First, they used the DMBQ cathode in combination with different sulfone-based solvents (sulfolane, ethyl-*i*-propyl sulfone and di-*n*-propyl sulfone) and $Mg(TFSI)_2$.[9] However, the passivating nature of these electrolytes meant that some reversibility could be achieved in a three electrode setup. Testing in a practical two-electrode cell resulted in a voltage hysteresis of more than 2 V. Afterwards, they carried out similar tests in $Mg(TFSI)_2$ electrolytes based on the different glyme solvents (monoglyme = DME, diglyme = DEG, triglyme and tetraglyme = TEG).[10] In these systems, the two-electrode cells showed better reversibility, but the voltage hysteresis still remained high.

The main motivation for the development of non-nucleophilic electrolytes was the demonstration of Mg–S battery electrochemical reversibility. This was realized in 2011 by the Muldoon group, who developed the first non-nucleophilic Mg electrolyte.[12] The non-nucleophilic nature of the electrolyte Mg_2Cl_3–$HMDSAlCl_3$ (HMDS = hexamethyldisilazane) in tetrahydrofuran (THF) enabled the demonstration of a proof-of-concept Mg–S battery and the electrolyte displayed a voltage stability of more than 3 V *vs.* Mg/Mg^{2+}. Our group combined this electrolyte with a poly(antraquinoyl sulfide) (PAQS) cathode and tested it in a two electrode battery setup containing Mg metal as a negative electrode.[11] PAQS is a good performance organic polymer, which contains anthraquinone electroactive moieties connected through S bridge atoms.[3] The Mg–PAQS battery displayed an initial capacity of 98 mA h g^{-1}, which later increased up to 105 mA h g^{-1}(Figure 9.2a) and then gradually faded down to 60 mA h g^{-1} after 80 cycles. Overpotential in the formation cycles was high and gradually decreased during cycling accompanied with an increase in capacity. Nevertheless, this system showed reasonable capacity stability and capacity retention. We need to emphasise that the maximum attainable capacity was lower than 50% of the theoretical capacity, although the same electrodes in combination with a lithium anode and corresponding electrolyte showed much higher capacity. We assume that such behaviour can be mainly attributed to the overpotential originating from the anode side of the battery. Thus, Mg foil was replaced with an in-house synthesized Mg powder anode. This battery setup resulted in a lower overpotential, and an initial capacity of 191 mA h g^{-1}. However, after more than 20 cycles the difference between the Mg–PAQS batteries with different anodes became

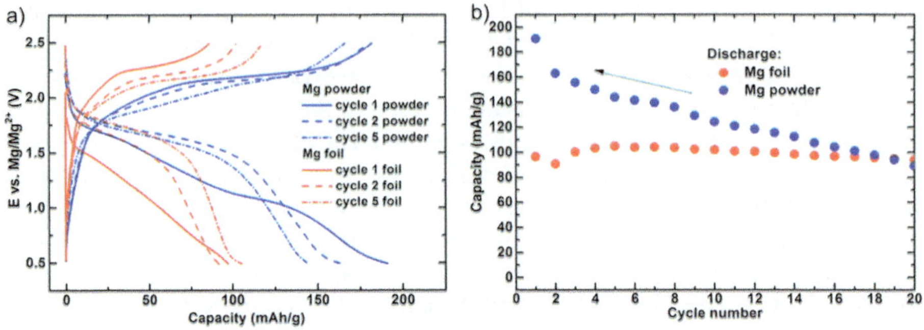

Figure 9.2 (a) Galvanostatic charge and discharge at 50 mA g^{-1} of a poly(anthraquinoyl sulfide) cathode with two different Mg anodes, synthesized Mg powder (blue) and Mg foil (red), in Mg_2Cl_3–HMDSAlCl_3 dissolved in THF. (b) Corresponding discharge capacities for the same two cells during the first 20 cycles. Adapted from ref. 11 with permission from John Wiley and Sons, © 2015 WILEY-VCH Verlag GmbH & Co. KGaA, Weinheim.

increasingly smaller and no long-term improvement could be observed (Figure 9.2b). Besides Mg_2Cl_3-HMDSAlCl_3 in THF, other electrolytes based on $MgCl_2$–$AlCl_3$ in THF and Mg$(TFSI)_2$–$MgCl_2$ in THF : DME = 2 : 3 vol% were also tested. Among these electrolytes, Mg$(TFSI)_2$–$MgCl_2$ in THF : DME = 2 : 3 vol% displayed the best performance both in terms of the maximum attainable capacity and rate capability performance. Later, the electrochemical mechanism of the PAQS cathode was investigated through application of *operando* infrared (IR) spectroscopy measurements through a silicon wafer window and the obtained results were compared with those of the cathode mechanism in Li battery.[13] Due to strong signals from the Si wafer and electrolyte, a methodology of subtraction of the spectra was utilized to give the difference spectra. Use of difference spectra enabled us to visualize all of the IR active changes that occurred during electrochemical cycling. The bands, which showed negative intensity in the difference spectra decreased during a specific step, while those with positive peaks increased in intensity. In both systems the reduction of PAQS proceeded through reduction of a C=O bond and formation of a –C–O^- bond, observed as a decrease in the intensity of the carbonyl bands at 1676 and 1651 cm^{-1} and the formation of a new enolate band. The difference between Li and Mg was in a small change in position of the –C–O^- IR band (1370 and 1376 cm^{-1}, respectively) and additional structural changes that were confirmed by density functional theory (DFT) calculations on a model PAQS trimer (Figure 9.3). Electrochemical testing in a different Mg electrolyte, Mg$(TFSI)_2$–$MgCl_2$ TEG : DOL = 1 : 1 vol% (DOL = dioxolane), displayed improved capacity retention. However, it has to be noted that the maximum capacities were 135 mA h g^{-1} at 0.1C and 72 mA h g^{-1} at 1C. Besides confirmation of the mechanism of redox active organic cathode materials through the reversible reduction of the quinone group and reoxidation of the enolate group, one can see also long term reversibility from Figure 9.3c. Even

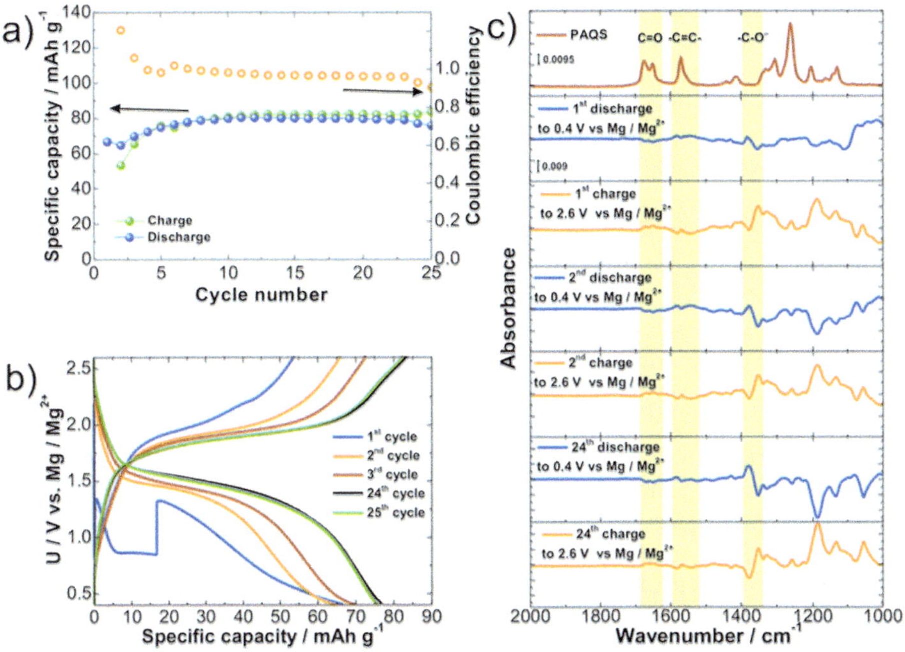

Figure 9.3 (a) Capacity and Coulombic efficiency of a Mg–PAQS battery cell at 50 mA g^{-1} in $Mg(TFSI)_2$–$MgCl_2$ in TEG : DOL = 1 : 1 vol% and (b) selected galvanostatic cycles. (c) ATR-IR difference spectra for the PAQS cathode during cycling. Reproduced from ref. 13, https://doi.org/10.1038/s41467-018-03114-1, under the terms of the CC BY 4.0 licence, http://creativecommons.org/licenses/by/4.0/.

after 25 cycles we were able to observe the same changes in the measured IR spectra. Another peculiarity of using PAQS in a Mg battery configuration is the interaction of enolate groups with magnesium cations, where two enolate groups from two PAQS monomers share one magnesium, while in a Li battery configuration each enolate group is coordinated by only one lithium cation. This causes changes in the ring vibrations between the Li and Mg systems, which was visualized through DFT calculations.

Application of an anthraquinone electroactive group was continued by Pan *et al.*, who tested the performance of polyanthraquinone (PAQ) in $Mg(HMDS)_2$–$4MgCl_2$ in THF.[14] PAQ is a next generation anthraquinone-based polymer with a higher theoretical capacity (260 mA h g^{-1}) due to the direct cross-coupling of anthraquinone groups without any linkers. It shows impressive performance in Li batteries, where only 1% drop in capacity is observed after 1000 cycles at 1C. Indeed, the capacity retention of PAQ cathode was unprecedented in the Mg battery setup and after 1000 cycles at 1C, more than 70% percent of the starting capacity was retained. Unfortunately, the starting capacity was well below half that of the theoretical one, at only 110 mA h g^{-1}. At a lower rate of 0.5C, a slightly higher capacity of 133 mA h

g^{-1} could be obtained, but still well below the theoretical one. In their work they did not test the practically attainable capacity of their PAQ in Li setup or test the PAQ cathode with different Mg electrolytes. This would help to understand whether a lower obtained capacity is an inherent limitation of the PAQ material used or due to the Mg electrolyte used. The electrochemical mechanism was investigated through *ex situ* XPS, where changes in the O1s spectra pointed to a reduction of the carbonyl bonds during the reduction of PAQ, similar to in the work on PAQS.

The anthraquinone electrochemical group has a relatively high molecular weight that is reflected in its low theoretical capacity (257 mA h g^{-1}). Additionally, its potential is only 1.5 V *vs.* Mg/Mg^{2+}, which further contributes to the low energy density of cathode compounds based on anthraquinone. Thus, one of the natural steps forward is the application of benzoquinone with a significantly higher capacity (496 mA h g^{-1}) and redox potential (2.0 V *vs.* Mg/Mg^{2+}). The application of this group was first attempted through the use of DMBQ, but use of Mg electrolytes with a high overpotential for Mg stripping/deposition resulted in poor cycling.[8–10] Pan *et al.* compared the performance of DMBQ in $Mg(TFSI)_2$ in DEG and $Mg(TFSI)_2$–$2MgCl_2$ in DME.[15] Firstly, they assessed the Mg stripping/deposition overpotential and demonstrated a significantly lower overpotential for $Mg(TFSI)_2$–$2MgCl_2$ in DME electrolyte. This was also later translated into two-electrode electrochemical cells, where the Mg–DMBQ battery displayed in $Mg(TFSI)_2$–$2MgCl_2$ in DME a discharged voltage of 2 V and could operate at elevated cycling rates of up to 2C (Figure 9.4a), while in $Mg(TFSI)_2$ in DEG the battery displayed two plateaus at 0.65 and 0.25 V (Figure 9.4b). This clearly demonstrates that use of Mg electrolyte with a low overpotential for Mg stripping/deposition is of the utmost importance for the realization of a practical Mg–organic battery or any other

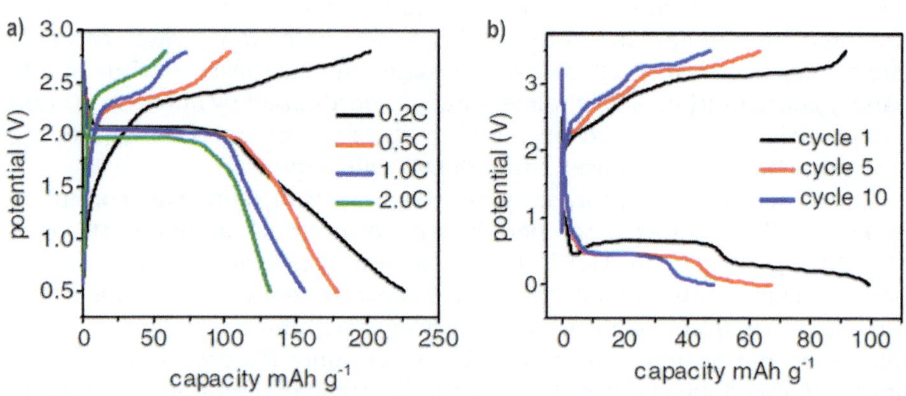

Figure 9.4 (a) First galvanostatic cycle of a Mg–DMBQ battery in $Mg(TFSI)_2$–$2MgCl_2$ DME electrolyte at different cycling rates. (b) Selected galvanostatic cycles of a Mg–DMBQ battery in $Mg(TFSI)_2$ DEG at a 1C cycling rate. Reproduced from ref. 15 with permission from The Electrochemical Society Copyright 2016.

practical Mg battery employing Mg metal as a negative electrode. The initial discharge capacity in $Mg(TFSI)_2$–$2MgCl_2$ in DME was around 226 mA h g^{-1}, which is more than 70% of the theoretical capacity. Later, the capacity faded quite rapidly due to dissolution of DMBQ into the electrolyte. The reversibility of DMBQ cycling was confirmed through *ex situ* XRD of the DMBQ electrodes.

A stabilization of the capacity retention could be achieved through the preparation of polymers. Unfortunately, preparation of benzoquinone-based polymers is much more difficult than preparation of their anthraquinone analogues. Although some benzoquinone-based polymers have already been prepared, their practical capacities are well below their theoretical capacities, even in Li electrolytes.[16] Through solvothermal reaction, a sulfur-bridged polymer was prepared containing both hydroquinone and benzoquinone species. The performance of this poly (hydroquinoyl–benzoquinoyl sulfide) PHBQS was tested in both Li and Mg battery setups.[17] In a Li battery it displayed a sloped voltage plateau with an average discharge voltage of 2.7 V with some initial capacity activation due to the gradual transformation of hydroquinone groups into benzoquinone ones. The maximum capacity was 240 mA h g^{-1} and after 340 cycles more than 160 mA h g^{-1} of the capacity was retained. Performance in a Mg battery was tested in two electrolytes both based on $Mg(TFSI)_2$–$MgCl_2$ salts, but in two different solvents: first in TEG : DOL = 1 : 1 vol% and second in TEG. In both Mg electrolytes, capacity activation could be observed similar to Li electrolyte, but TEG : DOL showed a higher maximum capacity of 158 mA h g^{-1}, while the maximum capacity in TEG-based electrolyte was 123 mA h g^{-1} (Figure 9.5). It has to be noted that the higher capacity in TEG : DOL came at the expense of a significantly lower Coulombic efficiency, which was lowered due to side reactions going on above 2.8 V *vs.* Mg/Mg^{2+}.

Again, this result clearly shows the importance of electrolyte selection and optimization when redox active organic cathodes are used in combination with metallic magnesium. The reasons can be manifold, from wetting and swelling properties at the interface between the electrolyte/polymer, the impact of electrolyte conductivity, passivation of the Mg surface and the nature of the passive film formation.

Another group of conjugated carbonyl compounds that was tested as potential cathode materials for Mg battery electrodes was the polyimides. These materials are typically synthesized through polycondensation of different dianhydrides and diamines. Their application has already been extensively tested in Li batteries as cathodes.[18] Interestingly, their first application was in aqueous electrolyte based on $MgSO_4$. Besides being used as cathodes, polyimides have also been applied as anodes in combination with a Prussian blue-type nickel hexacyanoferrate (PBN) cathode.[19] Testing of the polyimide formed through polycondensation of 1,4,5,8-naphtalenetetracarboxylic dianhydride and ethylene diamine (PNEDA) in aqueous electrolyte showed that the polyimide cathode could operate at high current densities of up to 10 A g^{-1} (Figure 9.6a) and be cycled for more than 2000 cycles with a small capacity drop of 30 mA h g^{-1} (Figure 9.6b). The full battery setup

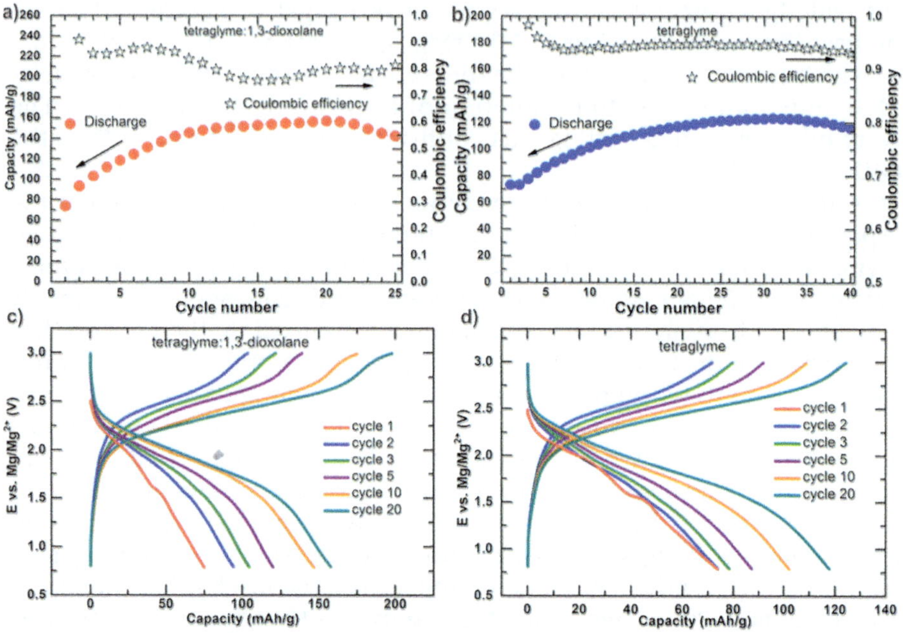

Figure 9.5 (a) Discharge capacity and Coulombic efficiency of a Mg–PHBQS battery at 50 mA g^{-1} in $Mg(TFSI)_2$–$MgCl_2$ electrolyte in TEG : DOL = 1 : 1 vol% and (b) TEG. (c) Selected galvanostatic cycles in $Mg(TFSI)_2$–$MgCl_2$ electrolyte in TEG : DOL = 1 : 1 vol% and (d) TEG. Adapted from ref. 17 with permission from Elsevier, Copyright 2016.

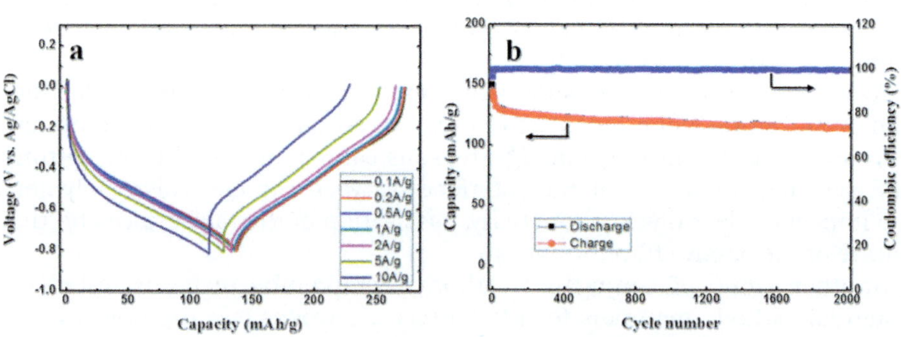

Figure 9.6 (a) Galvanostatic charge–discharge curves of a PNEDA electrode in aqueous $MgSO_4$ at different current rates from 0.1 to 10 A g^{-1}. (b) Capacities and Coulombic efficiency of the PNEDA electrode at 1 A g^{-1} for 2000 cycles. Adapted from ref. 19 with permission from American Chemical Society, Copyright 2017.

with polyimide–PBN had an average voltage of 1 V and delivered a capacity of 33 mA h g^{-1}, based on the weight of both active materials. This work was further extended by the application of polypyromellitic dianhydride ethylene diamine (PPMEDA) as an anode in aqueous $Mg(TFSI)_2$.[20] A high concentration of $Mg(TFSI)_2$ allowed the extension of the voltage window compared with the previous work based on $MgSO_4$. The PPMEDA anode had a sloping discharge plateau between 2.0 and 1.7 V *vs.* Mg/Mg^{2+}, the initial reversible capacity was 110 mA h g^{-1} and after 500 cycles at 100 mA g^{-1} (1C) more than 87% of the capacity was retained. Combination of this anode and a $Li_3V_2(PO_4)_3$ cathode gave a full cell with a capacity of 52 mA h g^{-1} that could operate at rates of as high as 60C.

A precursor for polyimide synthesis, 3,4,9,10-perylenetetracarboxyldianhydride, was tested again in aqueous electrolyte, but this time based on $Mg(NO_3)_2$ salt.[21] The reversible capacity of this cathode was 125 mA h g^{-1} at a current density of 20 mA g^{-1} and 75 mA h g^{-1} of the capacity was retained at 500 mA g^{-1}. An investigation of Mg^{2+} ion accommodation was carried out with *ex situ* IR, XRD and transmission electron microscopy. It was demonstrated that hosting one Mg^{2+} ion requires coordination from three PTCDA molecules in adjacent columns. This causes rotation of the columns and a reduction in the (011) spacing, but an increase in the (021) spacing. Additionally, it was demonstrated that PTCDA can even accommodate Ca^{2+} ions, although with a lower capacity. Later, polyimides were also tested in a Mg metal battery setup, where the polyimide prepared through polycondensation of 1,4,5,8-naphtalenetetracarboxylic dianhydride and hydrazine (PNH) was tested.[22] The electrochemical performance was tested in two different Mg electrolytes based on $Mg(TFSI)_2$–$MgCl_2$ in TEG : DOL = 1 : 1 vol% and TEG. In both electrolytes, the average discharge voltage was around 1.7 V, in good agreement with the potential obtained in a Li battery. PTNH displayed quite long capacity activation in TEG, as it took 53 cycles to achieve a maximum capacity of 86 mA h g^{-1}, while in TEG : DOL a maximum capacity of 85 mA h g^{-1} was achieved at the 97th cycle and seemed more stable. Voltage hysteresis between electrolytes was also different with TEG : DOL electrolyte displaying approximately 0.1 V smaller hysteresis. Rate capability tests showed that Mg–PTNH battery could be cycled at 2.5C and still deliver a capacity of around 50 mA h g^{-1}. The electrochemical mechanism of the PTNH cathode was investigated through *operando* IR spectroscopy and DFT calculations, which confirmed the reduction of a carbonyl group to an enolate anion in the form of a decrease in the intensity of a band at 1707 cm^{-1} and the appearance of a new IR band at 1516 cm^{-1}. Another study that tested the performance of a Mg metal–polyimide battery applied polyimide prepared from pyromellitic dianhydride and ethylene diamine (PPMEDA) in a composite with carbon nanotubes (CNT).[23] The PPMEDA/CNT composite showed good electrochemical performance with a starting capacity of more than 145 mA h g^{-1} and capacity retention of more than 120 mA h g^{-1} after 200 cycles at 150 mA g^{-1}. It has to be noted that the extracted cycles show mainly pseudocapacitance behaviour with a voltage slope extending from 1.7 to 0.5 V. Such behaviour could be caused by side reactions between the cathode and electrolyte, which are likely due to the fact that the $(PhMgCl)_2$–$AlCl_3$ electrolyte in THF used is based on Grignard compounds. The authors additionally showed that their Mg–organic battery could operate at varying current densities ranging from 0.5 to 20C (1C = 150 mA g^{-1}) and over an extended temperature range from −40 to 50 °C.

Recently, polyimide has also been used as an anode in non-aqueous Mg electrolyte in a type of organic battery, which employs a p-type organic material as a cathode and n-type organic material as an anode.[24] This battery operates on a dual ion mechanism, where electrolyte cations attaches/unattaches to the anode and anions with the cathode during charge/discharge, respectively. On the anode side, polyimide prepared from PTCDA and ethylene diamine (PPEDA) was employed, while on the cathode side polytriphenylamine (PTPA) from the class of conjugated amines was used. The electrolyte was $Mg(ClO_4)_2$ in acetonitrile. PTPA had two oxidation peaks at 3.26 and 3.38 V *vs.* Mg/Mg^{2+} and two reduction peaks at 3.13 and 3.26 V *vs.* Mg/Mg^{2+}. Polyimide PPEDA had one oxidation peak at 2.24 V *vs.* Mg/Mg^{2+} and two reduction peaks at 2.14 and 1.90 V *vs.* Mg/Mg^{2+}. The assembled full PPEDA–PTPA battery displayed a capacity of 90 mA h g^{-1} at a current density of 50 mA g^{-1} and a retained capacity of 73 mA h g^{-1} at a current density of 1000 mA g^{-1}. The cell suffered only minor capacity loss of 5 mA h g^{-1} during sustained long-term cycling at 1000 mA g^{-1} for 2000 cycles. Additionally, the cell could be cycled at different temperatures ranging from −20 to 25 °C. The capacity performance at low rates was quite comparable at lower temperatures, but greatly faded with increased current density. Additionally, the cell was prepared in a flexible pouch cell, demonstrating potential use for bendable energy storage devices.

Initial work employing nitroxyl radicals gave very poor results due to the use of unsuitable electrolytes.[7] Later, the same electrode material PTMA was used as one of the potential cathode materials for $Mg(TFSI)_2$ DME : DEG = 1 : 1 vol% electrolyte.[25] Although this electrolyte still suffered from a high overpotential for Mg deposition/stripping, the PTMA cathode displayed a well-defined discharge with a voltage of 2.6 V *vs.* Mg/Mg^{2+}. The capacity performance was very low, only 37.6 mA h g^{-1} could be obtained, but this can be attributed to the poor performance of the Mg anode in this electrolyte.

Among the conjugated hydrocarbons, only one material was tested. Fullerene C_{60} was tested in $(PhMgCl)_2$–$AlCl_3$ in THF and displayed two flat discharge plateaus at 1.4 and 1.1 V (Figure 9.7).[26] The total discharge capacity was quite low and reached only 50 mA h g^{-1}. The capacity fade was quite fast and at the lowest current density only 5 mA h g^{-1} could be obtained after 10 cycles due to dissolution of the active material. Although a 6 electron exchange was proposed for C_{60} (223 mA h g^{-1}) in the literature, practical experiments showed that only up to 2 electron exchange could be obtained and only the first plateau reached the theoretical capacity of 1 electron exchange (37 mA h g^{-1}).

Organics are an extremely variable group of materials and can be used as both anodes and cathodes in Mg batteries. Currently, most of the research has been carried out on n-type organics based on conjugated carbonyl compounds that can enable accommodation of Mg^{2+} ions. Application of these materials is the most realistic for practical batteries since it enables us to benefit from the use of a Mg metal anode with its high volumetric and gravimetric capacity. However, measured reversible capacities in the systems employing practical Mg electrolytes are still far from those obtained

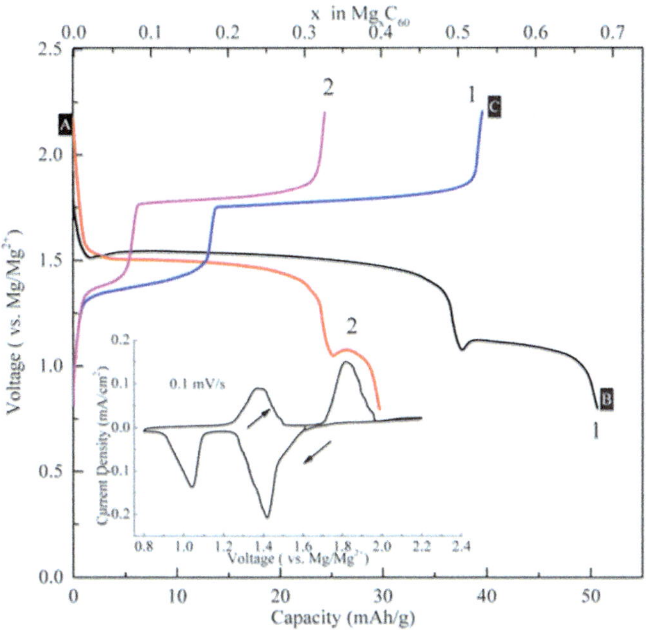

Figure 9.7 Galvanostatic cycling of a fullerene C_{60} cathode at a current density of 19 μA cm^{-2}. The inset shows the corresponding cyclic voltammogram curve. A, B, and C indicate the electrochemical states of the pristine, discharged and recharged electrodes, respectively. Reproduced from ref. 25 with permission from the Royal Society of Chemistry.

in Li electrolytes. This fact is mainly connected with the still developing field of Mg electrolytes, as it was shown during this chapter that electrolyte composition needs to be optimized for almost every combination of redox active organic material. Many studies show a long activation process before maximum capacity can be obtained but that the wetting and swelling properties of the as-prepared polymers are still not well understood. Cycling in aqueous Mg electrolytes has so far demonstrated that long-term cycling with good capacity retention is possible for several polyimide compounds. Furthermore, optimization of cathode compositions has not yet started since most reports show electrochemical performance with simple composite mixtures containing a large portion of carbon black as the conductive additive. Composites with only a few weight percent ratio of conductive additive are needed for commercial applications and these have to be developed for each interesting redox active organic material. One of the biggest challenges in the field of conjugated carbonyl compounds is the synthesis of benzoquinone-based polymers that can deliver capacities close to the theoretical ones. Combination of a benzoquinone polymer with magnesium metal can produce a Mg–organic battery with 2 V theoretical specific energy density of more than 600 W h kg^{-1}. Another option to increase the redox potentials of polymers is the preparation of polymers where the carbonyl groups are positioned in ortho positions instead of the more typical para position. Such proof of concept has recently demonstrated that it can lead to an increase of 400 mV in the redox potential of the electroactive group when preparing a polyphenanthrenequinone polymer instead of PAQ.[27] Other systems based on aqueous electrolytes or employing dual ion electrolytes offer good power performance, but are highly limited in terms of the energy density of the full cell. Practical application of these systems can be considered only in systems where the energy density is of secondary importance. A lot of room for future development also lies in the exploration of other groups of materials, where we currently have only isolated studies, but lack more complete and systematic surveys.

Although the first report on the use of organic cathodes appeared in the literature in 2007, it was an isolated report and was not followed up by the community. Recently, more groups have started investigating organic materials for use in Mg batteries and, correspondingly, significant progress has been achieved. When we compare the development and state-of-the-art of Mg batteries in recent years, we can conclude that organic materials have quickly caught up and in some cases even surpassed the performance of inorganic materials, especially when comparing practically attainable capacities and voltages in two-electrode battery setups with Mg metal anodes. Further progress is expected with the development of new non-nucleophilic electrolytes and the synthesis of new organic compounds. An important question is the behaviour of the Mg anode and the performance of Mg metal–organic cells in setups where the Mg anode is not present in a large excess. The commercialization of Mg–organic batteries will depend on the volumetric density of the electrode, areal loadings of the organic electrodes and the amount of electrolyte required for practical cells. From the point of view of volumetric energy density, despite the low gravimetric energy of polymers, the relatively high volumetric energy of magnesium metal promises cells with competitive volumetric energy density for applications where weight of the battery is not an issue. These are all topics that have not yet been discussed in the literature. Thus, future research efforts on the topic of organic materials should also address these questions. In the coming years we can expect reports about the first prototypes that will reflect the real value of this battery technology and its potential application for mobile or stationary applications.

References

1. P. Canepa, G. Sai Gautam, D. C. Hannah, R. Malik, M. Liu, K. G. Gallagher, K. A. Persson and G. Ceder, Odyssey of Multivalent Cathode Materials: Open Questions and Future Challenges, *Chem. Rev.*, 2017, **117**, 4287.
2. B. Genorio, K. Pirnat, R. Cerc-Korosec, R. Dominko and M. Gaberscek, Electroactive organic molecules immobilized onto solid nanoparticles as a cathode material for lithium-ion batteries, *Angew. Chem., Int. Ed.*, 2010, **49**, 7222.

3. Z. Song, H. Zhan and Y. Zhou, Anthraquinone based polymer as high performance cathode material for rechargeable lithium batteries, *Chem. Commun.*, 2009, 448.
4. H. Senoh, M. Yao, H. Sakaebe, K. Yasuda and Z. Siroma, A two-compartment cell for using soluble benzoquinone derivatives as active materials in lithium secondary batteries, *Electrochim. Acta*, 2011, **56**, 10145.
5. Y. NuLi, Z. Guo, H. Liu and J. Yang, A new class of cathode materials for rechargeable magnesium batteries: Organosulfur compounds based on sulfur–sulfur bonds, *Electrochem. Commun.*, 2007, **9**, 1913.
6. D. Lv, T. Xu, P. Saha, M. K. Datta, M. L. Gordin, A. Manivannaan, P. N. Kumta and D. Wang, A Scientific Study of Current Collectors for Mg Batteries in Mg(AlCl2EtBu)2/THF Electrolyte, *J. Electrochem. Soc.*, 2012, **160**, A351.
7. C. Qiang, Y.-N. Nuli, W. Guo, J. Yang, J.-L. Wang and Y.-G. Guo, PTMA/Graphene as a Novel Cathode Material for Rechargeable Magnesium Batteries, *Acta Phys.-Chim. Sin.*, 2013, **29**, 2295.
8. H. Sano, H. Senoh, M. Yao, H. Sakaebe and T. Kiyobayashi, Mg2+ Storage in Organic Positive-electrode Active Material Based on 2,5-Dimethoxy-1,4-benzoquinone, *Chem. Lett.*, 2012, **41**, 1594.
9. H. Senoh, H. Sakaebe, H. Sano, M. Yao, K. Kuratani, N. Takeichi and T. Kiyobayashi, Sulfone-Based Electrolyte Solutions for Rechargeable Magnesium Batteries Using 2,5-Dimethoxy-1,4-benzoquinone Positive Electrode, *J. Electrochem. Soc.*, 2014, **161**, A1315.
10. H. Senoh, H. Sakaebe, H. Tokiwa, M. Uchida, H. Sano, M. Yao and T. Kiyobayashi, Charge-Discharge Performance of Rechargeable Organic-Magnesium Batteries Using Glyme-Based Electrolytes, *ECS Trans.*, 2015, **69**, 33.
11. J. Bitenc, K. Pirnat, T. Bančič, M. Gaberšček, B. Genorio, A. Randon-Vitanova and R. Dominko, Anthraquinone-Based Polymer as Cathode in Rechargeable Magnesium Batteries, *ChemSusChem*, 2015, **8**, 4128.
12. H. S. Kim, T. S. Arthur, G. D. Allred, J. Zajicek, J. G. Newman, A. E. Rodnyansky, A. G. Oliver, W. C. Boggess and J. Muldoon, Structure and compatibility of a magnesium electrolyte with a sulphur cathode, *Nat. Commun.*, 2011, **2**, 427.
13. A. Vizintin, J. Bitenc, A. Kopač Lautar, K. Pirnat, J. Grdadolnik, J. Stare, A. Radon-Vitanova and R. Dominko, Probing electrochemical reactions in organic cathode materials *via* in operando infrared spectroscopy, *Nat. Commun.*, 2018, **9**, 661.
14. Z. Song, Y. Qian, M. L. Gordin, D. Tang, T. Xu, M. Otani, H. Zhan, H. Zhou and D. Wang, Polyanthraquinone as a Reliable Organic Electrode for Stable and Fast Lithium Storage, *Angew. Chem., Int. Ed.*, 2015, **54**, 13947.
15. B. Pan, D. Zhou, J. Huang, L. Zhang, A. K. Burrell, J. T. Vaughey, Z. Zhang and C. Liao, 2,5-Dimethoxy-1,4-Benzoquinone (DMBQ) as Organic Cathode for Rechargeable Magnesium-Ion Batteries, *J. Electrochem. Soc.*, 2016, **163**, A580.

16. Z. Song, Y. Qian, T. Zhang, M. Otani and H. Zhou, Poly(benzoquinonyl sulfide) as a High-Energy Organic Cathode for Rechargeable Li and Na Batteries, *Adv. Sci.*, 2015, **2**, 1500124.
17. J. Bitenc, K. Pirnat, G. Mali, B. Novosel, A. Randon Vitanova and R. Dominko, Poly(hydroquinoyl-benzoquinonyl sulfide) as an active material in Mg and Li organic batteries, *Electrochem. Commun.*, 2016, **69**, 1.
18. Z. Song, H. Zhan and Y. Zhou, Polyimides: Promising energy-storage materials, *Angew. Chem., Int. Ed.*, 2010, **49**, 8444.
19. L. Chen, J. L. Bao, X. Dong, D. G. Truhlar, Y. Wang, C. Wang and Y. Xia, Aqueous Mg-Ion Battery Based on Polyimide Anode and Prussian Blue Cathode, *ACS Energy Lett.*, 2017, **2**, 1115.
20. F. Wang, X. Fan, T. Gao, W. Sun, Z. Ma, C. Yang, F. Han, K. Xu and C. Wang, High-Voltage Aqueous Magnesium Ion Batteries, *ACS Cent. Sci.*, 2017, **3**, 1121.
21. I. A. Rodríguez-Pérez, Y. Yuan, C. Bommier, X. Wang, L. Ma, D. P. Leonard, M. M. Lerner, R. G. Carter, T. Wu, P. A. Greaney, J. Lu and X. Ji, Mg-Ion Battery Electrode: An Organic Solid's Herringbone Structure Squeezed upon Mg-Ion Insertion, *J. Am. Chem. Soc.*, 2017, **139**, 13031.
22. T. Bančič, J. Bitenc, K. Pirnat, A. Kopač Lautar, J. Grdadolnik, A. Randon Vitanova and R. Dominko, Electrochemical performance and redox mechanism of naphthalene-hydrazine diimide polymer as a cathode in magnesium battery, *J. Power Sources*, 2018, **395**, 25.
23. C. Wang, X. Fan, F. Wang, X. Ji, R. Wang, T. Gao, S. Hou, J. Chen, T. Deng, X. Li, L. Chen, C. Luo and L. Wang, A Universal Organic Cathode for Ultrafast Li- and Multivalent Metal Batteries, *Angew. Chem., Int. Ed.*, 2018, **57**, 7146.
24. D. Lu, H. Liu, T. Huang, Z. Xu, L. Ma, P. Yang, P. Qiang, F. Zhang and D. Wu, Magnesium ion based organic secondary batteries, *J. Mater. Chem. A*, 2018, **6**, 17297.
25. S.-Y. Ha, Y.-W. Lee, S. W. Woo, B. Koo, J.-S. Kim, J. Cho, K. T. Lee and N.-S. Choi, Magnesium(II) bis(trifluoromethane sulfonyl) imide-based electrolytes with wide electrochemical windows for rechargeable magnesium batteries, *ACS Appl. Mater. Interfaces*, 2014, **6**, 4063.
26. R. G. Zhang, F. Mizuno and C. Ling, Fullerenes: non-transition metal clusters as rechargeable magnesium battery cathodes, *Chem. Commun.*, 2015, **51**, 1108.
27. K. Pirnat, J. Bitenc, A. Vizintin, A. Krajnc and E. Tchernychova, Indirect Synthesis Route toward Cross-Coupled Polymers for High Voltage Organic Positive Electrodes, *Chem. Mater.*, 2018, **30**, 5726.

CHAPTER 10

Magnesium–Sulfur Batteries

Z. ZHAO-KARGER*

Helmholtz Institute Ulm (HIU), Helmholtzstr. 11, D-89081 Ulm, Germany
*E-mail: zhirong.zhao-karger@kit.edu

10.1 Introduction

Sulfur (S) is a low-cost, sustainable and environmentally benign material owing to its natural abundance and non-toxicity. Elemental sulfur can undergo a two-electron redox reaction and offers a theoretical specific capacity of 1672 mA h g^{-1}, which is approximately ten times more than those of the common intercalation cathodes in lithium-ion batteries (LIBs). Hence, sulfur is receiving intense interest as a promising cathode material. Due to the relatively low oxidation potential of sulfur, a metal anode would have to be employed to eventually achieve the high energy density of a sulfur-based battery because it possibly maximizes the electrode capacity and cell voltage. Metal–sulfur batteries are therefore particularly attractive for high-energy and low-cost electrochemical applications. Based on the Gibbs free energies of the reactions of sulfur with light metals including Li, Na, Mg and Al, the parameters of the battery systems are summarized in Table 10.1.

Among them, the lithium–sulfur (Li–S) battery has been regarded as a high-energy post Li-ion battery and has been extensively investigated over the past decade.[1] However, the unresolved technical problems associated with the chemistry of sulfur and Li metal hamper the commercialization of Li–S batteries.[2] One of the greatest challenges is how to deal with the

Energy and Environment Series No. 23
Magnesium Batteries: Research and Applications
Edited by Maximilian Fichtner

Published by the Royal Society of Chemistry, www.rsc.org

Table 10.1 Sulfur redox chemistries and the cell features in various metal-based systems.

Redox reaction	Δ_rG/kJ mol^{-1}	EMF/V	Energy density W h^{-1} kg^{-1}	Energy density W h^{-1} l^{-1}	Volume change/%
$2Li^+ + S + 2e^- \rightleftarrows Li_2S$	−439.0	2.27	2621	2907	80
$2Na^+ + S + 2e^- \rightleftarrows Na_2S$	−359.8	1.86	1275	1574	170
$Mg^{2+} + S + 2e^- \rightleftarrows MgS$	−341.8	1.77	1693	3227	28
$2Al^{3+} + 3S + 6e^- \rightleftarrows Al_2S_3$	−713.2	1.23	1319	2981	60

dendrite growth that occurs during the cycling of the batteries with a Li metal anode, which can induce internal short circuiting in the cells, leading to severe explosion risks. With a graphite anode, the Li–S system has a voltage penalty of ~0.3 V, which decreases the theoretical energy density of down to <1000 W h l^{-1}. Similarly, the development of room temperature sodium–sulfur (Na–S) batteries is also facing tremendous problems in the use of metallic Na as an anode due to its intrinsic high reactivity and severe dendritic growth. Therefore, before the metal–sulfur battery can become a viable technology, feasibility of a safe metal anode is a fundamental prerequisite. Additionally, the volume expansion of sulfur after being fully discharged can reach approximately 80% and as high as 170% in a Li–S and Na–S battery, respectively, which can lead to serious electrode degradation.

In contrast to Li and Na, Mg metal has inherent characteristics to form a homogeneous surface and is not plagued by dendritic deposition during repeated cycling.[3–5] The combination of a Mg anode with a sulfur cathode has therefore been suggested as an ideal electrochemical couple. The volume variation in the conversion between S and MgS is in a low range of approximately 28%. Owing to the relatively high density of Mg, a Mg–S battery can potentially offer the highest volumetric energy density among the metal–sulfur systems. The rich natural resources of both the Mg and S elements render Mg–S batteries one of the most economical options for practical applications in terms of cost per W h. Al–S batteries also show great promise as high energy, cost-effective and safe systems owing to the abundance and uniform plating properties of Al. However, the research on Al–S batteries is in a proof-of-concept stage and their redox mechanism in a complex ionic liquid electrolyte is still under investigation.[6–8] The key challenge in the realization of multivalent metal-based batteries is to develop a sulfur compatible electrolyte with efficient ion transport properties. In fact, substantive progress has been achieved in formulating effective Mg electrolytes, which has facilitated the research on Mg–S batteries.[9–12] In the following sections, we will mainly focus on recent development in Mg–S batteries with different electrolyte systems. Apart from that, the advances in cathode design and cell configuration are presented.

10.2 Features of a Mg–S Battery

A schematic of the components of a Mg–S cell and its operation features is shown in Figure 10.1a. A Mg–S cell is comprised of a sulfur-based composite cathode, magnesium metal anode, an organic liquid electrolyte and a separator between the two electrodes. When the sulfur is in a charged state, the cell operation starts with a discharge process. During discharge, the Mg metal is oxidized to Mg^{2+} ions at the negative electrode and releases electrons. The Mg ions migrate to the positive cathode through the electrolyte and separator, while the sulfur accepts the electrons through the external electrical circuit generating an electrical current being reduced to magnesium sulfide. Analogous to Li–S battery chemistry, the reduction of sulfur proceeds successively *via* various magnesium polysulfide (Mg–PS) intermediates during discharge, as shown in Figure 10.1b.[13] The most stable allotrope of elemental sulfur at room temperature is orthorhombic cyclic octasulfur α-S_8. During discharge, cyclic S_8 is first reduced to high-order long-chain polysulfides MgS_n ($6 < n \leq 8$), followed by the formation of lower-order magnesium polysulfides MgS_n ($2 < x \leq 6$) with the incorporation of more magnesium and finally to MgS. The end discharge products (MgS_2 and MgS) are insoluble in electrolyte. The charge reaction is generally reversed from MgS_2/MgS to S. The overall redox reactions of a Mg–S battery can be expressed as follows:

$$S_8 + 8Mg^{2+} + 16e^- \rightleftharpoons 8MgS \text{ (positive electrode)}$$

$$8Mg \rightleftharpoons 8Mg^{2+} + 16e^- \text{ (negative electrode)}$$

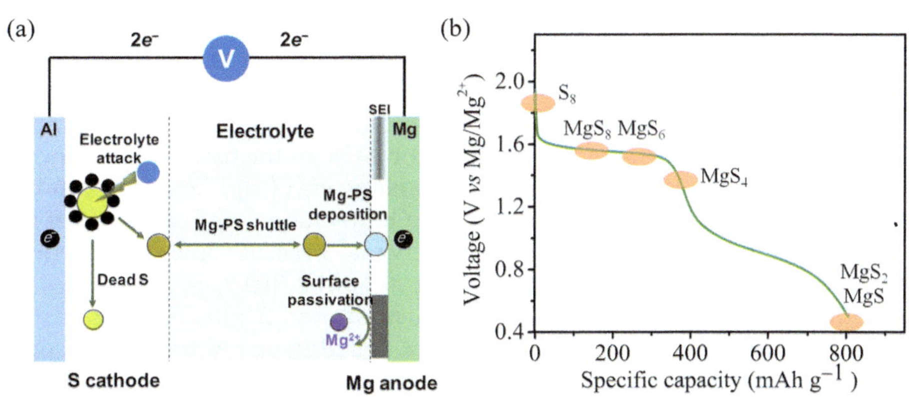

Figure 10.1 (a) Schematic illustration of the features of Mg–S batteries. Reproduced from ref. 12 with permission from John Wiley and Sons, © 2018 Zhengzhou University. (b) Electrochemical reduction mechanism of a sulfur cathode. Reproduced from ref. 13 with permission from John Wiley and Sons, © 2014 WILEY-VCH Verlag GmbH & Co. KGaA, Weinheim.

Because the composition and properties of the electrolyte have a strong impact on the sulfur conversion reactions, slightly different dominating intermediate products MgS_n have been identified in Mg–S systems with different electrolytes.[13–16] Nevertheless, it can be verified that the sulfur redox reaction in a Mg-based system proceeds through the multi-step formation of polysulfide intermediates. The high-order Mg–PSs are soluble in organic electrolyte and can lead to a loss of active material from the cathode and passivation of the Mg anode.

The Mg–S battery displays a number of similar features as those of the well-investigated Li–S chemistries, yet it has its own particularities. Due to the electrical insulating nature of both S and the final discharged products MgS_2 and MgS, rational design of the cathode architecture is required for fast electron and Mg ion transport to enable reversible electrochemical reactions between sulfur and Mg. A Mg–S battery also encounters polysulfide dissolution during electrochemical cycling as in a conventional Li–S battery. In particular, discovering a suitable electrolyte has been a key challenge for the R&D of a Mg–S battery. The electrolytes analogous to those commonly used for Li–S batteries cannot be employed for Mg-ion conduction. Substantial efforts have been devoted to the development of an efficient electrolyte that is capable of reversibly Mg plating/stripping and is chemically compatible with sulfur cathodes, Mg anodes, and other battery parts. In the following, we discuss the electrochemical performance of Mg–S batteries with special attention on the properties of the electrolytes and their correlations to the cell chemistry.

10.3 Electrolytes for Mg–S Batteries

The properties of an electrolyte play a crucial role in the cell performance. Electrolytes must, in general, fulfill multitude requirements for practical use in batteries, which include (i) a sufficiently broad electrochemical potential window that allows the redox reactions to proceed; (ii) good chemical compatibility with electrode materials and non-corrosive to the battery components such as the current collector and cell housing, *etc.*; (iii) high ionic conductivity and high ion mobility to ensure the high charge/discharge rate of the battery; (iv) thermal and hydrolysis stability may prevent some side reactions from occurring during cycling and thus extend the lifetimes of batteries. In addition, low cost and low toxicity are also desired features for a practical electrolyte. Due to the electrophilic nature of sulfur, an electrolyte with non-nucleophilic features is required for a metal–sulfur battery to prevent a parasitic chemical reaction from occurring between the cathode and electrolyte.

To date, the electrolytes explored for Mg–S batteries can be classified into two main groups: one class of electrolytes called "complex" electrolytes are typically generated from Lewis base–acid combinations such as $Mg(HMDS)_2$–$AlCl_3$,[17] $MgCl_2$–$AlCl_3$ (MACC),[18] and MgTFSI–$MgCl_2$.[19] Some recently reported reaction mixtures used as electrolytes are also included here in complex-type electrolytes.[20] Another category of electrolytes are solutions prepared by dissolving a Mg-ion conductive salt in organic solvents.[21] Next, the sulfur redox chemistry and electrochemical performance of Mg–S batteries in different electrolyte systems are discussed.

10.3.1 Complex Electrolytes

10.3.1.1 HMDS-based Electrolytes

Complex-type electrolytes generated from organomagnesium compounds cannot be employed for Mg–S batteries due to their nucleophilic nature and reductive chemical composition. A proof-of-concept Mg–S battery was demonstrated with a non-nucleophilic HMDSMgCl–$AlCl_3$ electrolyte in tetrahydrofuran (THF).[22] However, the cells exhibited a low discharge voltage of below 1 V and a large overpotential (>1 V) and were able to be operated only for few cycles with severe capacity fading and overcharging behavior. In addition, multistep reactions under strict conditions are required for the preparation of the HMDSMgCl–$AlCl_3$ electrolyte.[23] A convenient approach to non-nucleophilic electrolytes was established with the combination of $Mg(HMDS)_2$–$AlCl_3$.[17] Bis-amide-based electrolytes can be prepared in various solvents and show superior electrochemical characteristics in terms of high oxidative stability and high Coulombic efficiency (CE) for Mg cycling.[17] It was demonstrated that the Mg–S batteries using the bis-amide-based electrolytes in glymes or binary solvents considerably outperformed those in THF solution.[13] The Mg–S cell with the electrolyte in binary tetraglyme and ionic liquid (TEGIL) delivered an initial discharge capacity of 800 mA h g^{-1}, displaying two distinct potential plateaus in which the flat upper discharge voltage plateau at 1.65 V is close to the theoretical electromotive force (EMF) value, (Figure 10.2a).[13] The Mg–S batteries could retain a capacity of approximately 260 mA h g^{-1} after 20 cycles. Based on *ex situ* studies by means of ultraviolet–visible spectroscopy (UV-Vis) and X-ray photoelectron spectroscopy (XPS), a reaction mechanism was proposed for the Mg–S cell, where the redox chemistry of sulfur involves two main steps: a fast solid–liquid two phase reduction forming high-order polysulfides (MgS_8, MgS_6, and MgS_4), followed by liquid–solid reduction forming insoluble MgS_2 and MgS (Figure 10.1b).[13] However, critical issues of rapid capacity degradation, poor charging efficiency and large charge overpotential have been revealed, which can be ascribed to the dissolution of polysulfides into the electrolyte and its deleterious effect on the Mg deposition behavior. In addition, the cyclic voltammograms (CVs) indicate that the anodic peak associated with the oxidation of MgS/MgS_2 to sulfur is not as distinct as its cathodic counterpart, which reflects the high energy barrier to recharge the cathode in this system.

Gao *et al.* modified bis-amide electrolytes by introducing lithium bis(trifluoromethanesulfonyl)imide (LiTFSI) to enhance the reversibility of the Mg–S cell.[24] This hybrid system exhibited two discharge plateaus at 1.75 and 1.0 V and a stable capacity of approximately 1000 mA h g^{-1} after 30 cycles. It was proposed that Li^+ could undergo an ion displacement reaction with MgS and MgS_2, thus enhancing the re-oxidation kinetics of MgS_2/MgS (Figure 10.2c).[24]

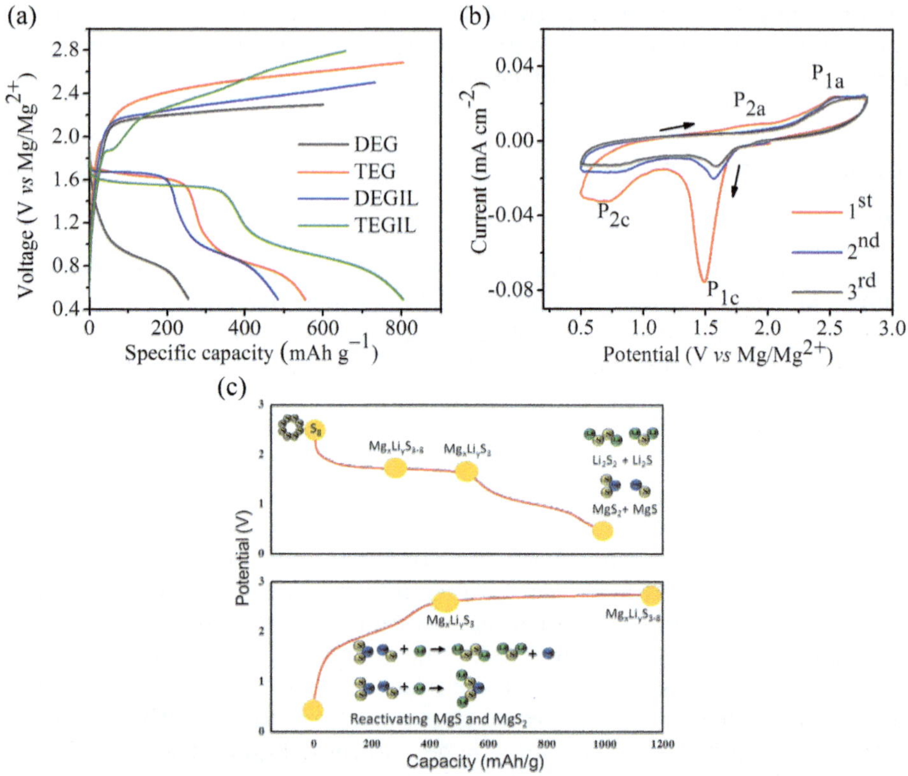

Figure 10.2 (a) Discharge/charge profiles of the Mg–S cells using $Mg(HMDS)_2$–$AlCl_3$ electrolytes in diglyme (DEG), tetraglyme (TEG), diglyme/PP14TFSI (DEGIL), and tetraglyme/PP14TFSI (TEGIL). (b) CVs of the sulfur cathode in the electrolyte TEGIL. Reproduced from ref. 13 with permission from John Wiley and Sons, © 2014 WILEY-VCH Verlag GmbH & Co. KGaA, Weinheim. (c) Discharge/charge profiles and illustration of the working mechanism of the Mg–S battery with $Mg(HMDS)_2$–$AlCl_3$–LiTFSI electrolyte. Reproduced from ref. 24 with permission from American Chemical Society, Copyright 2015.

The $Mg(HMDS)_2$–$AlCl_3$ electrolytes have also been used to study Mg–Se and Mg–SeS_2 batteries, respectively.[25] Benefiting from the intrinsic high electrical conductivity of elemental Se, the Se-based cathodes showed a superior rate capability. The Mg–Se cells could be cycled at a high rate of 2C, delivering a reversible capacity of 480 mA h cm^{-3} for 50 cycles. Moreover, SeS_2, which combines the merits of both selenium and sulfur, has also been demonstrated to be a potential high-capacity cathode material for use in Mg batteries.

10.3.1.2 $MgCl_2$ Combination-based Electrolytes

MACC electrolyte generated through the reaction of $MgCl_2$ with two equivalents of $AlCl_3$ is fully inorganic and chemically compatible with sulfur.[18] It is worth noting that a special conditioning process is needed to achieve reversible Mg deposition with the MACC electrolyte and its speciation is extremely complicated involving $[Mg_2Cl_3]^+$, $[AlCl_4]^-$, free Cl^- ions and various other species at chemical equilibria.[26,27] In addition, the presence of $[AlCl_4]^-$ anions causes the co-deposition of Al, resulting in the low columbic efficiency of the Mg anode.[18] A nitrogen-doped graphene–sulfur (NG–S) cathode was tested in Mg-based batteries using MACC electrolyte in the binary solvent of THF and an ionic liquid.[28] Unfortunately, the cells suffered from a low discharge potential (<1 V) and large overpotential (>1.5 V) and rapid capacity decline from 700 to 130 mA h g^{-1} after a few cycles. Recently, a new combination of $MgCl_2$ with yttrium chloride (YCl_3) was proposed to circumvent the co-deposition issue owing to the relatively low standard reductive potential of yttrium (Y) (−2.372 V for Y *vs.* −1.66 V for Al).[23] The Mg–S cells with a MgS_8 cathode using $MgCl_2$–YCl_3 electrolyte outperformed those in MACC electrolyte, delivering a discharge voltage of 1.2 V and a high discharge capacity of over 1000 mA h g^{-1} for more than 50 cycles at 0.05C.[23]

The Mg deposition performance of ethereal solutions of $Mg(TFSI)_2$ can be significantly improved with the addition of $MgCl_2$. However, further investigations indicated that highly reversible Mg deposition in this type of electrolyte could only be achieved through a chemical or electrochemical conditioning process.[29] Owing to the inorganic composition and ready commercial availabilities, this combination became the preferred electrolyte to use to study Mg batteries. The $Mg(TFSI)_2$–$2MgCl_2$ electrolyte in DME has been examined for Mg–S batteries.[30] The electrochemical performance could be enhanced by using more concentrated electrolyte, where a discharge capacity of about 600 mA h g^{-1} could be retained at 0.1C after 100 cycles (Figure 10.3).

To evaluate the applicability of an electrolyte, the interfacial properties between the electrolyte and electrode should be carefully examined. Studies on alterations to the anode have identified the decomposition of $TFSI^-$ anions and passivation on the Mg surface in $Mg(TFSI)_2$–$2MgCl_2$ electrolyte with repeated Mg plating at a practical current density of 1 mA cm^{-2}.[29,31] In metal–sulfur batteries, it is especially important to prevent the failure of

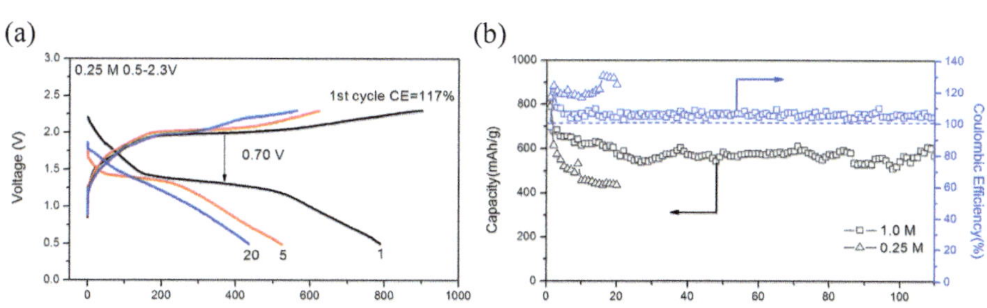

Figure 10.3 (a) Discharge/charge profiles and (b) cycling performance of the Mg–S cells in $Mg(TFSI)_2$–$MgCl_2$/DME electrolytes at 0.1C. Reproduced from ref. 30 with permission from John Wiley and Sons, © 2017 Wiley-VCH Verlag GmbH & Co. KGaA, Weinheim.

the metal anode due to dissolved polysulfides. Very recently, Salama *et al.* reported on the deleterious effects on a Mg anode due to the polysulfide species in $Mg(TFSI)_2$–$2MgCl_2$/DME electrolyte.[32]

10.3.1.3 Borate Derivative-based Electrolytes

Tris(2*H*-hexafluoroisopropyl)borate (THFPB) also called boric acid tris(2*H*-hexafluoroisopropyl) ester was recently used to formulate Mg electrolytes. Zhang *et al.* prepared a boron-centered anion-based magnesium (BCM) electrolyte by reacting THFPB with MgF_2 in a molar ratio of 20 : 1 in 1,2-dimethoxyethane (DME).[20] It was observed that both the current densities and the concentration of Mg ions in the solution substantially increased in the extended CV scans, which could be attributed to the chemical reaction between acidic THFPB and the Mg electrode. Such reaction mixture containing 0.05 M of the initial Mg source and an excessive amount of THFPB was used to study the electrolytic properties and performance of Mg–S and Mg–Se batteries.[20] Unfortunately, the chemical instability of the used electrolytes was not taken into account for the evaluation of the Mg–S batteries.

A modified borate-based electrolyte was prepared through the reaction of THFPB with $MgCl_2$ and additional Mg.[33] The crystal structures of $[Mg_4Cl_6(DME)_6]^+[B(HFP)_4]_2^-$ and $[Mg_2Cl_3(THF)_6]^+$ $[B(HFP)_4]^-$ were identified from the DME and THF solution, respectively. The Mg–S batteries were investigated with the reaction mixture as electrolyte and it was found that the sulfur–carbon nanotube (S–CNT) cathode could provide a relatively stable reversible capacity of about 400 mA h g^{-1} at a current rate of 500 mA g^{-1} for more than 100 cycles (Figure. 10.4).[33]

It needs to be mentioned that the use of a Cu current collector was shown to be crucial for the cell performance in this work.[33] The formation of copper sulfides $Cu(I)S_x$ and $Cu(II)S_x$ were identified in the cycled sulfur cathodes

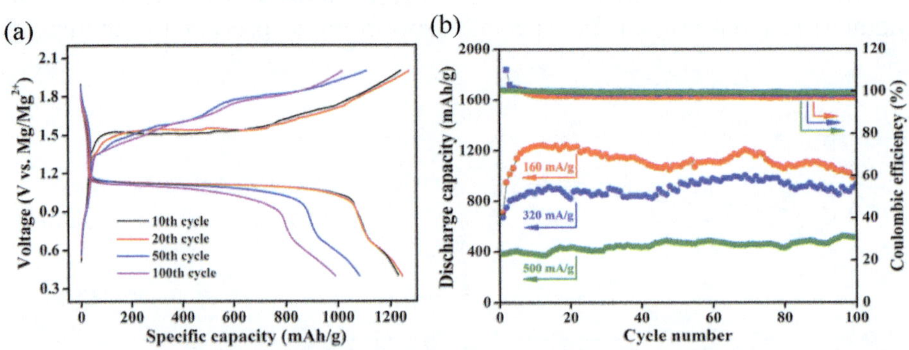

Figure 10.4 (a) Galvanostatic charge–discharge profiles of the Mg–S–CNT cell with the electrolyte $[B(HFP)_3]$–Mg–$MgCl_2$/DME at 0.1C. (b) Discharge capacities and Coulombic efficiencies at different current rates. Reproduced from ref. 33 with permission from the Royal Society of Chemistry.

by means of XPS, but surprisingly re-formation of elemental sulfur could not be confirmed. It seems that this system involves consecutive reactions between sulfur and the Cu current collector, and subsequently the redox reactions in the cell are between copper sulfides and Mg. The Mg–S cells in these studies exhibited a flat discharge voltage plateau at about 1.1 V,[20,33] which is lower than in the Mg–S cells in other reports.[13,15,30,34] Additionally, using the same sulfur cathode with common Al current collectors, the cell performance was dramatically inferior to those with Cu current collectors, where a discharge capacity of about 600 mA h g^{-1} at a low voltage of <0.7 V in the first cycle was measured and there was almost no reversible capacity from the second cycle.[33] A more fundamental study on the sulfur redox chemistry in this type of electrolyte is necessary to evaluate its suitability for Mg–S batteries.

In another approach, a reaction mixture of THFPB with $Mg(BH_4)_2$ was used as an electrolyte for Mg–S batteries.[35] THFPB was supposed to reduce the nucleophilicity of $Mg(BH_4)_2$, making the electrolyte chemically compatible with a sulfur cathode. However, the chemical characterization of the reaction mixture identified $[MgBH_4(Diglyme)_2]^+$ and a free BH_4^- species, which are actually nucleophiles. The compatibility of the THFPB/$Mg(BH_4)_2$ electrolyte with a sulfur cathode needs to be carefully examined. The Mg–S cells with THFPB/$Mg(BH_4)_2$ in diglyme exhibited an initial discharge capacity of 790 mA h g^{-1} at a current of 50 mA g^{-1}, but this decreased to 526 mA h g^{-1} after 30 cycles.[35]

To briefly summarize, initial progress has been made in developing non-nucleophilic electrolytes enabling Mg–S batteries. However, the cell performances with complex electrolytes indicate that a Mg–S battery is still far from to be regarded as a viable system. Besides the problems associated with the unfavorable features of sulfur and polysulfides, the inherent properties of the employed electrolyte should be a focus of attention. Comprehensive studies on complex electrolytes have revealed that the Mg electrochemical activities in the electrolytes generated through Lewis base–acid reactions strongly rely on the chemical equilibria of various complex ions and also the complicated interplay among the components in the bulk electrolyte.[27,36,37] Thus, the dissolved sulfur intermediates species could have severely deleterious effects on their electrochemical properties.[32,36,38] In addition, electrolytes containing Cl^- ions are corrosive to conventional current collectors and other battery components and may also negatively impact upon the battery performance.[39,40] Therefore, a non-corrosive and practical electrolyte with stable electrochemical properties is required for the realization of Mg battery technology.

10.3.2 Mg-ion Conductive Salt-based Electrolytes

With the progress made in achieving non-corrosive Mg electrolytes, the use of simple Mg-ion conductive salts has been explored for Mg–S batteries. $Mg(TFSI)_2$ was the one of the preferred Mg salts investigated for Mg electrolyte because of its molecular similarity to the commonly used Li-ion electrolyte salt LiTFSI and ready availability. However, the comprehensive studies verified the unsuitability of $Mg(TFSI)_2$ ethereal solutions to be used as an electrolyte for Mg batteries due to different factors including the extremely large overpotentials (~2 V *vs.* Mg/Mg^{2+}), low coulombic efficiencies, decomposition of the TFSI anion in contact with Mg and passivation of the Mg surface.[29,41] Consequently, the previous investigations on Mg–S batteries using $Mg(TFSI)_2$/glyme electrolytes revealed a poor performance with a very low discharge voltage (<0.5 V) and a large voltage hysteresis (>1.5 V) and almost no cycling capability.[41] Gao *et al.* studied the thermodynamics of the Mg–S cell with the solution of $Mg(TFSI)_2$ in dimethoxyethane (DME) at a low current rate.[14]

The Cl-free Mg ion-conductive salts magnesium tetrakis(hexafluoroisopropyloxy) borate $Mg[B(hfip)_4]_2$ (hfip = $OC(H)(CF_3)_2$) was synthesized through a dehydrogenation reaction of $Mg(BH_4)_2$ with hexafluoroisopropanol [HOC(H)$(CF_3)_2$] and was isolated as the solvate $Mg[B(hfip)_4]_2$·3DME with high purity.[21,34] The DME solutions of $Mg[B(hfip)_4]_2$ have been demonstrated to be state-of-the-art Mg electrolytes in terms of high oxidative stability (>4.3 V *vs.* Mg/Mg^{2+}), high conductivity(>10 mS cm^{-1}), low polarization (<0.1 V) and outstanding long-term cycling durability (>1000 h).[34] This ionic compound fulfills multiple requirements for a practical, non-corrosive and highly efficient electrolyte and is potentially compatible with any type of cathode material for Mg batteries. $Mg[B(hfip)_4]_2$/DME electrolytes have been employed to study Mg–S batteries. Using a sulfur cathode prepared with activated carbon cloth (ACC) and 0.4 M $Mg[B(hfip)_4]_2$/DME electrolyte, highly reversible sulfur redox chemistry in a Mg-based system has been demonstrated. The discharge/charge profiles indicated a discharge voltage of around 1.5 V for extensive cycling at a current rate of 0.1C. A initial discharge capacity of ~980 mA h g^{-1} was achieved and ~400 mA h g^{-1} was retained after over 30 cycles (Figure 10.6a).[34] The Mg–S cells could be cycled for at least 100 times, however, a capacity fade was observed (Figure 10.5b). The CV measurements with a three-electrode setup displayed well-defined redox pairs, indicating the high reversibility of sulfur. Moreover, the CV studies unveiled that the redox potential hysteresis stems from the half reaction at the Mg anode (Figure 10.5c), which could be caused by the dissolved polysulfides. Electrochemical and spectroscopic analyses verified the electrochemical reactions of the Mg–S system involving the transformation of a series of magnesium–polysulfide intermediates.[34] Additionally, the investigation on electrochemical Mg deposition in $Mg[B(hfip)_4]_2$/DME with additional Mg polysulfide verified the detrimental impacts of the soluble polysulfide on the electrolyte. To prevent the migration of the polysulfides within the Mg–S battery, a functional separator coated with activated carbon nanofiber (CNF) was used. The electrochemical performance of the Mg–S cells was substantially enhanced in terms of reversible capacity and mitigated potential hysteresis between discharge and charge (Figure 10.5d).

In comparison with the results of the Mg–S cells with complex electrolytes, the Mg-ion conductive salt $Mg[B(hfip)_4]_2$ is promising to be used as an

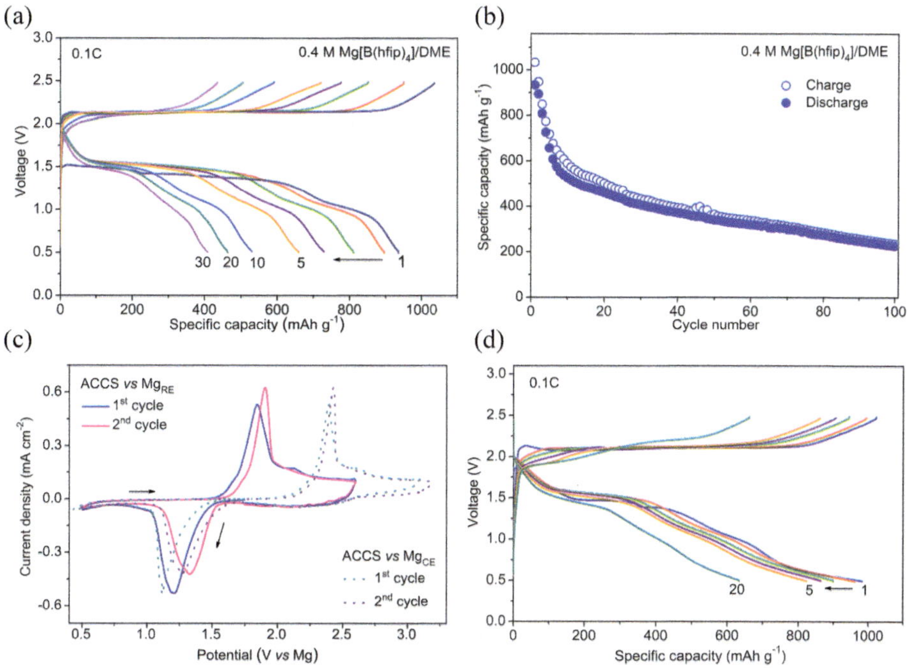

Figure 10.5 (a) Discharge/charge profiles and (b) cycling behavior of the Mg–ACCS cells in $Mg[B(hfip)_4]_2$/DME electrolyte at 0.1C. (d) Cyclic voltammograms of the ACC–S electrode using Mg as both counter and reference electrodes. (d) Discharge/charge profiles of the Mg–ACCS with CNF-coated separator. Reproduced from ref. 34 with permission from American Chemical Society, Copyright 2018.

efficient and practical electrolyte for a Mg–S battery. Yet, the polysulfide dissolution and the potential hysteresis between discharge and charge are the main issues to be resolved for further improvement.

10.4 Sulfur Cathodes and Cell Configuration

Similar to other metal–sulfur battery chemistries, sulfur redox reactions in Mg-based batteries also proceed through the formation of polysulfide intermediates. The design concepts of novel cathode architecture and cell configuration in Li–S batteries can be adopted in Mg–S batteries. In the research to develop suitable electrolytes for Mg–S batteries, advanced cathode materials and battery configurations have been explored in order to prevent the dissolution and migration of the polysulfides in the electrolytes to improve the battery performance.

Vinayan *et al.* reported on the incorporation of sulfur with a reduced graphene oxide (rGo) graphene–sulfur cathode to improve the cyclability and rate capability of Mg–S cells (Figure 10.6a).[42] Two-dimensional (2D)

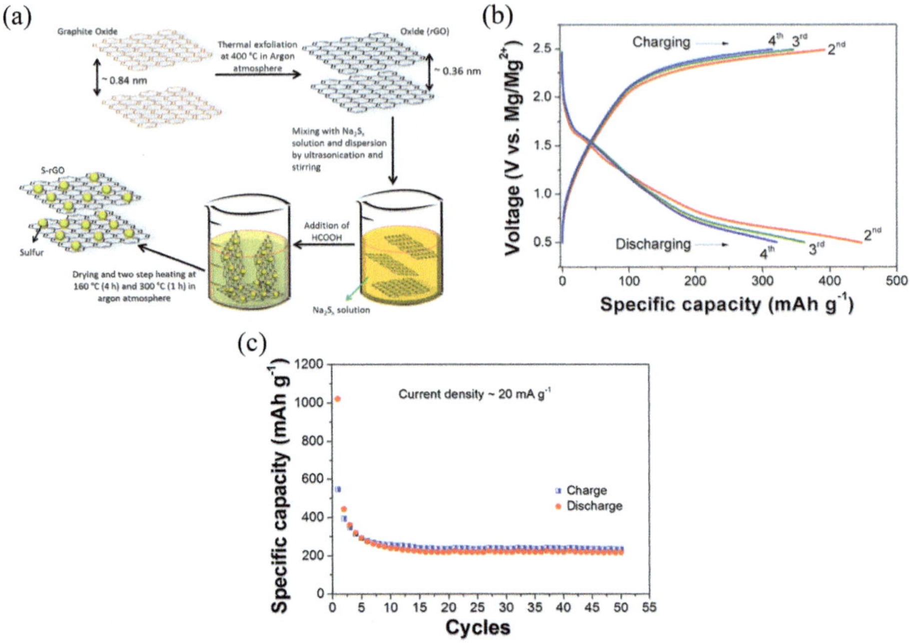

Figure 10.6 (a) Schematic diagram of the synthesis of a rGO–S nanocomposite. (b) Discharge/charge profiles and (c) cycling performance of an RGo–S cathode in $Mg(HMDS)_2$–$AlCl_3$ electrolyte. Reproduced from ref. 42 with permission from the Royal Society of Chemistry.

graphene-based materials possess unique characteristics of high electronical conductivity and large surface area for the dispersion of the active material. The oxygen functional groups over the surface of rGO could efficiently anchor the sulfide species. Additionally, rGO layers could buffer the volume changes of the sulfur cathode during discharge and charge. This approach enabled an initial discharge capacity of 448 mA h g^{-1} and a reversible capacity of 236 mA h g^{-1} after 50 cycles for Mg–S cells with bis-amide electrolytes (Figure 10.6b and c).

Graphdiyne (GDY) is a new 2D carbon allotrope that exhibits highly conjugated structures with linear acetylenic linkages, in which sp- and sp^2-hybridized carbon atoms are topologically ordered in a planar polymeric network.[43] 2D GDY carbon is a potentially promising electrode material owing to its semiconductor, uniform pore structure (5.42 Å) and large interlayer spacing (0.365 nm) properties. Du *et al.* demonstrated the use of GDY–sulfur composite cathodes in the study of both Li– and Mg–S batteries.[44] The composite was fabricated by heating GDY and elemental sulfur at 350 °C, where the short sulfur chains S_x, $(1 \le x \le 4)$, covalently bind with triple bonds (Figure 10.7a). As no highly soluble long-chain polysulfides were formed, the Mg–S batteries with APC electrolyte exhibited relatively stable cycling, achieving a reversible capacity of approximately 459 mA h g^{-1} after 36 cycles

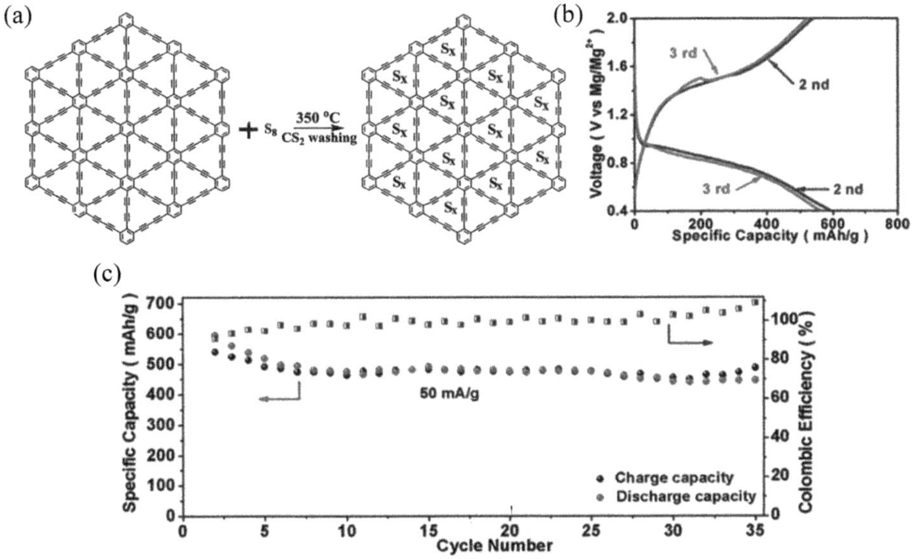

Figure 10.7 (a) Schematic illustration of the preparation of a SGDY nanocomposite. (b) Discharge/charge profiles and (c) cycling performance of the SGDY cathode in APC–LiCl electrolyte. Reproduced from ref. 44 with permission from John Wiley and Sons, © 2017 WILEY-VCH Verlag GmbH & Co. KGaA, Weinheim.

at a current of 50 mA g^{-1}. Compared to the superior electrochemical performance of Li-S cells with the same cathode, the Mg–S battery performance may be further enhanced by the use of a more suitable electrolyte.

Zhou *et al.* demonstrated a sulfur composite with a metal–organic framework (MOF)-derived carbon for use in Mg–S batteries.[45] A zeolitic imidazolate framework (ZIF-67) was used as a precursor to produce the host scaffolds with functional nitrogen (N) and cobalt (Co) moieties that can chemically bind polysulfides. Using a bis-amide-based electrolyte with LiTFSI as an additive, the MOF-based sulfur cathode exhibited a reversible capacity of about 400 mA h g^{-1} for more than 200 cycles at 1C and a capacity of about 300 mA h g^{-1} at a high rate of up to 5C (Figure 10.8). Compared to the previous study of Mg–S batteries with the same type of electrolytes, these remarkable results imply that MOF-derived carbon with an open framework structure and uniformly integrated functional groups could effectively control the dissolution of polysulfides and at the same time allow the Mg ions easy access to the active sulfur moieties. Hence, the rational design of functional host materials represents a viable solution to synergistically address the multiple issues associated with sulfur chemistry and Mg-ion transportation.

Modification of cell configuration can also be a feasible way to improve sulfur-based batteries. Manthiram's group initially established the concept of using a carbon interlayer between the separator and cathode in Li–S batteries and significantly enhanced the cell performance. The similar cell

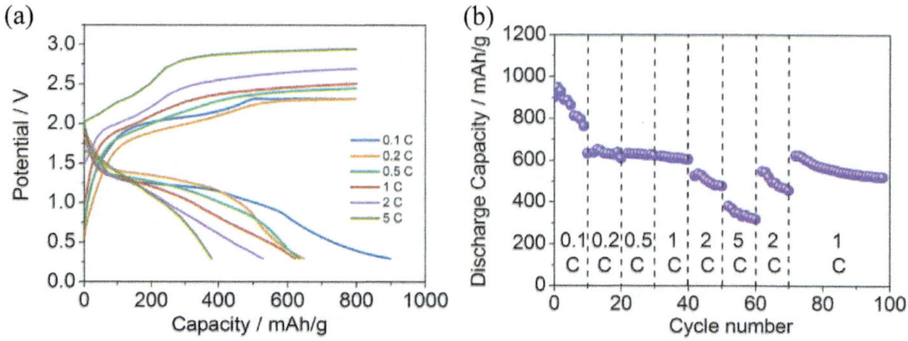

Figure 10.8 (a) Discharge/charge profiles and (b) rate performance of the ZIF–C–S cathode in $(HMDS)_2Mg$–$AlCl_3$–LiTFSI electrolyte. Reproduced from ref. 45 with permission from John Wiley and Sons, © 2018 WILEY-VCH Verlag GmbH & Co. KGaA, Weinheim.

configuration was applied to a Mg–S system, where the cathode was constructed by spreading sulfur on the activated carbon nanofiber (CNF) substrate and covered with a CNF-coated separator serving as both a polysulfide trapper and current collector for recovering the dissolved sulfur species (Figure 10.9a).[46] A high initial discharge capacity of ~1200 mA h g^{-1} and stable capacity retention after 20 cycles were demonstrated with this type of cell in combination with a bis-amide electrolyte (Figure 10.9b).

It is worth mentioning that these attempts to improve the cathode and cell configuration still cannot be properly evaluated because of the drawbacks of the used electrolytes, as discussed in Section 3.3. Further improvement and breakthroughs in Mg–S batteries could be realized by combination of the advanced cathode and cell design with a more efficient electrolyte.

10.5 Summary and Outlook

The electrochemical performances of Mg–S batteries have been investigated with different types of electrolytes. Extensive mechanistic studies have shown that the main challenges facing the development of Mg–S systems include large voltage hysteresis, capacity fade upon cycling, insufficient rate performance and coulombic efficiency. In comparison with other advanced metal–sulfur battery technologies, the dissolution of polysulfides could be the main cause of these critical issues. However, the particular features of complex Mg electrolytes should also seriously be taken into consideration for establishing viable approaches to overcome the challenges.

The recently achieved non-corrosive electrolytes based on simple Mg-ion conductive salts can alleviate the problematic side effects caused by the Cl-ion containing constituents. The preliminary results using $Mg[B(hfip)_4]_2$ electrolytes have indicated the significantly improved reversibility of sulfur

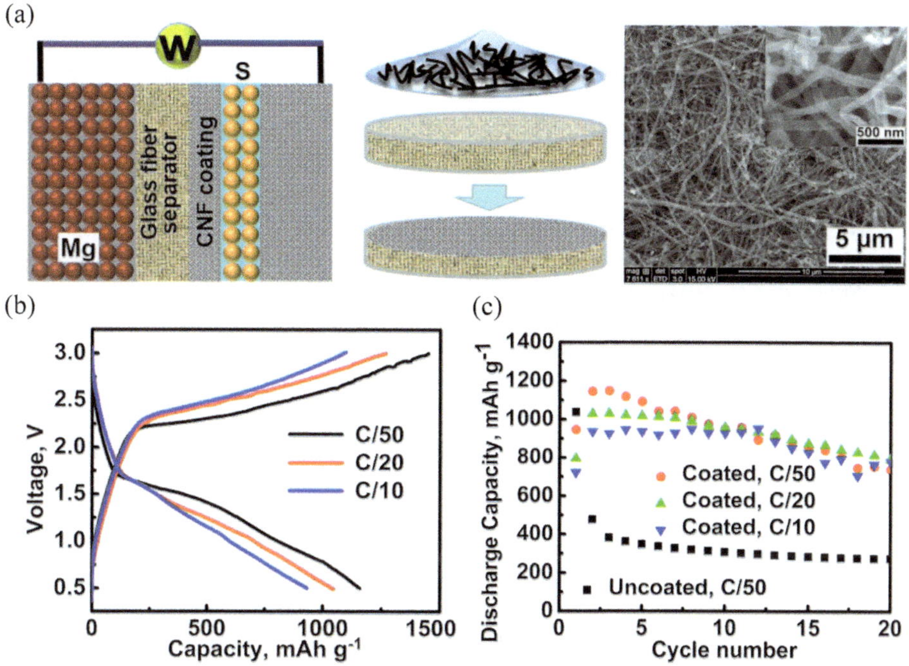

Figure 10.9 (a) Schematic of a Mg–S cell with activated-CNF coated separator(left), preparation of the CNF coated separator (middle), scanning electron microscopy image of the CNF(right). (b) Discharge/charge profiles and (c) discharge capacities of the Mg–S cells in $(HMDS)_2Mg-AlCl_3$ electrolytes at various C rates. Reproduced from ref. 46 with permission from American Chemical Society, Copyright 2016.

redox reactions. Moreover, the detrimental effects of the dissolved polysulfides on the Mg deposition behavior of the electrolyte have been identified, which could be attribute to the voltage hysteresis. Hence, effective strategies are needed to prevent the dissolution of the soluble sulfide species. In this respect, innovative concepts for cathode design and separator functionalization are believed to be feasible solutions to block the passage of polysulfide moieties to the anode side. An alternative option is to form active solid electrolyte interfaces (SEIs) on the Mg anode that allow the transport of Mg ions. To achieve this goal, more profound understanding of interfacial chemistries is essential in future research.

Acknowledgements

The support of the European H2020 FET Proactive project "E-Magic" (#824066) is acknowledged. This work is part of the newly founded platform CELEST Center for Electrochemical Storage Ulm-Karlsruhe.

References

1. A. Manthiram, Y. Fu, S. Chung, C. Zu and Y. Su, *Chem. Rev.*, 2014, **114**, 11751.
2. D. Lin, Y. Liu and Y. Cui, *Nat. Nanotechnol.*, 2017, **12**, 194.
3. D. Aurbach, Y. Cohen and M. Moshkovich, *Electrochem. Solid-State Lett.*, 2001, **4**, A113.
4. M. Jäckle and A. Groß, *J. Chem. Phys.*, 2014, 174710.
5. M. Jaeckle, K. Helmbrecht, M. Smits, D. Stottmeister and A. Gross, *Energy Environ. Sci.*, 2018, **11**, 3400.
6. G. Cohn, L. Ma and L. A. Archer, *J. Power Sources*, 2015, **283**, 416.
7. T. Gao, X. Li, X. Wang, J. Hu, F. Han, X. Fan, L. Suo, A. J. Pearse, S. B. Lee, G. W. Rubloff, K. J. Gaskell, M. Noked and C. Wang, *Angew. Chem., Int. Ed.*, 2016, **55**, 9898.
8. H. Yang, L. Yin, J. Liang, Z. Sun, Y. Wang, H. Li, K. He, L. Ma, Z. Peng, S. Qiu, C. Sun, H. M. Cheng and F. Li, *Angew. Chem., Int. Ed.*, 2018, **57**, 1898.
9. Z. Zhao-Karger and M. Fichtner, *MRS Commun.*, 2017, **7**, 770.
10. Z. Zhao-Karger and M. Fichtner, *Front. Chem.*, 2019, **6**, 656.
11. Z. Zhang, S. Dong, Z. Cui, A. Du, G. Li and G. Cui, *Small Methods*, 2018, **2**, 1800020.
12. L. Kong, C. Yan, J. Huang, M. Zhao and M. Titirici, *Energy Environ. Mater.*, 2018, **1**, 100.
13. Z. Zhao-Karger, X. Zhao, D. Wang, T. Diemant, R. J. Behm and M. Fichtner, *Adv. Energy Mater.*, 2015, **5**, 1.
14. T. Gao, X. Ji, S. Hou, X. Fan, X. Li, C. Yang, F. Han, F. Wang, J. Jiang, K. Xu and C. Wang, *Adv. Mater.*, 2018, **30**, 2.
15. A. Robba, A. Vizintin, J. Bitenc, G. Mali, I. Arčon, M. Kavčič, M. Žitnik, K. Bučar, G. Aquilanti, C. Martineau-Corcos, A. Randon-Vitanova and R. Dominko, *Chem. Mater.*, 2017, **29**, 9555.
16. Z. Zhao-Karger, R. Liu, W. Dai, Z. Li, T. Diemant, B. P. Vinayan, C. Bonatto Minella, X. Yu, A. Manthiram, R. J. Behm, M. Ruben and M. Fichtner, *ACS Energy Lett.*, 2018, 2005.
17. Z. Zhao-Karger, X. Zhao, O. Fuhr and M. Fichtner, *RSC Adv.*, 2013, **3**, 16330.
18. F. Wang, Y. Guo, J. Yang and S. Hirano, *Chem. Commun.*, 2012, **48**, 10763.
19. A. Robert, E. Doe, M. A. Us, G. H. Lane, S. Helens, R. E. Jilek, I. Robert, E. Doe, M. A. Us, G. H. Lane and S. Helens, *US Pat. Appl. Publ.*, 2013/0252112 A1, 2013.
20. Z. Zhang, Z. Cui, L. Qiao, J. Guan, H. Xu, X. Wang, P. Hu, H. Du, S. Li, X. Zhou, S. Dong, Z. Liu, G. Cui and L. Chen, *Adv. Energy Mater.*, 2017, **7**, 1602055.
21. Z. Zhao-Karger, M. E. Gil Bardaji, O. Fuhr and M. Fichtner, *J. Mater. Chem. A*, 2017, **5**, 10815.
22. H. S. Kim, T. S. Arthur, G. D. Allred, J. Zajicek, J. G. Newman, A. E. Rodnyansky, A. G. Oliver, W. C. Boggess and J. Muldoon, *Nat. Commun.*, 2011, **2**, 426.

23. Y. Xu, G. Zhou, S. Zhao, W. Li, F. Shi, J. Li and J. Feng, *Adv. Sci.*, 2019, **6**, 1800981.
24. T. Gao, M. Noked, A. J. Pearse, E. Gillette, X. Fan, Y. Zhu, C. Luo, L. Suo, M. A. Schroeder, K. Xu, S. B. Lee, G. W. Rubloff and C. Wang, *J. Am. Chem. Soc.*, 2015, **137**, 12388.
25. Z. Zhao-Karger, X. M. Lin, C. Bonatto Minella, D. Wang, T. Diemant, R. J. Behm and M. Fichtner, *J. Power Sources*, 2016, **323**, 213.
26. C. J. Barile, R. G. Nuzzo and A. A. Gewirth, *J. Phys. Chem. C*, 2015, **119**, 13524.
27. K. A. See, K. W. Chapman, L. Zhu, K. M. Wiaderek, O. J. Borkiewicz, C. J. Barile, P. J. Chupas and A. A. Gewirth, *J. Am. Chem. Soc.*, 2016, **138**, 328.
28. W. Li, S. Cheng, J. Wang, Y. Qiu, Z. Zheng, H. Lin, S. Nanda, Q. Ma, Y. Xu, F. Ye, M. Liu, L. Zhou and Y. Zhang, *Angew. Chem., Int. Ed.*, 2016, **55**, 6406.
29. I. Shterenberg, M. Salama, H. D. Yoo, Y. Gofer, J.-B. Park, Y.-K. Sun and D. Aurbach, *J. Electrochem. Soc.*, 2015, **162**, A7118.
30. T. Gao, S. Hou, F. Wang, Z. Ma, X. Li, K. Xu and C. Wang, *Angew. Chem., Int. Ed.*, 2017, **129**, 13711.
31. H. D. Yoo, S. D. Han, I. L. Bolotin, G. M. Nolis, R. D. Bayliss, A. K. Burrell, J. T. Vaughey and J. Cabana, *Langmuir*, 2017, **33**, 9398.
32. M. Salama, R. Attias, B. Hirsch, R. Yemini, Y. Gofer, M. Noked and D. Aurbach, *ACS Appl. Mater. Interfaces*, 2018, **10**, 36910.
33. A. Du, Z. Zhang, H. Qu, Z. Cui, L. Qiao, L. Wang, J. Chai, T. Lu, S. Dong, T. Dong, H. Xu, X. Zhou and G. Cui, *Energy Environ. Sci.*, 2017, **10**, 2616.
34. Z. Zhao-Karger, R. Liu, W. Dai, Z. Li, T. Diemant, B. P. Vinayan, C. Bonatto Minella, X. Yu, A. Manthiram, R. J. Behm, M. Ruben and M. Fichtner, *ACS Energy Lett.*, 2018, **3**, 2005.
35. H. Xu, Z. Zhang, J. Li, L. Qiao, C. Lu, K. Tang, S. Dong, J. Ma, Y. Liu, X. Zhou and G. Cui, *ACS Appl. Mater. Interfaces*, 2018, **10**, 23757.
36. Z. Zhao-Karger, J. E. Mueller, X. Zhao, O. Fuhr, T. Jacob and M. Fichtner, *RSC Adv.*, 2014, **4**, 26924.
37. Y. Cheng, R. M. Stolley, K. S. Han, Y. Shao, B. W. Arey, N. M. Washton, K. T. Mueller, M. L. Helm, V. L. Sprenkle, J. Liu and G. Li, *Phys. Chem. Chem. Phys.*, 2015, **17**, 13307.
38. N. Pour, Y. Gofer, D. T. Major and D. Aurbach, *J. Am. Chem. Soc.*, 2011, **133**, 6270.
39. C. Qiang, N. Yan-Na, Y. Jun, K. Kailibinuer and W. Jiu-Lin, *Acta Phys.-Chim. Sin.*, 2012, **28**, 2625.
40. C. Wall, Z. Zhao-Karger and M. Fichtner, *ECS Electrochem. Lett.*, 2014, **4**, C8.
41. S. Y. Ha, Y. W. Lee, S. W. Woo, B. Koo, J. S. Kim, J. Cho, K. T. Lee and N. S. Choi, *ACS Appl. Mater. Interfaces*, 2014, **6**, 4063.
42. B. P. Vinayan, Z. Zhao-Karger, T. Diemant, V. S. K. Chakravadhanula, N. I. Schwarzburger, M. A. Cambaz, R. J. Behm, C. Kübel and M. Fichtner, *Nanoscale*, 2016, **8**, 3296.

43. N. Wang, Z. Zuo, H. Liu, Y. Xue, Y. Li, C. Huang and Y. Li, *Chem. Rev.*, 2018, **118**, 7744.
44. Z. Yang, G. Cui, Z. Zhang, J. Ma, J. He, J. Chai, H. Du, C. Huang and Z. Cui, *Small*, 2017, **13**, 1702277.
45. X. Zhou, J. Tian, J. Hu and C. Li, *Adv. Mater.*, 2018, **30**, 1.
46. X. Yu and A. Manthiram, *ACS Energy Lett.*, 2016, **1**, 431.

CHAPTER 11

Mg–Li Dual-cation Batteries

HONGYI LI*[a], TETSU ICHITSUBO*[a] AND EIICHIRO MATSUBARA*[b]

[a]Tohoku University, Institute for Materials Research, 2-1-1 Katahira, Sendai 980-8577, Japan; [b]Kyoto University, Department of Materials Science and Engineering, Yoshidahonmachi, Sakyo-ku, Kyoto 606-8501, Japan
*E-mail: li.hongyi@imr.tohoku.ac.jp, tichi@imr.tohoku.ac.jp, matsubara.eiichiro.6z@kyoto-u.ac.jp

11.1 Introduction

Magnesium battery technologies have attracted much attention as promising candidates for next-generation energy storage systems. Among the several superior characteristics of Mg, its low redox potential is a superior property for use as an anode, and in particular, dangerous dendritic formation can be avoided on the electrodeposition of Mg in contrast to Li, which ensures the safety of the batteries. Besides this, since Mg ions are divalent, Mg metal has a higher volumetric capacity compared to that of Li metal. In addition, Mg as an element is relatively abundant, which is helpful for reducing production costs. Owing to these advantages, Mg ions are appropriate for use as carrier ions in future battery systems.

However, toward the practical use of magnesium batteries, several serious problems are raised. One of the major problems is the passivation of Mg metal in commonly used electrolytes. The passivated surface usually causes a large overpotential for anodic deposition, and as a result, reversible

Energy and Environment Series No. 23
Magnesium Batteries: Research and Applications
Edited by Maximilian Fichtner

Published by the Royal Society of Chemistry, www.rsc.org

Mg deposition/dissolution may be hard obtain. An early study reported an effective approach to enhance the reversibility of Mg deposition, in which a Lewis base and Lewis acid are employed with ether solvents.[1] Based on this approach, a series of Mg electrolytes, such as PhMgCl–$AlCl_3$ (APC),[2] $MgHMDS_2$–$AlCl_3$,[3] $MgCl_2$–$AlCl_3$ (MACC),[4] and $MgTFSA_2$–$MgCl_2$,[5] was developed. In these electrolytes, it was revealed that Mg–Cl complexes, such as $MgCl^+$ and $Mg_2Cl_3^+$, are formed in the reactions between the Lewis bases and Lewis acids, which leads to an improvement in the electrolyte performance.[2–7] Although the Mg–Cl complexes can effectively suppress the passivation, the presence of highly reactive Cl anions also causes considerable corrosiveness and low anodic stability of the electrolyte.[8,9] To avoid these problems, Cl-free electrolytes, such as $Mg(BH_4)_2$,[10] $MgTFSA_2$,[11] $Mg[B(hfip)_4]_2$,[12] were developed as alternative choices. Although many efforts are still required to realize high-performance Mg electrolytes, there is no doubt that Mg passivation would be gradually overcome and no longer an essential issue hindering the development of Mg batteries.

Another crucial problem is the well-known sluggish diffusion of Mg ions inside cathode hosts. For intercalation cathode materials, carrier ions must be inserted and extracted reversibly during battery reactions. Compared with Li ions, divalent Mg ions have much stronger coulombic interactions with the cathode hosts, because the Mg ion has a similar ionic radius to that of the monovalent Li ion. Such a strong interaction usually makes Mg migration in the solid phase considerably difficult, it retards the intercalation reactions, and, consequently, causes remarkable capacity loss and poor cyclability of the cathode materials. For example, the Chevrel compound Mo_6S_8, reported by Aurbach *et al.*,[13] has a theoretical capacity of ~122 mA h g^{-1}, whereas, at room temperature and at a moderate current, a significant amount (20–30%) of the capacity is irreversible.[14] The previous studies indicate that the inevitable capacity loss is caused by a kinetic issue in which inserted Mg ions are partially "trapped" in the cathode host, which can be improved remarkably by elevating the operation temperature up to around 60 °C.[15]

As an approach to overcome the problems associated with Mg batteries, we proposed a new battery system, which has been referred to as Mg–Li dual-ion batteries (also originally called Mg–Li dual-salt batteries).[16–18] In this battery system, both Mg and Li ions are employed as carrier ions. As shown in Figure 11.1, two typical combinations can be considered. One is a Daniell-type combination, where the Li ions work as carriers for the cathode, while the Mg ions participate in an anodic reaction. The other is a (quasi-)rocking-chair combination, where both the Mg and Li ions simultaneously participate in cathode and anode reactions.

In this chapter, we introduce fundamental chemistry and recent progress on the two types of Mg–Li dual-ion batteries. Then, we focus on the intercalation behavior of Mg–Li dual ions with cathode hosts in the rocking-chair combination and reveal a facilitating mechanism of Mg solid-phase diffusion based on a concerted interaction between the Mg and Li ions. The aim

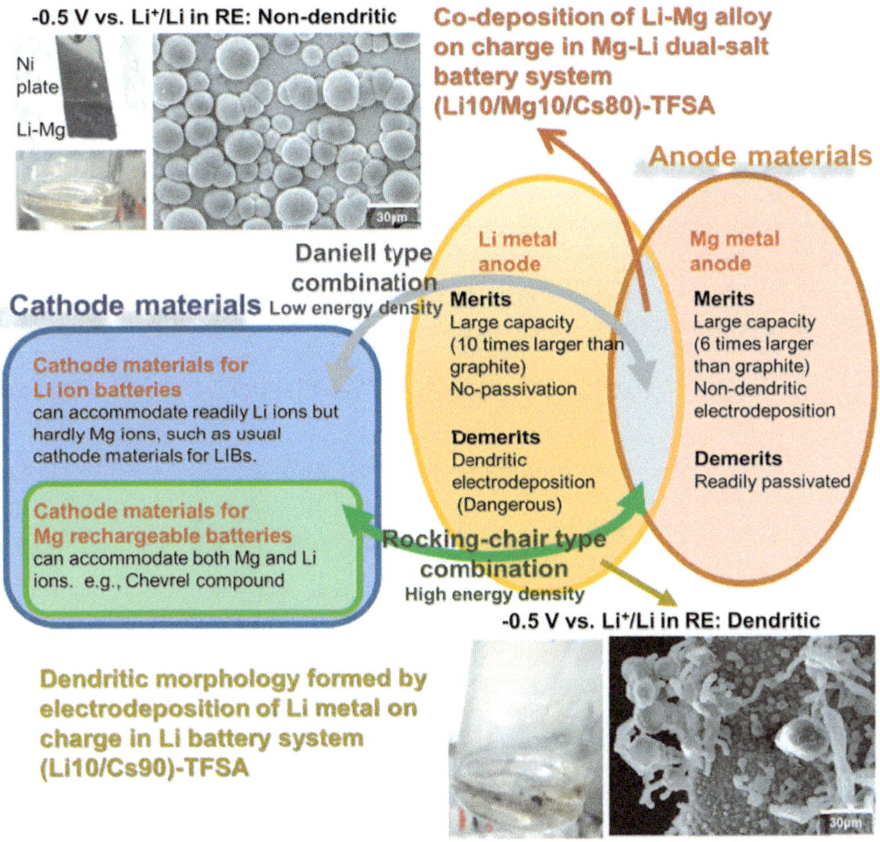

Figure 11.1 Concept of the Mg–Li dual-ion battery system based on the anode characteristics. The respective merits and demerits of Li and Mg batteries are illustrated with scanning electron and optical microscopy images of the electrodeposited Li metal in (Li10Cs90)–TFSA (lower) and Mg–Li alloy (Li10/Mg10/Cs80)–TFSA (upper) at 150 °C. The notation, such as (Li10Cs90)–TFSA, indicates an ionic liquid of 10 at.% LiTFSA and 90 at.% CsTFSA. Reproduced from ref. 17 with permission from the Royal Society of Chemistry.

of this work is to provide a comprehensive understanding of Mg–Li dual-ion battery technologies, which presents a new perspective on building high-performance batteries with multivalent ions.

11.2 Mg–Li Dual-ion Batteries: Daniell-type

Daniell-type Mg–Li dual-ion batteries consist of a Mg metal anode and a Li-intercalation cathode. By combining the advantages of Mg- and Li-ion batteries, the intrinsic problems of the single-ion systems, (*i.e.* dendritic growth of Li on the anode and sluggish diffusion of Mg ions in the cathode host) are successfully circumvented.

11.2.1 Battery Reactions

Figures 11.2a and b show structures of classical Cu–Zn Daniell batteries and Daniell-type Mg–Li dual-ion batteries, respectively. In the classical Daniell batteries, the Cu cathode and Zn anode are divided by a septum in the cell to prevent the displacement deposition of a more noble metal (Cu) on the less-noble electrode (Zn). In contrast, there is no need to prepare a septum for Daniell-type Mg–Li dual-ion batteries, because each carrier is preferred in each electrode reaction. Here, as an example, we deal with $FePO_4$ as a cathode material for the dual-ion battery. The expected battery reactions with the $FePO_4$ cathode are shown in eqn (11.1)–(11.3).

$$\text{Cathode: } FePO_4 + Li^+ + e^- \leftrightarrow LiFePO_4 \tag{11.1}$$

$$\text{Anode: } Mg \leftrightarrow Mg^{2+} + 2e^- \tag{11.2}$$

$$\text{Total: } FePO_4 + Li^+ + \frac{1}{2}Mg \leftrightarrow LiFePO_4 + \frac{1}{2}Mg^{2+} \tag{11.3}$$

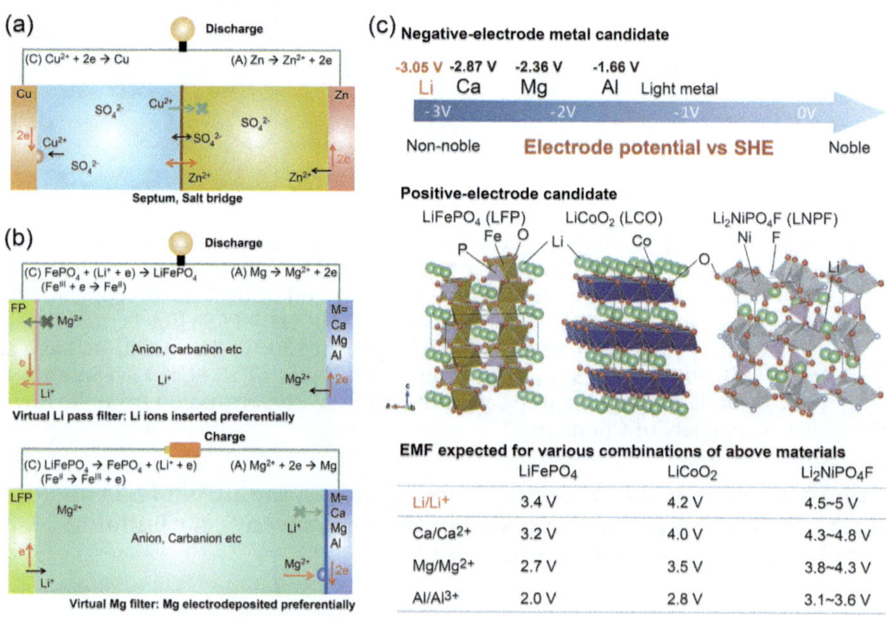

Figure 11.2 Examples of Daniell-type Mg–Li dual-ion batteries. Schematic figures illustrate (a) the classical Daniell battery and (b) Daniell-type Mg–Li dual-ion batteries. Other metal anodes, such as Ca and Al, are also shown as alternative choices. (c) Examples of anode and cathode materials for Daniell-type dual-ion batteries, and expected emf values for various combinations are given. Reproduced from ref. 16 with permission from the Royal Society of Chemistry.

In the cathode reaction, Mg intercalation is hard to achieve because of the high diffusion barrier and instability of the Mg-inserted $FePO_4$ host.[19] Thus, the cathode host plays the role of a "Li pass filter" during discharge. In the anode reaction, more noble Mg is preferentially deposited during charge at a moderate constant current. Therefore, if enough carriers exist in the electrolyte, each carrier would mainly participate in the half-cell reaction and a septum is not required for this battery system.

As shown in Figure 11.2, the concept of the Daniell-type dual-ion batteries can also be extended to Ca–Li, Al–Li systems, *etc.* Since the concept was first proposed in 2013,[16] many subsequent studies have been reported, which include not only Mg–Li,[20,21] but also Mg–Na,[22,23] Al–Li,[24] Al–Na,[25] systems. Besides dual metal cations, dual-ion batteries of cation–anion pairs, such as Mg–I[26] and Al–$AlCl_4$,[27] have also developed in recent years, which have similar structures to the Daniell-type battery.

11.2.2 Example of a Practical System

To examine the feasibility of the battery reactions, we investigated the electrode properties of the $FePO_4$ cathode and Mg metal anode experimentally. For the electrolyte, PhMgCl–$AlCl_3$/THF (APC),[2] was employed due to its excellent reversibility of Mg deposition. As the Li-ion salt, $LiBF_4$ salt was added into the APC electrolyte. Figures 11.3a and b show the cyclic

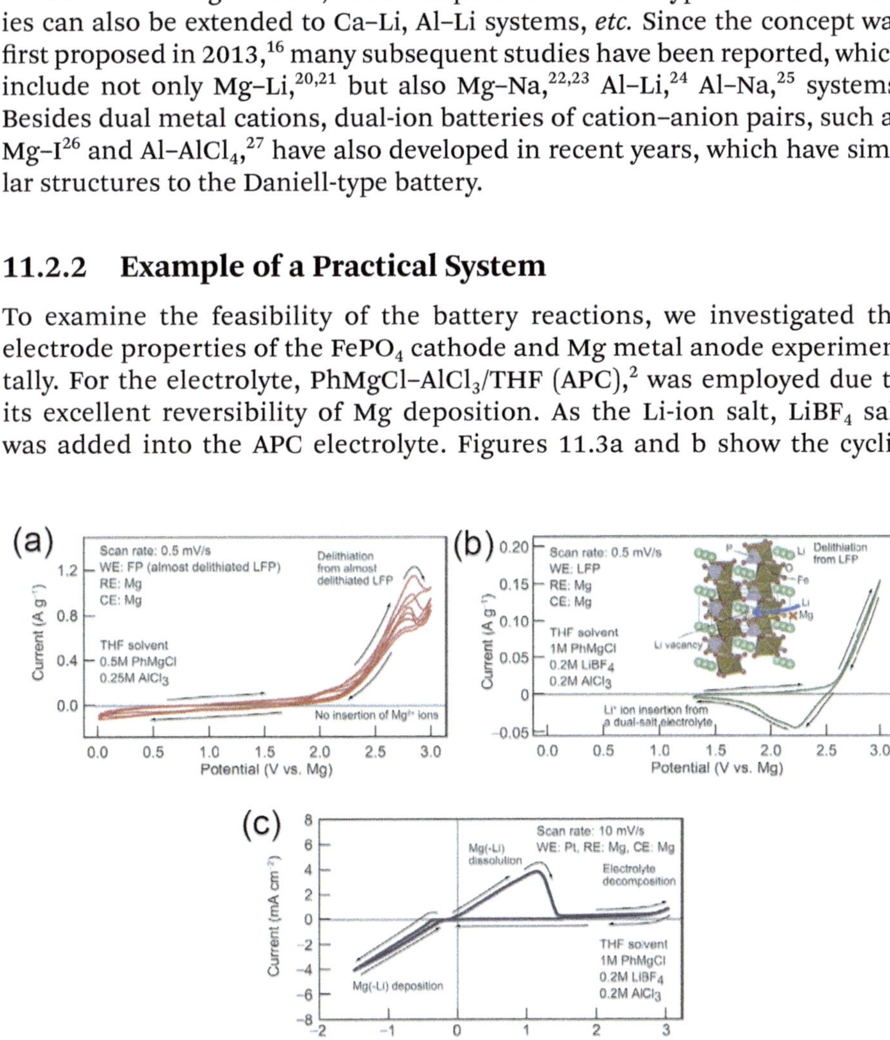

Figure 11.3 Cyclic voltammetry (CV) profiles obtained in (a) PhMgCl–$AlCl_3$/THF, (b) and (c) PhMgCl–$LiBF_4$–$AlCl_3$/THF, using $FePO_4$ (almost delithiated $LiFePO_4$), $LiFePO_4$, and Pt as the working electrode, respectively. Reproduced from ref. 16 with permission from the Royal Society of Chemistry.

voltammograms (CVs) of the $FePO_4$ (delithiated $LiFePO_4$) and $LiFePO_4$ cathode in Mg single-ion electrolyte and Mg–Li dual-ion electrolyte, respectively. In the Mg single-ion electrolyte, a distinct cathodic peak (*i.e.* meaning insertion of Mg ions) cannot be observed. This is consistent with our conjecture that it is difficult to intercalate Mg ions into the $FePO_4$ host. The rise in the anodic current on the high potential side would be caused by decomposition of the electrolyte. In contrast, a cathodic peak is clearly observed in the electrolyte containing both Mg and Li ions. Thus, Li ions are preferentially intercalated into the $FePO_4$ host, which confirms that the $FePO_4$ plays the role of a Li pass filter, as expected above. In the present case, the instability of the APC electrolyte (anodic limit of the electrochemical window) restricts the operating voltage of the cathode. This problem can be improved by employing advanced electrolytes, which has been confirmed in the several following studies.[28]

Regarding the reaction of the Mg anode, as shown in Figure 11.3c (a Pt electrode is used as the working electrode), reversible deposition and dissolution are observed at around 0 V *vs.* Mg. To confirm the morphology on the metal anode during charge, deposition experiments were carried out at constant potentials of −0.5 and −1.5 V *vs.* Mg (electrical charge: 3 C cm^{-2}). As shown in Figure 11.4, the deposits have a smooth and flat morphology, even at the latter potential (−1.5 V *vs.* Mg), being far below the electrodeposition

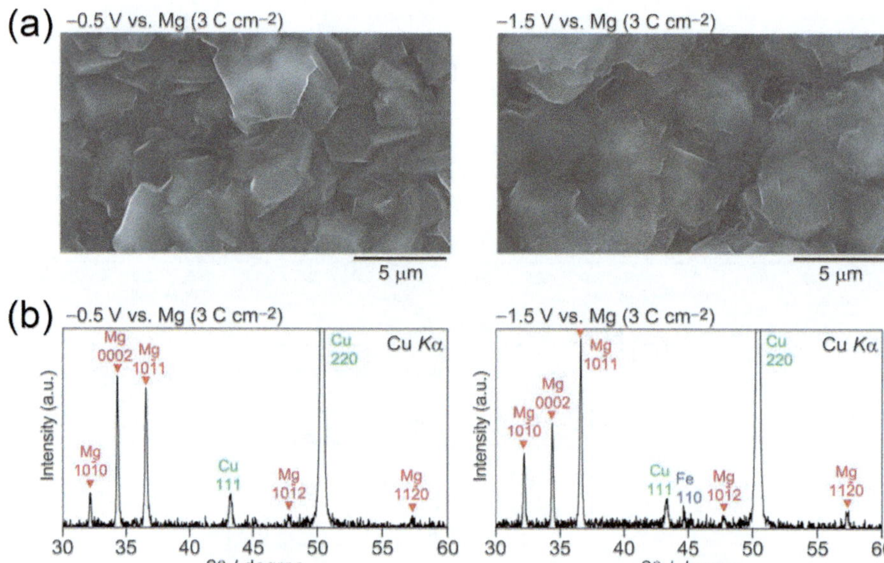

Figure 11.4 (a) SEM images and (b) XRD profiles of the deposit obtained in PhMgCl–$LiBF_4$–$AlCl_3$/THF at −0.5 and −1.5 V *vs.* Mg (electrical charge: 3 C cm^{-2}). The peaks of Cu and Fe correspond to Cu foil (current collector) and the stage of the XRD apparatus. Reproduced from ref. 16 with permission from the Royal Society of Chemistry.

potential of Li. Since the deposits show no dendritic morphology even when Mg and Li ions are co-electrodeposited, the metal (alloy) anode can therefore be safely used in storage batteries. This is one of the most important findings in the study of Mg–Li dual ion battery systems.

11.2.3 Toward High Energy Density Dual-ion Batteries

As demonstrated experimentally, the Daniell-type combination of Mg–Li dual ions can work well. However, there are some critical issues that must be overcome to achieve high energy-density batteries. For Daniell-type batteries, either type of carrier must be accommodated in the electrolyte (all Mg ions after discharge and all Li ions after charge). To accommodate the carriers, an adequate amount of electrolyte is required in the batteries, which consequently markedly diminishes the energy density. This issue would be improved if a saturated electrolyte was used, where the precipitated salts in the saturated electrolyte can work as a reservoir and source of the carriers. Nevertheless, Daniell-type dual-ion batteries hardly provide higher energy densities compared to those of commonly used Li-ion batteries.

11.3 Mg–Li Dual-ion Batteries: Rocking-chair Type

To reduce the amount of electrolyte required for Daniell-type dual-ion batteries, a rocking-chair type structure is necessary, in which the carriers are accommodated in either of the electrodes during battery reactions. Since the electrolyte does not need to accommodate the carrier ions and works as a mediator for the transfer of carrier ions between the electrodes, rocking-chair type batteries only require a minimum amount of electrolyte, which allows high energy densities to be realized.

11.3.1 Ideal Charge and Discharge Processes

In rocking-chair-type Mg–Li dual-ion batteries, both Mg and Li ions must simultaneously participate in the electrode reactions. Figure 11.5a shows the ideal charge–discharge processes. Note that the electrolyte amount is assumed to be minimal and almost no extra carriers are contained in the electrolyte. In the ideal charge process, Mg and Li ions are deintercalated from the cathode and at the same time co-electrodeposited on the anode to form a Mg–Li alloy, while in the ideal discharge process, the Mg–Li alloy would be dissolved from the anode (presumably, Li would be dissolved preferentially from the Mg–Li alloy, because Li is less noble than Mg) and the dissolved Li and Mg ions are (co-)intercalated to the cathode. The presumed composition change of electrodes during battery reactions is shown in Figure 11.5b.

Since these types of battery reactions have not been studied so far, the feasibility was explored through the preparation of prototype batteries with

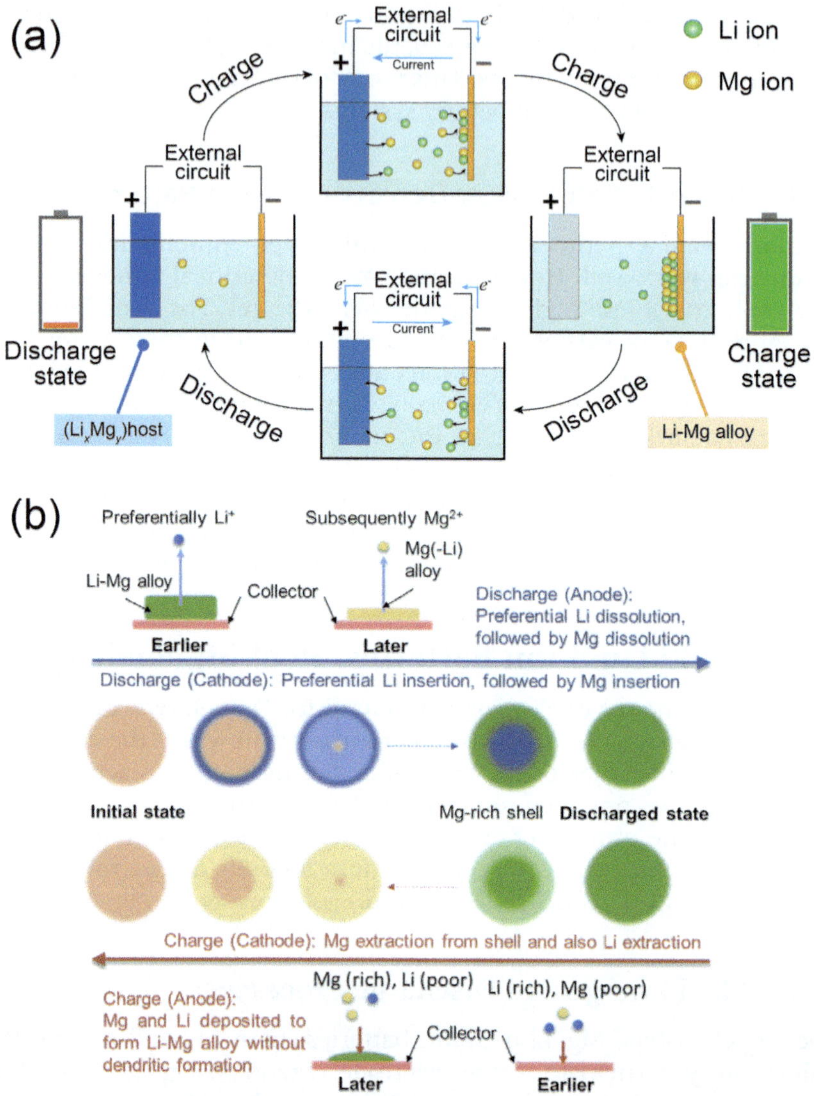

Figure 11.5 (a) Schematic figure of ideal charge–discharge processes of rocking-chair-type Mg–Li dual-ion batteries. Reproduced from ref. 18 with permission from The Royal Society of Chemistry. (b) Presumable composition change of the electrodes during battery reactions. Reproduced from ref. 17 with permission from the Royal Society of Chemistry.

typical cathode materials and electrolytes, then focusing on two key points in the charge–discharge processes. One is the co-electrodeposition morphology of the Mg–Li alloy during the charge process, and the other is to judge whether Mg and Li ions can be reversibly and concomitantly intercalated in the cathode materials during the battery reactions.

11.3.2 Prototype Battery System

In the prototype batteries, the Chevrel compound Mo_6S_8 was employed as the cathode material. Mo_6S_8 is a unique material that allows reversible intercalation of Mg ions with high cyclability at room temperature, among all the candidate cathode materials for Mg batteries that have been reported to date.[29] The host structure and electrode properties of Mo_6S_8 have been well-investigated in Mg or Li single-ion systems, and therefore Mo_6S_8 is a suitable example to use to investigate the intercalation behavior in the new Mg–Li dual-ion systems. Regarding the electrolytes, two typical electrolytes were employed, which can be classified by as to whether or not they contained Cl anions. As mentioned in the introduction, Cl anions can significantly affect the solvation structures of Mg ions in electrolytes. Since the solvation environments would be an important and interesting factor for Mg–Li dual-ion systems, both Cl-containing and Cl-free electrolytes were employed in the prototype. The Cl-containing electrolyte was $MgHMDS_2$–LiHMDS–$AlCl_3$/G3, in which a $Mg_2Cl_3^+$ complex is generated.[3] The Cl-free electrolyte was $MgTFSA_2$–LiTFSA/G3. In this electrolyte, it is reported that Mg ions are mainly solvated with G3 solvent molecules, due to the weak bonding between small Mg cations and very large TFSA anions.[30] When investigating the electrode properties, the concentration of each salt was unified to 0.5 M (mol L^{-1}) to make it easy to compare the results in different electrolytes.

11.3.3 Anode Properties of a Mg–Li Alloy

During the charge–discharge processes, the deposition and dissolution of a Mg–Li alloy occurs on the anode. Since Mg and Li have different redox potentials, the electrode potential of a Mg–Li alloy anode varies depending on the ratio of Mg and Li ions participating in the reactions. Moreover, because Li is also deposited during charging, morphology of the Mg–Li alloy is an important issue that needs to be carefully examined.

11.3.3.1 Thermodynamic Analysis for the Alloy Anode

As reported in a previous study,[20] the electrode potential of a Mg–Li alloy can be thermodynamically estimated with regular solution approximation. Here, we follow the treatment to obtain the electrode potential. Figure 11.6a shows the Mg–Li binary phase diagram; pure Li and Mg metal have bcc and hcp structures at room temperature, respectively. For the Mg–Li alloy, Mg and Li form a substitutional solid solution in any component ratio. Based on the phase diagram, the Mg–Li alloy has a bcc single-phase structure until the Mg fraction comes up to ~68% of the atomic ratio. If the Mg fraction is higher than ~83%, the alloy has a hcp single-phase structure. Between the bcc and hcp single-phase regions, a eutectic alloy is formed containing both bcc and hcp phases.

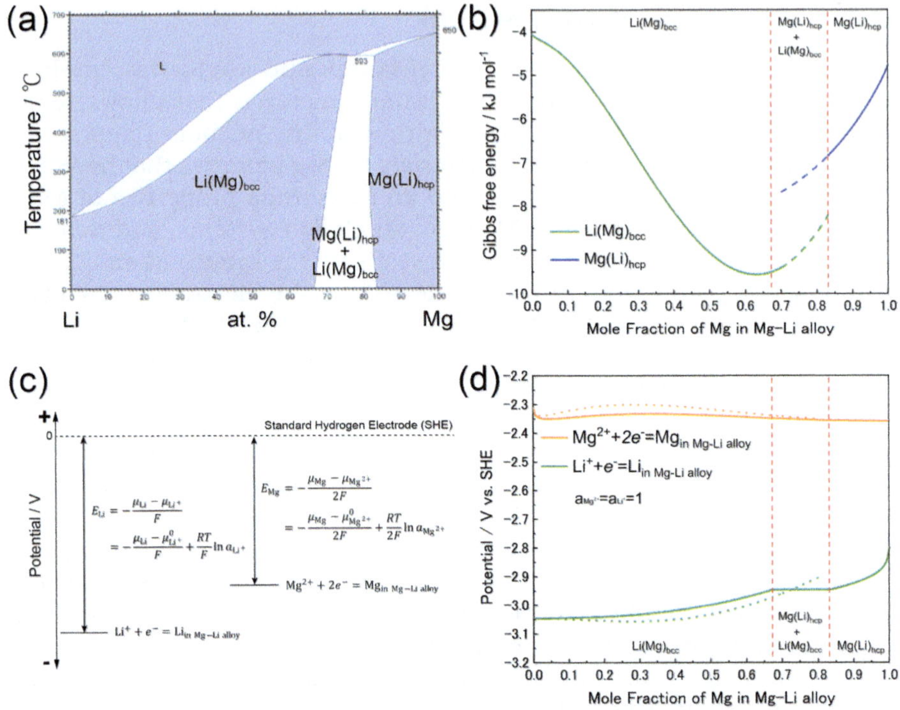

Figure 11.6 (a) Li–Mg binary phase diagram. (b) Calculated Gibbs free energy of a Mg–Li alloy based on the regular solution model. (c) Relationship between the redox potential and the chemical potential of Li and Mg in the Mg–Li alloy. (d) Calculated redox potential of Li and Mg in the Mg–Li alloy, where the activity of Li and Mg ions in the electrolyte is assumed as 1. Solid lines are based on eqn (11.11) and (11.12). Dotted lines are derived from eqn (11.9) and (11.10). Note that, theoretically, the redox potential in the eutectic region should be constant because there is no change in the chemical potential of Mg and Li in the alloy. For this reason, the solid lines suggest more reasonable results of the redox potentials. The vertical dashed lines in (b) and (d) indicate the regions of different phases corresponding to the phase diagram.

Here we consider the equilibrium dissolution potential of each element of the Mg–Li alloy, and we can reasonably estimate the electrodeposition potential of the alloy based on the dissolution potential. The electrochemical deposition/dissolution reactions of Mg and Li in the alloy can be written as follows:

$$Mg^{2+} + 2e^- = Mg_{\text{in Mg–Li alloy}} \tag{11.4}$$

$$Li^+ + e^- = Li_{\text{in Mg–Li alloy}} \tag{11.5}$$

Based on the Nernst equation with the following relations associated with the activities, $\mu_M = \mu_M^0 + RT \ln a_M$ (in the alloy) and $\mu_{M^{n+}} = \mu_{M^{n+}}^0 + RT \ln a_{M^{n+}} = \mu_M^0 + nFE_M^0 + RT \ln a_{M^{n+}}$ (in the electrolyte), the redox potential of the reactions shown in eqn (11.4) and (11.5) can be determined by:

$$E_{Mg} = -\frac{\mu_{Mg} - \mu_{Mg^{2+}}}{2F} = -\frac{\mu_{Mg}^0 - \mu_{Mg^{2+}}^0}{2F} + \frac{RT}{2F}\ln\frac{a_{Mg^{2+}}}{a_{Mg}} = E_{Mg}^0 + \frac{RT}{2F}\ln\frac{a_{Mg^{2+}}}{a_{Mg}} \tag{11.6}$$

$$E_{Li} = -\frac{\mu_{Li} - \mu_{Li^+}}{F} = -\frac{\mu_{Li}^0 - \mu_{Li^+}^0}{F} + \frac{RT}{F}\ln\frac{a_{Li^+}}{a_{Li}} = E_{Li}^0 + \frac{RT}{F}\ln\frac{a_{Li^+}}{a_{Li}} \tag{11.7}$$

where F is the Faraday constant, μ_{Mg}, μ_{Li} are the chemical potentials of Mg and Li in the Mg–Li alloy, and similarly $\mu_{Mg^{2+}}$, μ_{Li^+} are the chemical potentials of $Mg^{2+}$and Li^+ in the Mg–Li dual-ion electrolyte. μ_{Mg}^0, μ_{Li}^0 and $\mu_{Mg^{2+}}^0$, $\mu_{Li^+}^0$ are the chemical potentials of pure Mg, Li metal and Mg^{2+}, Li^+ in the electrolyte at standard state (298 K, 1 atm), respectively. E_{Mg}^0, E_{Li}^0 are the standard electrode potentials of Mg and Li. a_{Mg}, a_{Li} and $a_{Mg^{2+}}$, a_{Li^+} are the activities of Mg and Li in the alloy and $Mg^{2+}$and Li^+ in the electrolyte, respectively. Based on eqn (11.6) and (11.7), if $a_{Mg^{2+}}$ and a_{Li^+} are fixed at certain values, which means that the concentrations of $Mg^{2+}$and Li^+ in the electrolyte are constant, then the redox potential of each element would only depend on their chemical potential in the Mg–Li alloy at a certain temperature. For the sake of simplicity, we here employ the regular solution model to calculate the chemical potential of each element in the Mg–Li alloy. Based on the regular solution model, the molar free energy of the alloy is expressed as:

$$G_{alloy} = x\mu_{Mg}^0 + (1-x)\mu_{Li}^0 + \Omega x(1-x) + RT[x \ln x + (1-x)\ln(1-x)] \tag{11.8}$$

where R is the gas constant, T is the temperature (298 K in this work), x is the molar fraction of Mg in the Mg–Li alloy, and Ω is the interaction parameter. Here, we adopt the value of Ω derived in a previous study.[31] μ_{Mg}^0, μ_{Li}^0 are adopted from the Scientific Group Thermodata Europe (SGTE) data.[32] Figure 11.6b shows the calculated molar free energy of the Mg_xLi_{1-x} alloy. The chemical potential of Mg and Li, μ_{Mg} and μ_{Li}, thus can be evaluated using the following equations:

$$\mu_{Mg} = G(x) + \frac{\partial G}{\partial x}(1-x) \tag{11.9}$$

$$\mu_{Li} = G(x) - \frac{\partial G}{\partial x}x \tag{11.10}$$

Alternatively, based on the regular solution model, if the interaction parameter Ω is not a function of the composition x, the following equations can also be used:

$$\mu_{Mg} = \mu_{Mg}^0 + \Omega(1-x)^2 + RT\ln x \tag{11.11}$$

$$\mu_{Li} = \mu_{Li}^0 + \Omega x^2 + RT\ln(1-x) \tag{11.12}$$

Substituting the μ_{Mg} and μ_{Li} into eqn (11.6) and (11.7), we can calculate the redox potentials. Here, we assume that the activity of Mg^{2+} and Li^+ in the electrolyte is 1 and adopt the standard electrode potentials of Mg^{2+}/Mg and Li^+/Li as E_{Mg}^0 and E_{Li}^0, where the Standard Hydrogen Electrode (SHE) is employed as the reference electrode. Of course, the E^0 values depend on the solvation structure and are modified when electrolytes other than aqueous liquid are used, but empirically we know that the difference in these values ($E_{Mg}^0 - E_{Li}^0 \approx 0.7$ V) is almost unchanged even for various electrolytes. The relationship between the chemical potential and the redox potential is illustrated in Figure 11.6c. Consequently, the redox potentials of Mg and Li in the alloy are calculated and shown in Figure 11.6d.

Based on the results, the redox potentials of each element in the Mg–Li alloy are basically higher than the standard electrode potential due to the decrease in the chemical potential caused by the alloying. Compared to the redox potential of Mg, Li always has a lower redox potential at any composition. For this reason, Li would be dissolved preferentially in the discharge process from a thermodynamic point of view (or very low constant current rates). In the charge process, Li and Mg ions are released from the cathode host, and they are co-electrodeposited on the anode due to the overpotential effects at fast current rates, even though the redox potential of Mg is higher than that of Li. Namely, in order to enable co-electrodeposition, the electrode potential of the anode must be lower than the calculated value by increasing the current rates. Besides this, apart from thermodynamic aspects, the practical properties of a Mg–Li alloy, such as diffusion in a substitutional solid solution, solid electrolyte interphase (SEI), *etc.*, should also be taken into account.

11.3.3.2 Cyclic Voltammetry Experiments in Three-electrode Beaker Cells

To explore the practical properties of a Mg–Li alloy anode, cyclic voltammetry (CV) experiments were conducted in Mg–Li dual-ion electrolytes with three-electrode beaker cells. Figures 11.7a and b show the CV profiles in $MgHMDS_2$–LiHMDS–$AlCl_3$/G3, and $MgTFSA_2$–LiTFSA/G3 electrolytes, respectively. As mentioned above, the concentration of each salt was unified to 0.5 M (mol L^{-1}). In $TFSA^-$/G3 electrolytes, a Li-metal reference electrode was employed, where a Li foil was immersed in 0.5 M LiTFSA/G3 and separated from the external electrolyte with a ceramic filter; hereafter we refer to this reference as "Li RE". To investigate the influence of the Li salts, CV experiments in Mg single-ion electrolytes were also conducted and the results are shown in Figure 11.7c and d as a reference. In $MgHMDS_2$–$AlCl_3$/G3 electrolyte (Figure 11.7c), reversible Mg deposition was observed with high coulombic efficiency, which suggests that the Mg passivation was significantly overcome. In contrast, wide plateaus were observed between the cathodic and anodic peaks in $MgTFSA_2$/G3 electrolyte (Figure 11.7d), indicating marked Mg passivation. In the Mg–Li dual-ion electrolytes, since the potential scan was done to values lower than the potential of Li deposition, the co-electrodeposition of both Mg and Li ions should occur during CV experiments. In the $MgHMDS_2$–LiHMDS-$AlCl_3$/G3 electrolyte (Figure 11.7a), the current density was enhanced as the number of cycles increased, and one anodic peak was mainly observed. Varying the cathodic curves indicates that the active species gradually changed with the number of cycles, hinting at complicated deposition and dissolution processes. In contrast, the CV profile in $MgTFSA_2$–LiTFSA/G3 (Figure 11.7b) shows more simple shapes, which suggests a relatively simple mechanism. Comparing Figure 11.7b and

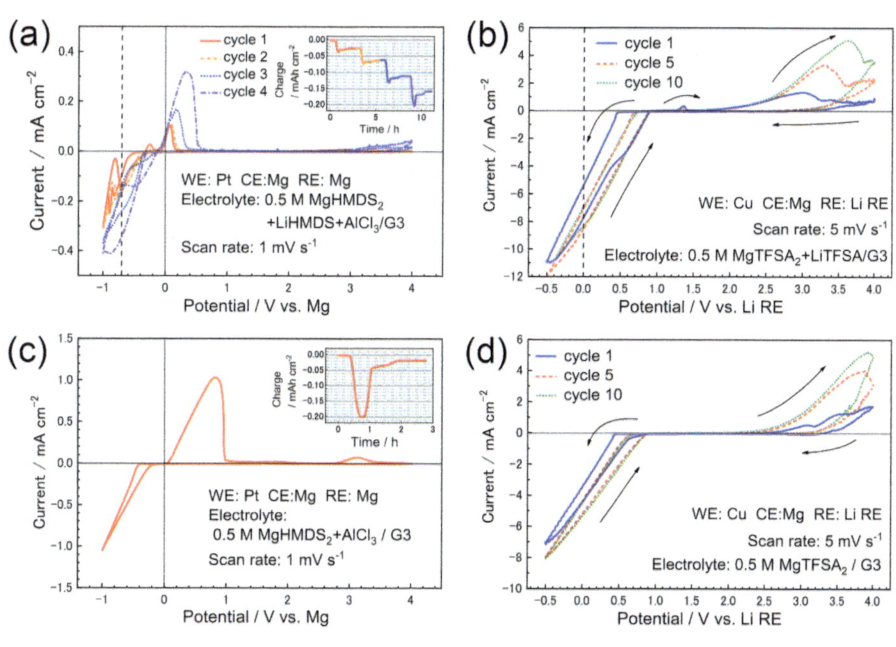

Figure 11.7 CV profiles of anodes in two types of electrolytes in beaker cells. (a) $MgHMDS_2$–LiHMDS–$AlCl_3$/G3, (b) $MgTFSA_2$–LiTFSA/G3, (c) $MgHMDS_2$–$AlCl_3$/G3, and (d) $MgTFSA_2$/G3. Pt foil was used as a current collector in (a) and (c), and Cu foil was used in (b) and (d). The redox potential of Li is shown as a dashed vertical line in (a) and (b).

d, the rise in the anodic current starts from a lower potential in the dual-ion electrolyte, which suggests that introducing Li salt may be helpful to improving the dissolution of Mg metal.

Because a much larger number of carrier ions are accommodated in the electrolyte of the beaker cells compared to that practically participating in the electrode reactions, the concentration of each carrier was almost constant in the electrolyte during the CV experiments. Therefore, the beaker cells cannot completely simulate the designed rocking-chair-type dual-ion batteries, where charge carriers in the electrolyte should be limited and the concentration of Mg and Li ions should be determined by the progress of the electrode reactions. Further investigations are therefore still needed, although fundamental knowledge provided from the results in the beaker cells is helpful for understanding the practical properties of the Mg–Li alloy anode.

11.3.3.3 Co-electrodeposition Morphology

Since Li ions can also be deposited during the charge process, it is necessary to investigate the electrodeposition morphology in Mg–Li dual-ion electrolytes in order to check whether or not dendrites are formed. Figure 11.8 compares the morphology of electrodeposits obtained in Li single-ion electrolyte with Mg–Li dual-ion electrolytes. Cu or Ni foil was used as the working electrode (WE) to collect the deposits. The experiments started from a fast potential scan rate at 50 mV s^{-1} from the open circuit potential of the Cu or Ni (about 3 V *vs.* Li RE) to −0.5 V *vs.* Li RE, where co-electrodeposition of Mg and Li should occur. Then, the potential was kept at −0.5 V for several minutes and the electrodeposits were subsequently observed by scanning electron microscope (SEM). As shown in Figure 11.8b, the electrodeposit in Li single-ion electrolyte clearly shows a rod-like, dendritic morphology. In contrast, electrodeposits in both two types of Mg–Li dual-ion electrolytes show a much smoother, sphere-like morphology (Figure 11.8c and d). In $MgTFSA_2$–LiTFSA/G3 and LiTFSA/G3, the deposit shows a smooth morphology even at a higher current density than in Li single-ion electrolyte. These results suggest that the co-electrodeposition of Mg and Li can successfully suppress the dendritic growth of Li, confirming the safety of Mg–Li alloy anodes.

11.3.4 Cathode Properties

In the above sections, it has been described that Mg–Li alloy can be used as a safe metal anode, if co-electrodeposition of Mg–Li dual ions occurs during the charge process. Because there are almost no extra carrier ions in the electrolyte, co-(de)intercalation of Mg and Li ions should occur in the cathode reaction to enable co-electrodeposition on the anode. In this section, the intercalation behavior in Mg–Li dual-ion systems is investigated with the Chevrel compound Mo_6S_8 cathode.

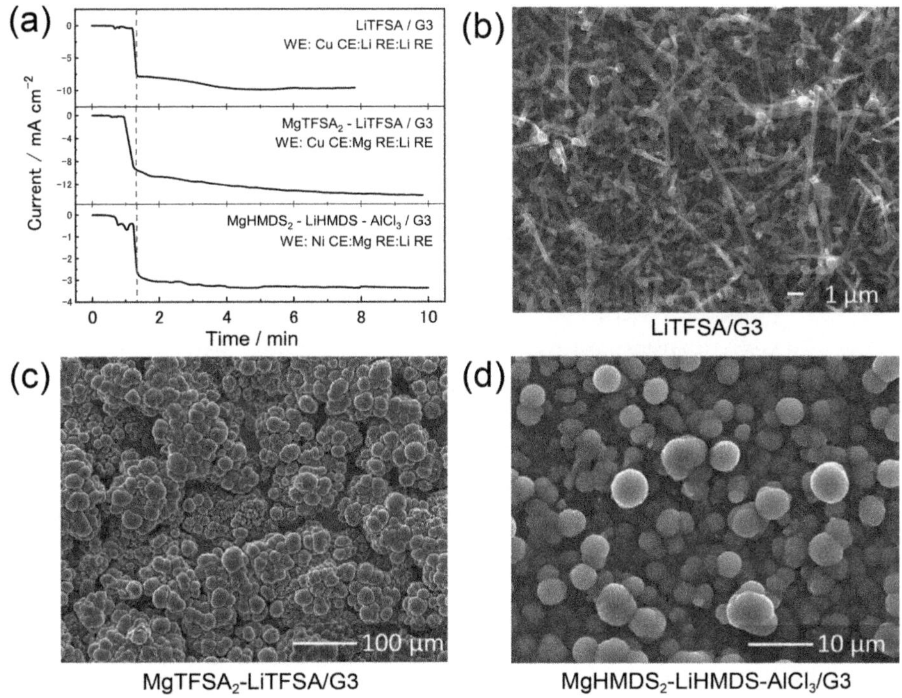

Figure 11.8 Deposition morphology of a Mg–Li alloy in beaker cells. (a) Current densities during deposition experiments in LiTFSA/G3, $MgTFSA_2$–LiTFSA/G3, and $MgHMDS_2$–LiHMDS–$AlCl_3$/G3 electrolytes. The left side of the dashed vertical line corresponds to the potential scan from open-circuit potential to −0.5 V *vs.* Li RE. The right side of the dashed vertical line corresponds to the constant potential electrodeposition at −0.5 V *vs.* Li RE. SEM images of the deposit are shown in (b), (c) and (d), respectively.

11.3.4.1 *Cyclic Voltammetry Experiments in Different Types of Electrolytes*

CV profiles of a Mo_6S_8 cathode in the two types of Mg–Li dual-ion electrolytes, $MgHMDS_2$–LiHMDS–$AlCl_3$/G3 and $MgTFSA_2$–LiTFSA/G3 are shown in Figure 11.9, accompanied by CV profiles of Mo_6S_8 in corresponding Mg or Li single-ion electrolytes. To facilitate the comparison of the intercalation potentials in different electrolytes, a Li-metal reference electrode (Li RE) was employed in both types of electrolytes, in which LiHMDS–$AlCl_3$/G3 or LiTFSA/G3 were used as internal electrolytes for Li RE, being the same type as the external bulk electrolytes; it has been confirmed that Li metal in the two types of internal electrolytes has an equal electrode potential. As with the Li single-ion electrolytes (Figure 11.9a and b), Li metal was immersed into the bulk electrolyte directly without separation.

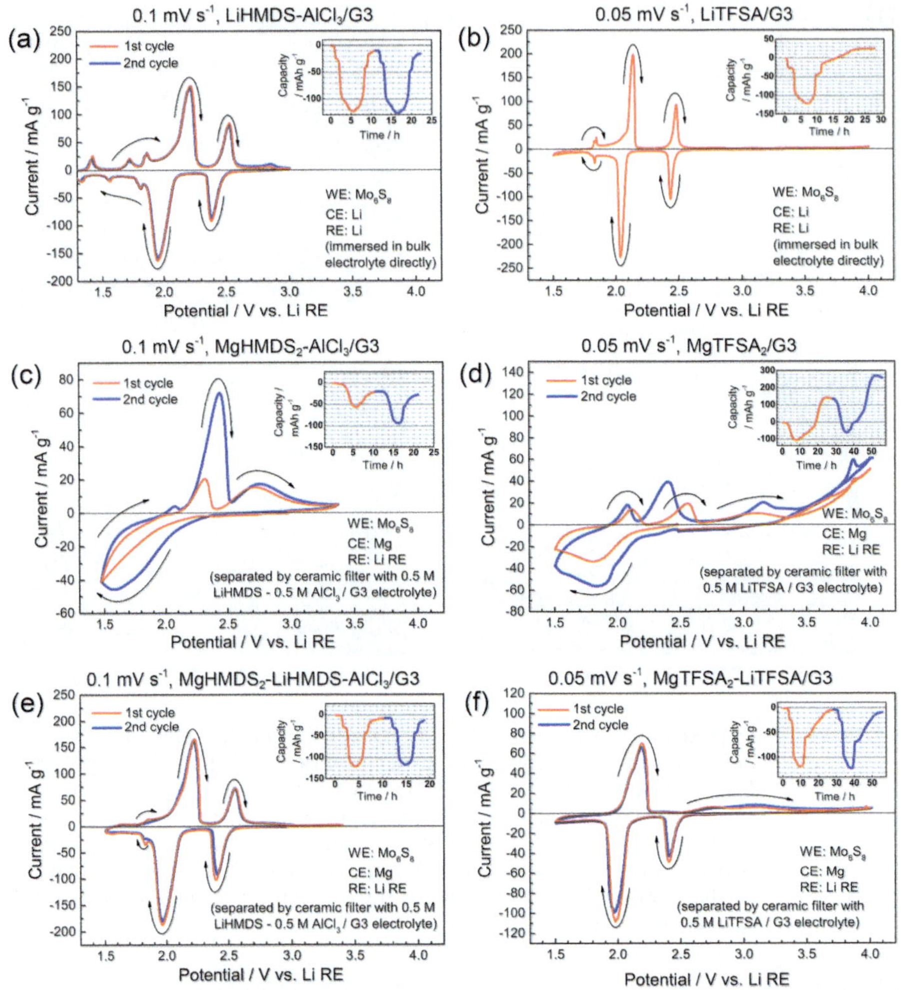

Figure 11.9 Cyclic voltammograms of Mo_6S_8 in (a) LiHMDS–$AlCl_3$/G3, (b) LiTFSA/G3, (c) $MgHMDS_2$–$AlCl_3$/G3, (d) $MgTFSA_2$/G3, (e) $MgHMDS_2$–LiHMDS–$AlCl_3$/G3, and (f) $MgTFSA_2$–LiTFSA/G3 electrolytes. The concentrations of all of the salts were set to 0.5 M. The insets show the capacities of each cycle. The scan rate of each experiment is shown above each figure; The scan rate in $HMDS^-$–$AlCl_3$/G3 was two times faster than that in $TFSA^-$/G3, due to the relatively poor chemical stability of $HMDS^-$–$AlCl_3$/G3. Reproduced from ref. 18 with permission from the Royal Society of Chemistry.

The intercalation and deintercalation peaks in the Li single-ion electrolytes show good symmetry with almost full theoretical capacities. Since the reference electrodes were properly prepared as mentioned above, the Li intercalation potentials in Figure 11.9a and b are substantially equal, as is expected. Thus, we confirm that the intercalation potential of Li ions with the

Mo_6S_8 host is not influenced by the different types of electrolytes. Based on this setup, we can easily compare the intercalation behavior of Mo_6S_8 even in different kinds of electrolytes.

Unlike the results in Li single-ion electrolytes, the CV profiles in Mg single-ion electrolytes (Figure 11.9c and d) show that the intercalation–deintercalation peaks in the two types of Mg single-ion electrolytes are broad and lack symmetry. Because the intercalation reactions of Li or Mg ions with the Mo_6S_8 host are mainly two-phase reactions,[33,34] the broad peaks are attributed to a kinetic issue, *i.e.* sluggish Mg diffusion in contrast to agile Li ions in the cathode host.

In $MgHMDS_2$–LiHMDS–$AlCl_3$/G3 (Figure 11.9e), the CV profile of Mo_6S_8 is a similar shape to that in the Li single-ion electrolyte (Figure 11.9a), which indicates that mainly Li ions are intercalated into the Mo_6S_8 host. On the other hand, in $MgTFSA_2$–LiTFSA/G3 (Figure 11.9f), the CV profile shows a unique shape that is different from the results either in the Li or Mg single-ion electrolytes (Figure 11.9b and d). There are two pairs of intercalation/deintercalation peaks at around 2.5 and 2.1 V *vs.* Li RE. Although the redox peaks at around 2.1 V appear to be normal sharp peaks, the redox peaks at around 2.5 V consist of a sharp intercalation peak and a broad deintercalation peak, showing features of intercalation of Li and deintercalation of Mg, respectively. The unique shape suggests that Li and Mg ions have complicated intercalation behavior in $MgTFSA_2$–LiTFSA/G3 electrolyte. Although it is still unclear as to what causes the difference in the two types of dual-ion electrolytes, we believe it should be mainly attributed to the different solvation structures in the electrolyte; strong Mg–Cl bonding may affect the intercalation behavior of Mg ions into the cathode material and be one of the reasons for the different results. Eventually, the co-intercalation of Mg–Li dual ions is required to realize the co-electrodeposition on the anode, and therefore Cl-free electrolytes would be suitable for use in rocking-chair-type Mg–Li dual-ion batteries.

11.3.4.2 Concomitant Intercalation of Mg–Li Dual Ions

The unique shape of the CV profile in the $MgTFSA_2$–LiTFSA/G3 electrolyte suggests that both Mg and Li ions are intercalated into the cathode host. To clarify the intercalation behavior of Mg and Li ions, stepwise CV experiments were carried out. First, the potential of the Mo_6S_8 cathode was scanned in the range of 2.3 to 4 V, corresponding to the higher-potential (about 2.5 V) intercalation reaction. As seen in Figure 11.10a, both the cathodic and anodic peaks are very sharp and are almost the same as the peaks in the Li single-ion electrolyte (Figure 11.9b). Then, the potential scan was extended to in the range of 1.5 to 4 V, in which the lower potential reaction near to 2.1 V should also occur. As shown in Figure 11.10b, the deintercalation peak at 2.5 V experiences an obvious change in shape from sharp to broad after undergoing the intercalation reaction at the lower potential (about 2.1 V). The broad

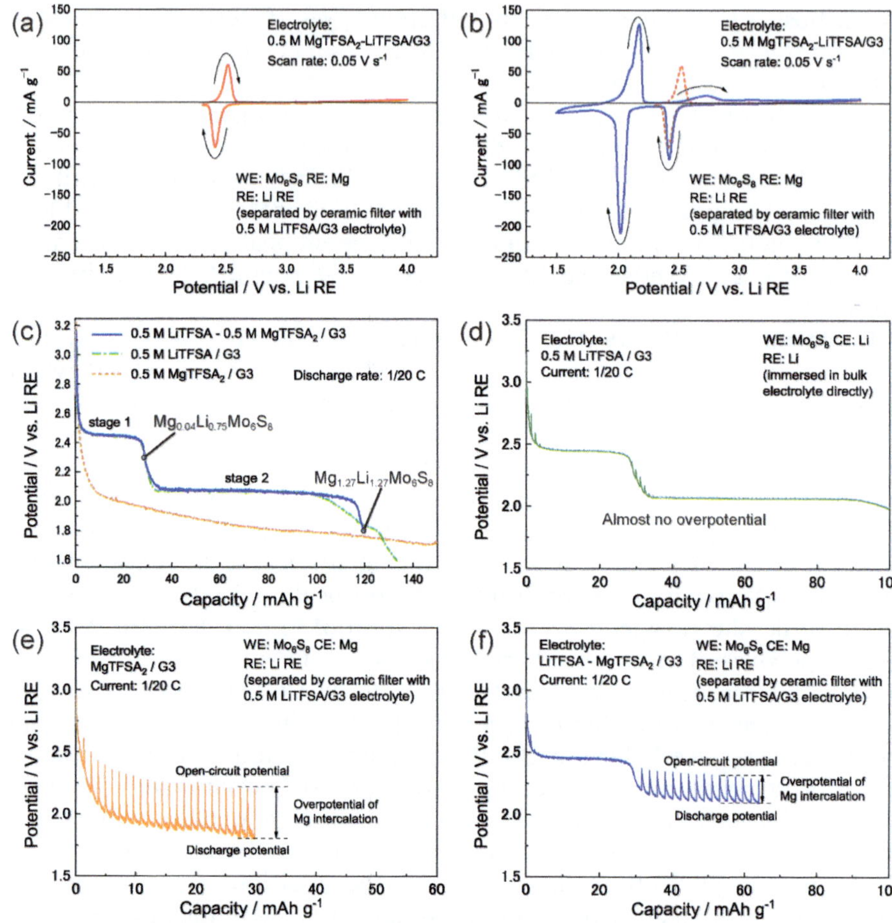

Figure 11.10 Cathode properties of Mo_6S_8 in $TFSA^-$/G3 electrolytes. CV profiles in $MgTFSA_2$–LiTFSA/G3 with a potential scan range of (a) 2.3–4 V, and (b) 1.5–4 V. (b) Galvanostatic discharge profiles in Li or Mg single-ion electrolytes and Mg–Li dual-ion electrolyte. Composition of the discharged Mo_6S_8 electrode after each stage in dual-ion electrolyte was measured by inductively coupled plasma (ICP) analysis and is also displayed. Galvanostatic intermittent titration technique (GITT) experiments on the Mo_6S_8 cathode in (d) Li single-ion electrolyte, (e) Mg single-ion electrolyte, and (f) Mg–Li dual-ion electrolyte were conducted to estimate the overpotential during discharge. Figure (a)–(c) are reproduced from ref. 18 with permission from the Royal Society of Chemistry.

deintercalation peak indicates the typical features of the deintercalation of Mg, as seen in Figure 11.9c and d. This result strongly indicates that a certain amount of Mg ions was intercalated into Mo_6S_8, accompanied by Li ions at around 2.1 V.

To understand the intercalation behavior quantitively, galvanostatic discharge experiments were conducted, and compositions of the discharged cathodes after each reaction were checked by inductively coupled plasma (ICP) analysis. After the first plateau (stage 1 in Figure 11.10c) at about 2.5 V, the composition of the cathode was $Mg_{0.04}Li_{0.75}Mo_6S_8$. This result indicates that mainly Li ions participate in the intercalation reaction, which is reasonable from the viewpoint of the stepwise CV profiles. After the second plateau (stage 2 in Figure 11.10c), the intercalation of Mg ions is remarkably improved. In the fully discharged Mo_6S_8 host, almost the same amount of Mg and Li ions were inserted ($Mg_{1.27}Li_{1.27}Mo_6S_8$, $1.27 \times 2 + 1.27 \approx 4$ electrons).

The discharge profiles of the Mo_6S_8 cathode in Mg or Li single-ion electrolytes are also shown in Figure 11.10c. Compared with the discharge profile in the Mg single-ion electrolyte, it is obvious that Mg intercalation occurs at a higher potential in Mg–Li dual-ion electrolyte. This phenomenon suggests that introducing Li ions into the electrolyte can reduce the overpotential of Mg intercalation that arises from the sluggish diffusion of Mg ions in the cathode host. To investigate the overpotential for Mg intercalation, galvanostatic intermittent titration technique (GITT) experiments were carried out at a current density of 1/20C. In Li single-ion electrolyte (Figure 11.10d), almost no potential difference could be observed between the open-circuit and discharge potentials, which indicates the very fast Li intercalation and host relaxation processes. On the other hand, as shown in Figure 11.10e, a large potential difference of more than 0.4 V was observed in the GITT profile in the Mg single-ion electrolyte. The large overpotential can be attributed to the sluggish Mg diffusion in the cathode host. In the Mg–Li dual-ion electrolyte (Figure 11.10f), the higher-potential reaction (~2.5 V) shows a similar feature to the Li single-ion case, which is consistent with the ICP results in that mainly Li ions are inserted at this stage. However, in the low potential reaction (~2.1 V), a relatively small difference of about 0.2 V appears in the overpotential (between the discharge potential and open-circuit potential), which is distinctively different from the results in Li single-ion electrolyte (Figure 11.10d). The potential difference arises from the structural relaxation of the Mo_6S_8 host after the Mg intercalation (specifically, the interchange of Li and Mg), which indicates that there is a thermodynamic preference for Mg ions to be inserted into the Mo_6S_8 host than Li ions at this stage. Compared to the Mg single-ion case, the overpotential of Mg intercalation is markedly decreased after certain Li ions are pre-inserted into the Mo_6S_8 host. This is a certain kind of concerted interaction between cations, and further discussion of this phenomenon is presented in Section 11.4.

11.3.5 Charge Tests Using Coin Cells

In the above section, it is revealed that Mo_6S_8 can accommodate and release both Mg and Li ions concomitantly. Thus, practical cells of Mg–Li dual-ion batteries can be constructed with a Mo_6S_8 cathode. However, since the Mg passivation cannot be suppressed in $TFSA^-$/G3, it is still difficult to obtain good discharge performance of practical cells. Here, we focus on the charging process to investigate whether a non-dendritic morphology can be obtained on the anode in practical cells. Figure 11.11a shows the structure of two-electrode coin cells. The coin cells need a very slight amount of electrolyte, which enables battery tests under similar conditions to those of the designed rocking-chair-type structure. To compare the deposited morphologies, both a Li cell (cell 1: $LiFePO_4$ was used for the cathode) and a Li–Mg dual-ion cell (cell 2: $Mg_{1.27}Li_{1.27}Mo_6S_8$ was used for the cathode) were prepared. The $Mg_{1.27}Li_{1.27}Mo_6S_8$ cathode used in cell 2 was prepared by discharging a Mo_6S_8 cathode in the $MgTFSA_2$–LiTFSA/G3 electrolyte. In both cells, a Cu foil was used as the anode to collect electrodeposits, and only the LiTFSA/G3 electrolyte was used to guarantee that Mg ions were supplied only from the cathode material. The two coin-cells were then fully charged to 4 V, as seen in Figure 11.11b. The electrodeposited morphologies on the anodes were subsequently observed by a SEM equipped with an inert transfer vessel. The images are shown in Figure 11.11c and d. Electrodeposits in the Li-ion

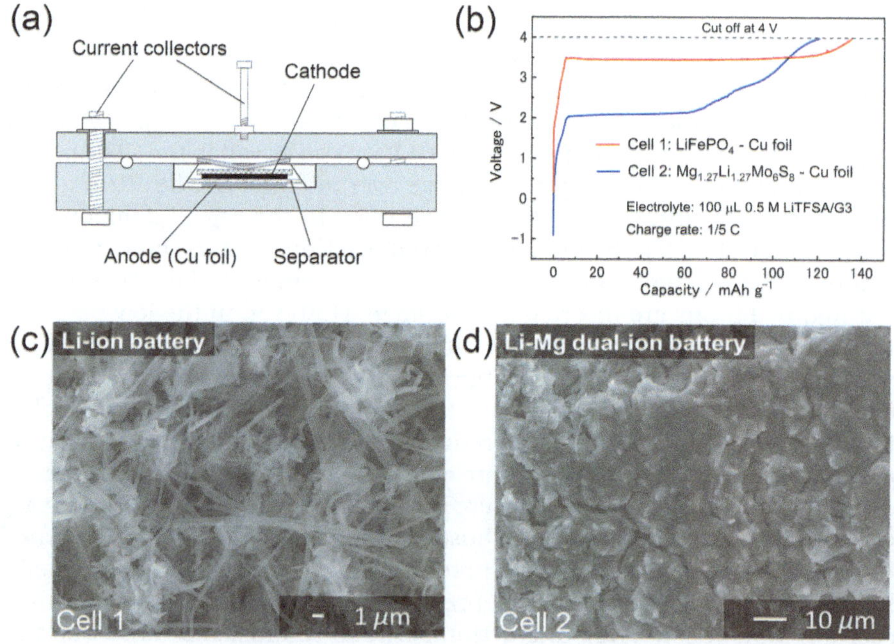

Figure 11.11 (a) Structure of two-electrode coin cells. (b) Galvanostatic charge profiles of the two-electrode coin cells with 100 mL of 0.25 M LiTFSA/G3 electrolyte, a Cu foil (anode/negative electrode), and a cathode (positive electrode) of (red) $LiFePO_4$ or (blue) $Mg_{1.27}Li_{1.27}Mo_6S_8$; the latter cathode was obtained by discharging the Mo_6S_8 electrode in $MgTFSA_2$–LiTFSA/G3 electrolyte. After charging, electrodeposits on the Cu foil in the two-electrode coin cells were observed by SEM, which are shown in (c) and (d), corresponding to cell 1 and cell 2 in (b), respectively. Reproduced from ref. 18 with permission from the Royal Society of Chemistry.

cell showed a typical dendritic morphology, whereas, in the Mg–Li dual-ion cell, a much smoother morphology was observed, which demonstrates that co-electrodeposition can effectively suppress dendritic growth. Thus, we confirm that the co-electrodeposition of Mg and Li can be realized in the practical charge process, and a safe metal anode can be used because of the smooth morphology, which is a key point in constructing rocking-chair-type Mg–Li dual-ion batteries.

11.4 Facilitating Mechanism of Mg Diffusion

In this section, our special focus is put on the improvement of Mg intercalation that is observed in dual-ion electrolyte; see Figure 11.10. To investigate this phenomenon in detail, *ab initio* computational models were used to simulate the Mo_6S_8 hosts at different discharge states. The activation energies of Mg migration in the hosts were then estimated by the nudged band elastic (NEB) method,[35] implemented by first-principles calculations.[36,37]

11.4.1 Structure and Diffusion Path in the Mo_6S_8 Host

Figure 11.12 shows the crystal structure of the Mo_6S_8 host and diffusion paths inside it. Interstitial sites in Mo_6S_8 are usually divided into inner and outer sites.[15,34] Site occupancy adopted from the Inorganic Crystal Structure Database (ICSD) is shown in Table 11.1. The inner sites are more stable for both Li and Mg ions, and one cation usually inserts in the inner ring of 6 equivalent sites (inner-site ring). The outer sites surrounding the inner-site ring are less stable and are occupied by the successive insertion of further cations.

As shown in Figure 11.12, the inner-site rings are connected to the nearest neighbour outer sites. Thus, the inserted ions at the inner sites need to pass through the neighbouring outer sites to hop to adjacent inner-site rings. For this reason, the inserted ions must pass though the inner site space (blue in Figure 11.12) and outer site space (green in Figure 11.12) alternately during the diffusion process.

11.4.2 Single Ion Migration in a Dilute Mo_6S_8 Host

Because of the structure of the Mo_6S_8 host, two basic cation migration processes can be considered: (i) migration between inner sites in the same ring (Figure 11.13a), and (ii) migration from an inner site to another inner site through neighbouring outer sites (Figure 11.13b). To compare the activation energies of the two basic processes, single Mg or Li migration in Mo_6S_8 host were calculated using the NEB method. Each Mg or Li ion was placed in every three Mo_6S_8 unit cells and the NEB calculations were conducted using the Mo_6S_8 supercells that correspond to the concentrations of $Mg_{0.33}Mo_6S_8$ and $Li_{0.33}Mo_6S_8$, respectively. The computational results of

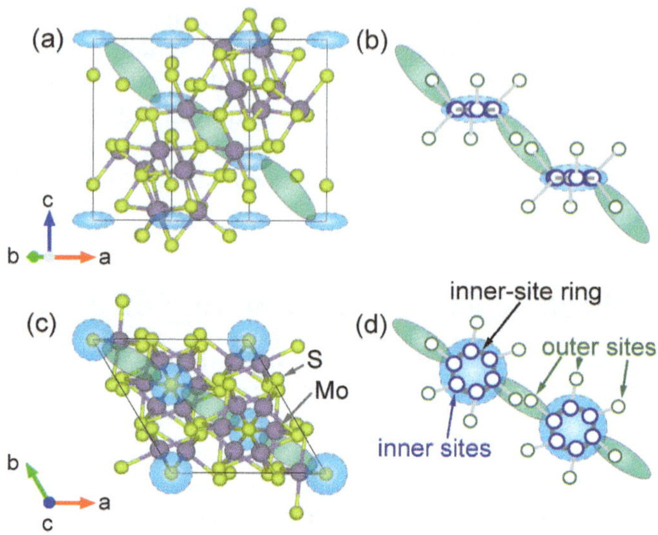

Figure 11.12 Crystal structure and internal ion diffusion paths of Mo_6S_8. (a) and (c) show the structure of a hexagonal Mo_6S_8 cell from different projection directions. Mo and S atoms are colored violet and yellow, respectively. (b) and (d) show the interstitial sites of Li and Mg ions in Mo_6S_8 corresponding to (a) and (c), respectively. The positions of interstitial sites indicate the relationship between them, but do not represent the exact distances. Blue ellipses (inner sites) and green circles (outer sites) indicate the spaces of the interstitial sites inside the unit cell and the connections between them. Note that only the outer-site spaces along the diagonal of the unit cell are displayed with green ellipses and the other equivalent spaces are omitted for clarity. Reproduced from ref. 41 with permission from John Wiley & Sons, © 2018 WILEY-VCH Verlag GmbH & Co. KGaA, Weinheim.

Table 11.1 Site occupancy of Li- and Mg-inserted Mo_6S_8.

	Occupancy of cation site	
	Inner site	Outer site
$LiMo_6S_8$	0.167	0
$Li_3Mo_6S_8$	0.188	0.312
$Li_4Mo_6S_8$	0.190	0.477
$MgMo_6S_8$	0.167	0
$Mg_2Mo_6S_8$	0.167	0.167

Mg and Li migration are shown in Figure 11.13c and d, respectively. Since migration in an inner-site ring shows very slight activation energies (~20 meV), this process would not significantly affect the diffusion in Mo_6S_8 hosts. In contrast, passing through the outer sites shows a remarkable energy barrier of several hundred meV, which indicates that passing through the outer sites would be the rate-controlling step during solid-phase diffusion.

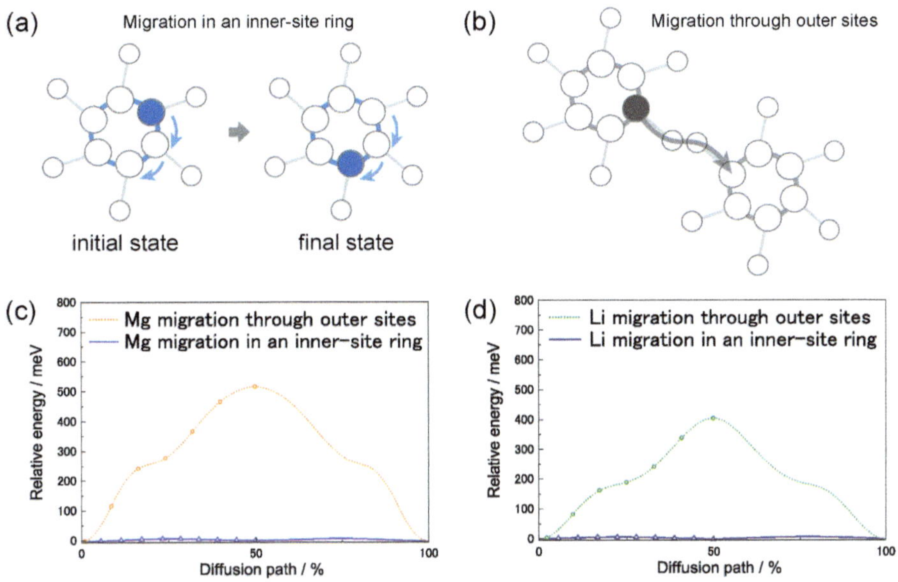

Figure 11.13 Migration process of inserted Li and Mg ions in the Mo_6S_8 host. Schematic figures illustrate (a) ion migration inside an inner-site ring, and (b) ion migration between neighbouring inner-site rings through outer sites. The activation energies of (c) Mg migration and (d) Li migration were computed using the NEB method. Nodes along the curves indicate interpolated images used in the NEB calculations. The right halves of the curves were inverted from the left halves based on the symmetry of the diffusion processes.

Comparing the results of Mg and Li, it was found that the activation energy of Mg migration in the empty Mo_6S_8 host is much higher than that of Li. The peak values of the activation energies are in good agreement with those in previous reports.[14,38] On the activation energy curves, shoulders can be observed on both sides of the peaks. According to the crystal structure in Figure 11.12, the shoulders indicate the position of the outer sites on the diffusion path. There are no local minima (around the outer sites) in the diffusion path, which indicates that the outer sites are not stable in the Mo_6S_8 host for a very low concentration of cations, which is consistent with the experimental results that Mg or Li ions cannot occupy the outer sites until the inner sites are occupied first.[15,34]

11.4.3 Mg Migration in Mg–Li Dual-ion Systems

To investigate what improves the Mg intercalation in dual-ion systems, a $LiMo_6S_8$ host was prepared to simulate the Mo_6S_8 cathode discharged after stage 1 (see Figure 11.10c). Then, one Mg ion was placed into the $LiMo_6S_8$ host, and the activation energy of the Mg migration between the neighbouring inner-site rings was computed. To determine the most stable configurations for the initial and final states for the NEB calculations, the total energies of the Mg-inserted $LiMo_6S_8$ host were computed by static calculations. The lowest total energy could be realized when the Mg ion was located in the inner site and pushes out a pre-inserted Li ion to an adjacent outer site. Based on this result, the stable initial and final states of the computational model are as shown in Figure 11.14c(i) and c(v), respectively.

Prior to consideration of concerted motion, the energy barrier of a single hopping process was first calculated for the case where the Mg ion at the inner site solely hops to the next inner-site ring, whereas the Li ion was allowed to relax only around the outer site (Li1 in Figure 11.14c(i)). The result is indicated by a red 'X' marked in Figure 11.14a. Next, a concerted motion was considered as follows. As a Mg ion hops to the next inner-site ring from its initial inner site, an outer-site Li ion (Li1 in Figure 11.14c(i)) simultaneously enters a neighbouring inner site. Then, another Li ion (Li2 in Figure 11.14c(i)) is pushed out to a neighbouring outer site from the initial inner site, as shown in Figure 11.14c(v). The Li ion pushed out to the outer site will hop further to the next inner-site ring. As a result, concerted motion occurs successively along the diffusion path. The activation energy computed for this process is shown in Figure 11.14a (blue curve). The activation energy for the Mg hopping in a $LiMo_6S_8$ host decreases significantly relative to that for single Mg hopping in a Mo_6S_8 host and is even lower than that of a Mg ion hopping without concerted motion, as indicated by 'X'. Furthermore, to explore what decreases the activation energy, the distances between the Li and Mg ions (Li1&Mg from 0 to 50% and Mg & Li2 from 50 to 100% of the diffusion path) during the diffusion process are shown as the grey curve in Figure 11.14a above the energy curves. Interestingly, the Mg and Li ions in an inner-site ring and neighbouring outer site always prefer to keep a distance in the range of 3.5–4 Å rather than hopping individually, which strongly reflects the concerted motion of both Li and Mg dual ions in a $LiMo_6S_8$ host. As a reference, the farthest distance (~5.78 Å, as drawn in Figure 11.14c(iii)) that Li and Mg can reach during the diffusion process is indicated by a dashed line in Figure 11.14a. Consequently, we can conclude that the facilitated Mg intercalation observed in the Mg–Li dual-ion system is attributable to the remarkable lowering of the activation energy of Mg-ion migration after a certain amount of Li ions are inserted into the electrode prior to Mg insertion.

11.4.4 Concerted Motion in Single-ion Systems

From the above results, it is also reasonable to consider that the same phenomenon will also occur in single-ion systems, once the inserted ions reach a certain concentration in a Mo_6S_8 host. Computational models were prepared by putting an additional Mg or Li ion into the $MgMo_6S_8$ and $LiMo_6S_8$ hosts, respectively, which corresponds to concentrations of $Mg_{1.33}Mo_6S_8$ and $Li_{1.33}Mo_6S_8$. Initial and final states of the diffusion processes are shown in

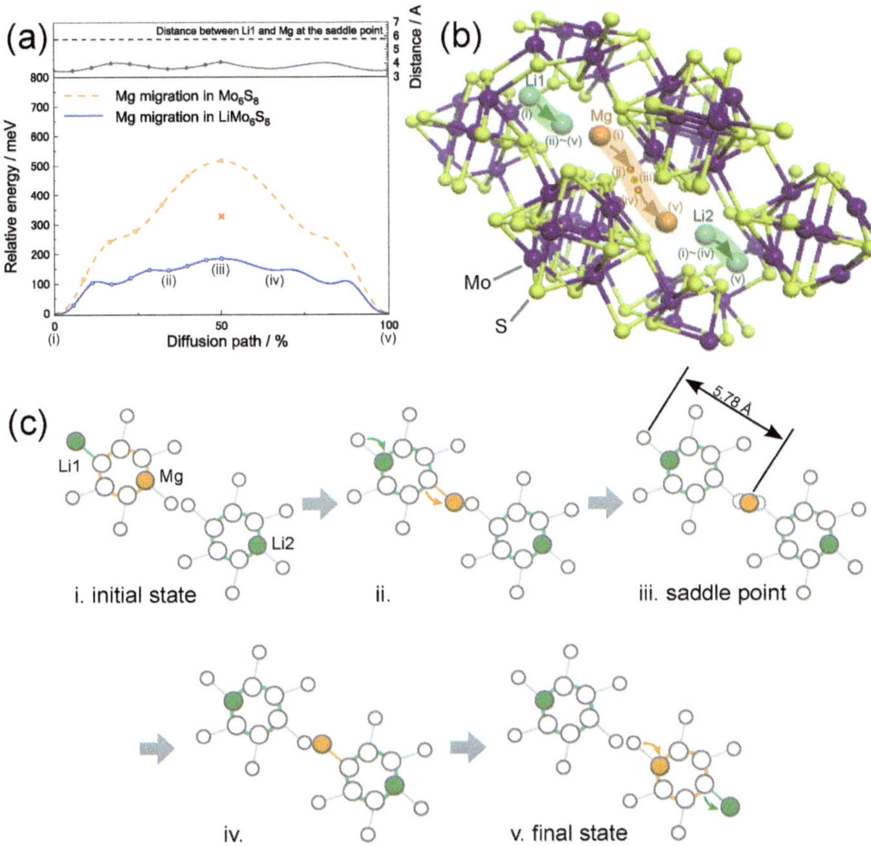

Figure 11.14 Minimum energy paths of Mg diffusion in a Mo_6S_8 host. (a) Comparison of the activation energies computed using the NEB method for the diffusion processes of (orange) a Mg ion in a Mo_6S_8 host, and (blue) a Mg ion in a $LiMo_6S_8$ host. Nodes along the curves indicate the energies of the interpolated images used in the NEB calculations. The right halves of the curves were inverted from the left halves based on the symmetry of the diffusion processes. The red 'X' mark indicates the energy barrier of the Mg migration in a $LiMo_6S_8$ host, where the Mg ion hops between inner-site rings without the concerted motion of Li ions. (b) Schematic figure illustrating the diffusion process of a Mg ion in a $LiMo_6S_8$ host. Mg and Li ions are colored orange and green, respectively. (c) Figure illustrates the initial and final states of the NEB calculations and the atomic arrangements of inserted ions in several important configurations. Distances between concerted hopping Mg and Li ions during the diffusion process; Li1&Mg from 0 to 50% and Mg&Li2 from 50 to 100% of the diffusion path are plotted in the upper part of (a). The dashed line in (a) indicates the distance between the initial position of Li1 with Mg at the saddle point. Reproduced from ref. 41 with permission from John Wiley & Sons, © 2018 WILEY-VCH Verlag GmbH & Co. KGaA, Weinheim.

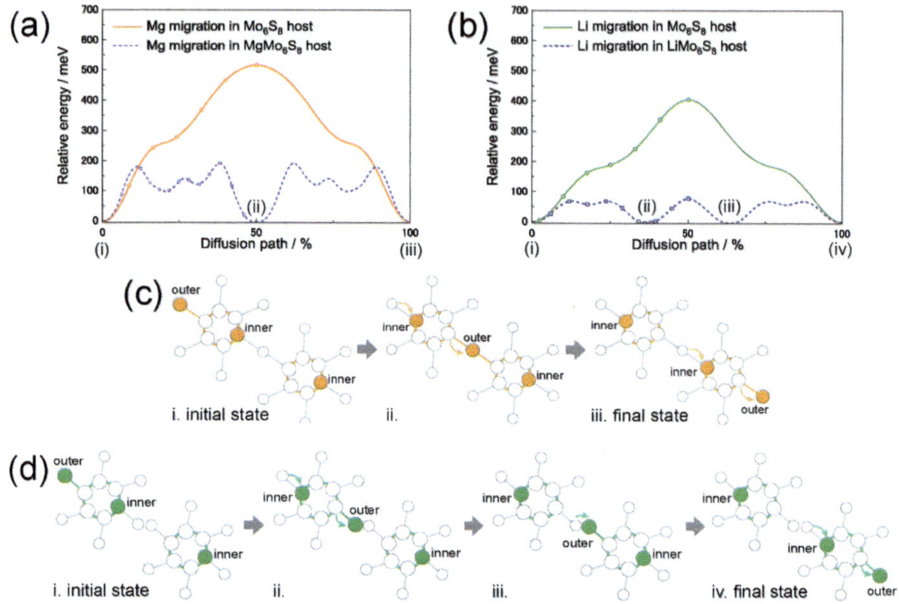

Figure 11.15 Concerted motion in Li/Mg single-ion systems: (a) and (b) show the calculated energy barriers in the Mg and Li single-ion systems, respectively. Nodes along the curves indicate energies of the interpolated images used in the NEB calculations. The right halves of the curves were inverted from the left halves based on the symmetry of the diffusion processes. (c) and (d) show the initial and final states of the calculations, where Mg ions are colored orange and Li ions are colored green. Atomic arrangements of the inserted ions in several important configurations are also illustrated. Reproduced from ref. 41 with permission from John Wiley & Sons, © 2018 WILEY-VCH Verlag GmbH & Co. KGaA, Weinheim.

Figure 11.15c and d, respectively. It should be noted that, especially in the case of the $MgMo_6S_8$ host with an additional Mg ion (corresponding to the composition $Mg_{1.33}Mo_6S_8$), the Mg ion intended to occupy an outer site is located in the middle of the neighbouring inner-site rings in the relaxed structure, that is, the two outer sites neighbouring each other are merged into one site in this special composition.

From the computational results (Figure 11.15a and b), it is found that the activation energies are also much lower than those in the Mo_6S_8 host, and local minima can be observed in the diffusion path in the Mg and Li single-ion systems. Unlike the case of Mg ion diffusion in a $LiMo_6S_8$ host, the relative energy decreases to zero at several points along the diffusion path. By checking the atomic arrangements, it is found that the configurations at these zero points are equivalent to the initial and final states (Figure 11.15c and d). The results indicate that the concerted motion of the inserted ions is a common phenomenon by which a much higher ion mobility can be obtained in host materials with relatively high concentrations of inserted ions. Since the activation energies are reduced to a very low level (lower than 200 meV) in all three cases, the present results suggest that, if such a concerted motion works well, solid-phase diffusion would be significantly improved and no longer a significant kinetic rate-controlling factor on the intercalation reaction.

11.4.5 Facilitating Intercalation in Mg–Li Dual-ion Systems

Based on the experimental and computational results, we discuss the essential differences between Li and Mg ions in intercalation reactions and, eventually, show a novel way to facilitate the diffusion of multivalent ions in solid phases using dual-ion systems.

Solid-phase diffusion in electrode materials during intercalation reactions can be roughly divided into two stages, "individual hopping motion" and "concerted motion", based on the concentration of the inserted ions. Figure 11.16 illustrates the activation energy of basic diffusion processes and the corresponding intercalation behavior in the active electrode materials for Li single-ion systems, Mg single-ion systems and Li–Mg dual-ion systems. Strongly diluted in host materials, the inserted ions hop individually, and the interaction between inserted ions is barely effective because of the relatively far distances and random directions among the inserted ions. The energy barriers of the Li and Mg ions show essential differences in dilute hosts (Figure 11.16a and b). In the case of Li ions, intercalation progresses rapidly because of the relatively low energy barrier. On the other hand, in the case of divalent Mg ions, the relatively high energy barrier usually becomes the rate-controlling factor, and consequently, the intercalation is retarded at or around the surface of the particles. However, if the carrier-ion concentration reaches a certain degree, the distance between inserted ions is shortened, and eventually the activation energies would be significantly lowered by the concerted interaction effect. Consequently, by virtue of the concerted interaction, the activation energy along the diffusion path diminishes, and the solid-phase diffusion is considerably facilitated.

Experimentally, the concerted motion would hardly be detectable in single-ion systems of multivalent ions because the individual hopping motion in a dilute host usually determines whether intercalation electrode materials can work practically or not. However, in Mg–Li dual-ion systems, the concerted motion is a key factor by which the sluggish solid-phase diffusion of divalent Mg ions can markedly be modified, and fast Mg diffusion can be realized intentionally by introducing Li ions. As shown in Figure 11.16c, in a dilute host, Li intercalation precedes preferentially at the individual motion stage, because the relatively high energy barrier of individual motion kinetically retards Mg intercalation. However, after Li ions hop into the host, Mg intercalation is facilitated because the concerted interaction of Mg–Li dual ions significantly decreases the energy barrier of Mg diffusion to a degree comparable with that of Li ions. Consequently, the dual ions can enter the host material without any obvious loss in electrode performance.

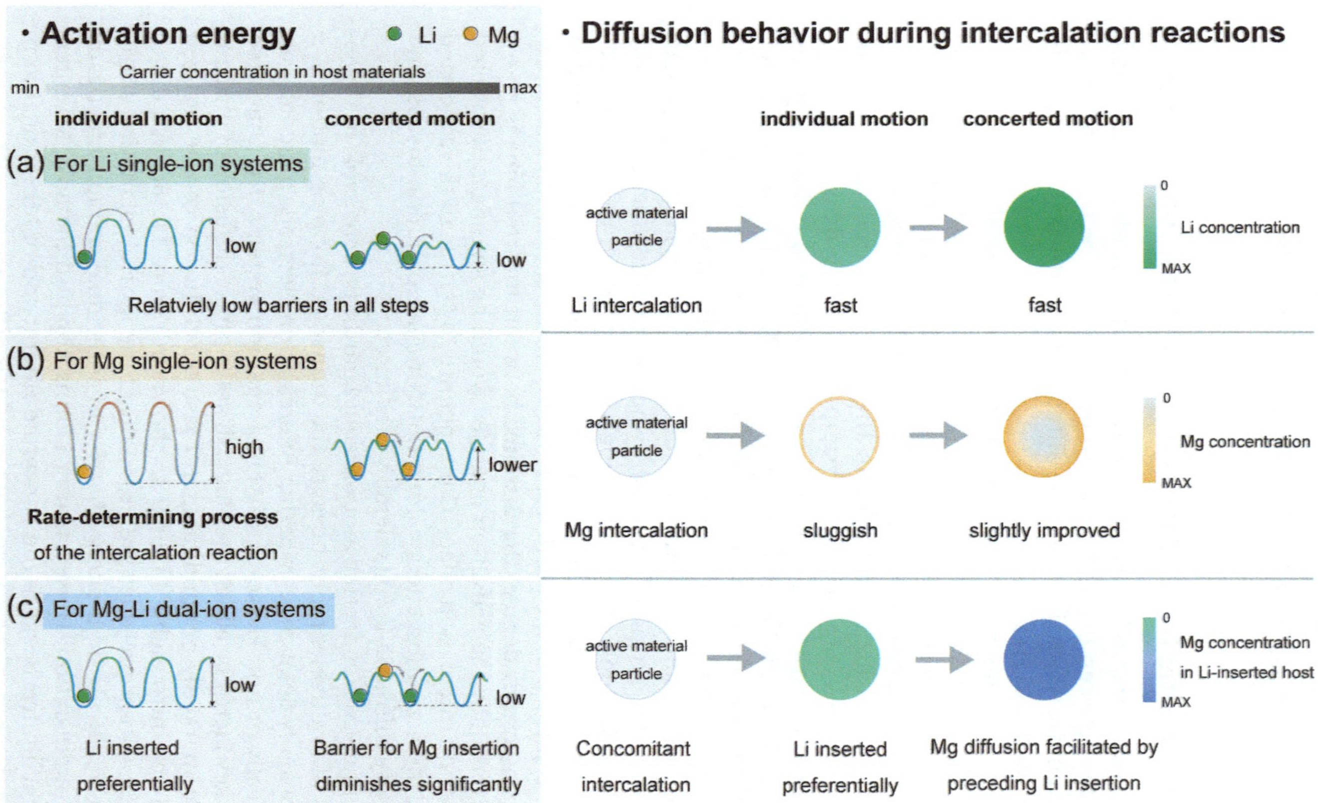

Figure 11.16 Solid-phase diffusion processes during the intercalation reactions: activation energy of basic solid-phase diffusion steps and diffusion behavior with active material articles during intercalation reactions in (a) Li single-ion systems, (b) Mg single-ion systems, and (c) Li–Mg dual-ion systems. Reproduced from ref. 41 with permission from John Wiley & Sons, © 2018 WILEY-VCH Verlag GmbH & Co. KGaA, Weinheim.

11.4.6 Versatility of the Facilitating Mechanism

The computational results for Mo_6S_8 cathode indicate that the sluggish solid-phase diffusion of Mg ions can be improved remarkably by the concerted interaction with agile Li ions in Mg–Li dual-ion systems. In this part, versatility of the concerted motion phenomena is examined with spinel oxide materials, which are promising high-performance cathode materials of Mg batteries.

11.4.6.1 Concomitant Intercalation in Oxide Hosts

Compared to the Chevrel compounds, spinel oxides are considered as high performance cathodes of Mg batteries, because of high theoretical capacities and operating potentials.[39,40] However, considerable loss of cathode performance practically occurs, due to the high energy barriers of Mg migration,[40,41] and instability of Mg-inserted oxide hosts.[39,42] Many techniques have been employed to improve the cathode performance, such as grain refinement (shortening diffusion paths), and elevating the operating temperature (increasing the jumping frequency). For example, by elevating the operation temperature to around 150 °C, reversible Mg intercalation was successfully activated in $MgCo_2O_4$ cathode (Figure 11.17c).[39] As shown in Figure 11.17a and b, inserted Mg ions would push out cations at tetrahedral 8a sites to octahedral 16c sites, which transforms the spinel host ($MgCo_2O_4$) to a rock-salt structure ($Mg_{1+x}Co_2O_4$); a similar phase transition also occurs when Li ions are inserted into the $MgCo_2O_4$ host. Since both Mg and Li ions are possible to insert, the cathode properties of the $MgCo_2O_4$ host were tested in a Mg–Li dual-ion electrolyte ($MgTFSA_2$–LiTFSA–CsTFSA ionic liquid with a molar ratio of 10 : 10 : 80) at 150 °C. Figure 11.17d shows that intercalation of both Mg and Li ions occurs in the $MgCo_2O_4$ host, which indicates that concomitant intercalation can be also obtained with these spinel type oxide cathodes.

11.4.6.2 Facilitating Diffusion in Spinel Oxide at Room Temperature

To further study the versatility of concerted motion phenomenon in spinel oxide hosts, λ-Mn_2O_4 was employed as another cathode material. Compared to the unstable spinel Co_2O_4,[39] λ-Mn_2O_4 can be easily prepared by delithiation of $LiMn_2O_4$. λ-Mn_2O_4 is one of the most well-known oxide cathode materials for Li-ion batteries. However, in a Mg single-ion system, intercalation of Mg ions is generally difficult and shows a significantly high overpotential in non-aqueous electrolytes.[39] Here, λ-Mn_2O_4 hosts with typical concentrations of pre-inserted Li ions were prepared by charging commercial $LiMn_2O_4$ cathodes (Sigma-Aldrich, <0.5 μm particle size) to the corresponding cut-off potentials in a Li single-ion electrolyte (Figure 11.18a). The prepared λ-Mn_2O_4 and $Li_{0.5}Mn_2O_4$ cathodes were subsequently discharged in a

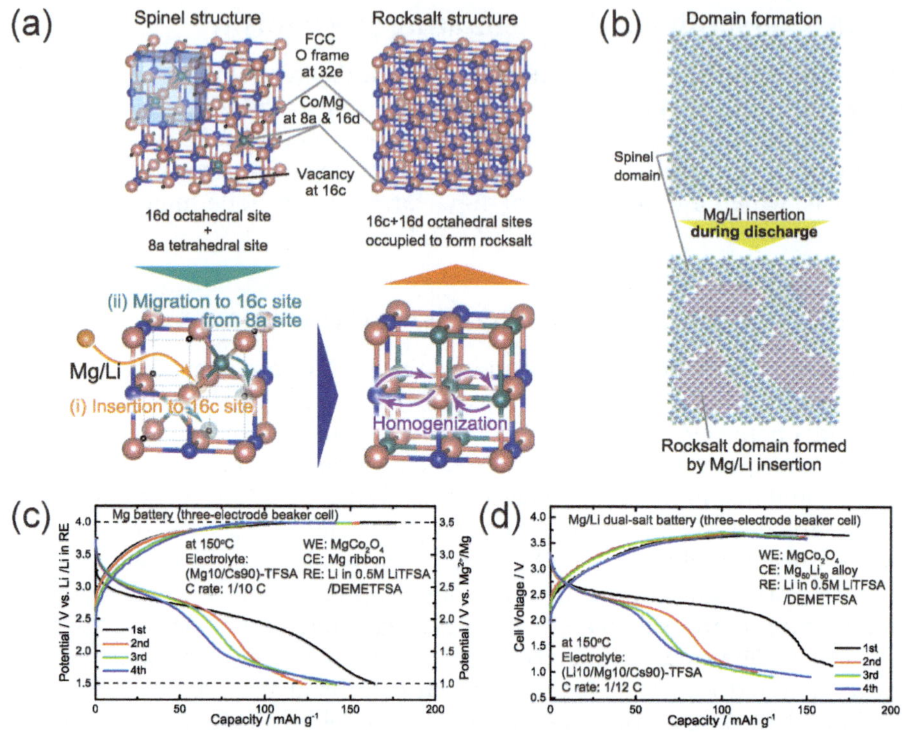

Figure 11.17 (a) Schematic illustration showing the cation insertion and transformation process in the $MgCo_2O_4$ cathode. In the spinel structure, where a 32e site is set at the origin, 8a, 16c, 16d, and 32e indicate the Wyckoff positions in $Fd\bar{3}m$ (space group no. 227) and usually 16c sites (marked by small black circles) are vacant in spinel structures. After a Mg or Li cation is inserted into a 16c site in the spinel structure, the original cation located in its neighbouring 8a site moves to an adjacent 16c site due to the electrostatic repulsion between the cations, to form a rock-salt structure. This fundamental process occurs within the framework of the spinel structure, and the phase change occurs coherently. (b) After continuous insertion of cations, atomic-level rock-salt domains are formed in the spinel structure, resulting in coherent phase separation. Battery cycle tests (c) for the Mg battery system, and (d) for the Mg–Li dual-ion battery system, were conducted in three beaker cells. Cut-off conditions in the battery tests were made based on the WE potential. In the cathode active material after the battery tests, the insertion-amount ratio of Mg and Li was measured to be Mg/Li = 7/5 by ICP analysis. Reproduced from ref. 17 with permission from the Royal Society of Chemistry.

Mg single-ion electrolyte; the discharge profiles are shown in Figure 11.18b. Apparently, the discharge potential and capacity of Mg intercalation with the $Li_{0.5}Mn_2O_4$ cathode are much higher than those of the empty λ-Mn_2O_4 host, which shows the opposite trend and is significantly different from the cases of Li-insertion into the λ-Mn_2O_4 and $Li_{0.5}Mn_2O_4$ hosts (dashed lines in Figure

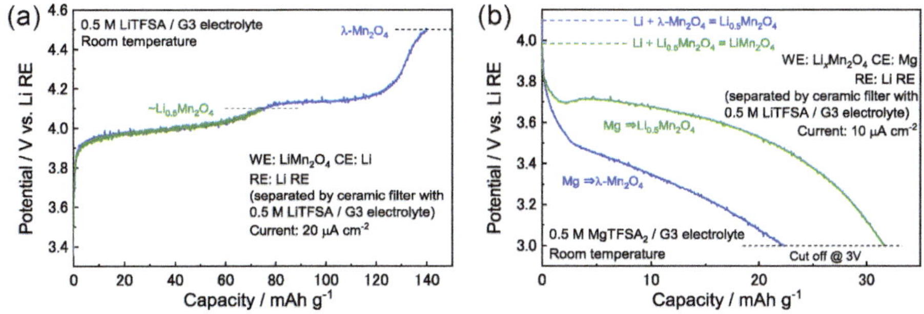

Figure 11.18 Galvanostatic charge and discharge profiles of spinel $Li_xMn_2O_4$ (x = 1–0) cathodes. (a) Galvanostatic charge profiles of commercial $LiMn_2O_4$ cathodes in 0.5 M LiTFSA/G3 electrolyte. Cut-off energies were set to 4.5 and 4.1 V (*vs.* Li RE), by which typical compositions of λ-Mn_2O_4 and $Li_{0.5}Mn_2O_4$ hosts were obtained. (b) Galvanostatic discharge profiles of λ-Mn_2O_4 and $Li_{0.5}Mn_2O_4$ cathodes in 0.5 M $MgTFSA_2$/G3 electrolyte. The discharge curve of each cathode is colored corresponding to the charge curves in (a). The discharge potentials of Li intercalation into the λ-Mn_2O_4 and $Li_{0.5}Mn_2O_4$ cathodes are indicated with dashed lines for comparison. Reproduced from ref. 41 with permission from John Wiley & Sons, © 2018 WILEY-VCH Verlag GmbH & Co. KGaA, Weinheim.

11.18b). The higher discharge potential and capacity of the $Li_{0.5}Mn_2O_4$ host indicate that the overpotential for Mg intercalation is drastically decreased owing to the preceding intercalation of Li ions, which indicates that the concerted interaction of Mg–Li dual ions is effective in improving Mg intercalation behavior.

11.5 Conclusions and Remarks

Novel rechargeable battery systems employing Mg–Li dual ions have been reviewed in this chapter. Daniell-type Mg–Li dual-ion batteries are an effective approach to combine a safe Mg metal anode and high-performance Li-intercalation cathode. Although the demand for a large amount of electrolyte causes the difficulty in achieving a higher energy density than in Li-ion batteries, the high safety makes the Daniell-type Mg–Li dual-ion batteries a promising alternative for stationary energy storage systems.

To achieve a high energy density, rocking-chair type Mg–Li dual-ion batteries are proposed, which consist of a Mg–Li alloy anode and a Mg–Li co-intercalation cathode. Since both Mg and Li ions participate in each electrode reaction, almost all charge carriers can be accommodated at the electrode, which means the volume of electrolyte can be reduced to a minimum. In the study of the Mg–Li alloy anode, it is confirmed that dendritic growth of Li can be successfully suppressed by the co-electrodeposition of Mg–Li dual ions, which means that a Mg–Li alloy anode can be used safely.

Moreover, regarding the cathode, concomitant intercalation of Mg–Li dual ions is observed with Chevrel compound and spinel oxide cathodes, where Mg intercalation is found to be significantly improved in Mg–Li dual-ion electrolytes. Further investigation of this phenomenon with first-principles calculations revealed that concerted interaction between preceding Li ions and following Mg ions can remarkably reduce the activation energy of Mg migration in the cathode host, which effectively facilitates Mg diffusion during intercalation reactions.

Since there are two kinds of cations that participate in the intercalation reactions simultaneously, the differences in the thermodynamic and kinetic properties associated with the dual ions in the host materials determine whether concerted intercalation can occur or not. Based on the knowledge obtained from the Mo_6S_8 and spinel oxide hosts, we point out several necessary conditions to facilitate the solid-phase diffusion of multivalent ions in dual-ion systems:

(i) Stable solid solutions containing a certain range of dual ions must exist in the host material;
(ii) Sluggish multivalent ions, *e.g.* Mg ions in Mg–Li dual-ion systems, are thermodynamically preferred to be inserted into hosts;
(iii) More than two cation sites of different stabilities must be connected with each other alternately along the diffusion paths.

In addition, carrier ions with similar ion radii,[43] such as Li–Mg, Li–Zn, Mg–Zn, and Na–Ca dual-ion systems, would be suitable for concerted motion, because the similar ion radii of dual ions usually need relatively smaller structural changes in host materials.

Rocking-chair type dual-ion batteries are still in the developmental stages and not yet ready for practical use, but the electrode properties and phenomena investigated here provide intriguing perspectives and inspiration for realizing high-performance future rechargeable batteries with multivalent ions.

Acknowledgements

The authors would like to express their sincere thanks to Dr Yu Kumagai and Prof. Fumiyasu Oba (Tokyo Institute of Technology, Japan) for the kind guidance and useful discussions on the computational methods. The authors also would like to thank Kohei Shimokawa (Tohoku University, Japan) for the kind advice on the manuscript. The first-principles calculations in this work were performed under the inter-university cooperative research program of the Center for Computational Materials Science, Institute for Materials Research, Tohoku University (Proposal No. 18S0405). This work is supported by a Grand-in-Aid for Scientific Research (S) no. 18H05249, and a Grand-in-Aid for JSPS Research Fellow no. 18J11696, commissioned by the Japan Society for the Promotion of Science (JSPS).

References

1. T. D. Gregory, *J. Electrochem. Soc.*, 1990, **137**, 775.
2. O. Mizrahi, N. Amir, E. Pollak, O. Chusid, V. Marks, H. Gottlieb, L. Larush, E. Zinigrad and D. Aurbach, *J. Electrochem. Soc.*, 2008, **155**, A103.
3. Z. Zhao-Karger, X. Zhao, O. Fuhr and M. Fichtner, *RSC Adv.*, 2013, **3**, 16330.
4. R. E. Doe, R. Han, J. Hwang, A. J. Gmitter, I. Shterenberg, H. D. Yoo, N. Pour and D. Aurbach, *Chem. Commun.*, 2014, **50**, 243.
5. I. Shterenberg, M. Salama, H. D. Yoo, Y. Gofer, J.-B. Park, Y.-K. Sun and D. Aurbach, *J. Electrochem. Soc.*, 2015, **162**, A7118.
6. P. Canepa, S. Jayaraman, L. Cheng, N. N. Rajput, W. D. Richards, G. S. Gautam, L. A. Curtiss, K. A. Persson and G. Ceder, *Energy Environ. Sci.*, 2015, **8**, 3718.
7. N. Sa, B. Pan, A. Saha-Shah, A. A. Hubaud, J. T. Vaughey, L. A. Baker, C. Liao and A. K. Burrell, *ACS Appl. Mater. Interfaces*, 2016, **8**, 16002.
8. J. Muldoon, C. B. Bucur, A. G. Oliver, J. Zajicek, G. D. Allred and W. C. Boggess, *Energy Environ. Sci.*, 2013, **6**, 482.
9. K. Shimokawa, H. Matsumoto and T. Ichitsubo, *J. Phys. Chem. Lett.*, 2018, **9**, 4732.
10. R. Mohtadi, M. Matsui, T. S. Arthur and S.-J. Hwang, *Angew. Chem., Int. Ed.*, 2012, **51**, 9780.
11. T. Fukutsuka, K. Asaka, A. Inoo, R. Yasui, K. Miyazaki, T. Abe, K. Nishio and Y. Uchimoto, *Chem. Lett.*, 2014, **43**, 1788.
12. Z. Zhao-Karger, M. E. Gil Bardaji, O. Fuhr and M. Fichtner, *J. Mater. Chem. A*, 2017, **5**, 10815.
13. D. Aurbach, Z. Lu, A. Schechter, Y. Gofer, H. Gizbar, R. Turgeman, Y. Cohen, M. Moshkovich and E. Levi, *Nature*, 2000, **407**, 724.
14. C. Ling and K. Suto, *Chem. Mater.*, 2017, **29**, 3731.
15. E. Levi, G. Gershinsky, D. Aurbach, O. Isnard and G. Ceder, *Chem. Mater.*, 2009, **21**, 1390.
16. S. Yagi, T. Ichitsubo, Y. Shirai, S. Yanai, T. Doi, K. Murase and E. Matsubara, *J. Mater. Chem. A*, 2014, **2**, 1144.
17. T. Ichitsubo, S. Okamoto, T. Kawaguchi, Y. Kumagai, F. Oba, S. Yagi, N. Goto, T. Doi and E. Matsubara, *J. Mater. Chem. A*, 2015, **3**, 10188.
18. H. Li, T. Ichitsubo, S. Yagi and E. Matsubara, *J. Mater. Chem. A*, 2017, **5**, 3534.
19. Z. Zhang, H. Xu, Z. Cui, P. Hu, J. Chai, H. Du, J. He, J. Zhang, X. Zhou, P. Han, G. Cui and L. Chen, *J. Mater. Chem. A*, 2016, 2277.
20. J.-H. Cho, M. Aykol, S. Kim, J.-H. Ha, C. Wolverton, K. Y. Chung, K.-B. Kim and B.-W. Cho, *J. Am. Chem. Soc.*, 2014, **136**, 16116.
21. T. Gao, F. Han, Y. Zhu, L. Suo, C. Luo, K. Xu and C. Wang, *Adv. Energy Mater.*, 2015, **5**, 1401507.
22. M. Walter, K. V. Kravchyk, M. Ibáñez and M. V. Kovalenko, *Chem. Mater.*, 2015, **27**, 7452.

23. H. Dong, Y. Li, Y. Liang, G. Li, C.-J. Sun, Y. Ren, Y. Lu and Y. Yao, *Chem. Commun.*, 2016, **52**, 8263.
24. X.-G. Sun, Z. Bi, H. Liu, Y. Fang, C. A. Bridges, M. P. Paranthaman, S. Dai and G. M. Brown, *Chem. Commun.*, 2016, **52**, 1713.
25. X.-G. Sun, Z. Zhang, H. Y. Guan, C. A. Bridges, Y. Fang, Y.-S. Hu, G. M. Veith and S. Dai, *J. Mater. Chem. A*, 2017, **5**, 6589.
26. H. Tian, T. Gao, X. Li, X. Wang, C. Luo, X. Fan, C. Yang, L. Suo, Z. Ma, W. Han and C. Wang, *Nat. Commun.*, 2017, **8**, 14083.
27. M. C. Lin, M. Gong, B. Lu, Y. Wu, D. Y. Wang, M. Guan, M. Angell, C. Chen, J. Yang, B. J. Hwang and H. Dai, *Nature*, 2015, **520**, 325.
28. Y. Cheng, D. Choi, K. S. Han, K. T. Mueller, J. Zhang, V. L. Sprenkle, J. Liu and G. Li, *Chem. Commun.*, 2016, **52**, 5379.
29. D. Aurbach, Z. Lu, A. Schechter, Y. Gofer, H. Gizbar, R. Turgeman, Y. Cohen, M. Moshkovich and E. Levi, *Nature*, 2000, **407**, 724.
30. T. Kimura, K. Fujii, Y. Sato, M. Morita and N. Yoshimoto, *J. Phys. Chem. C*, 2015, **119**, 18911.
31. W. Gasior, Z. Moser, W. Zakulski and G. Schwitzgebel, *Metall. Mater. Trans. A*, 1996, **27**, 2419.
32. P. Press, *Elements*, 1991, **15**, 317.
33. E. Levi, A. Mitelman, O. Isnard, M. Brunelli and D. Aurbach, *Inorg. Chem.*, 2008, **47**, 1975.
34. C. Ritter, E. Gocke, C. Fischer and R. Schöllhorn, *Mater. Res. Bull.*, 1992, **27**, 1217.
35. G. Henkelman, B. P. Uberuaga and H. Jónsson, *J. Chem. Phys.*, 2000, **113**, 9901.
36. P. E. Blöchl, *Phys. Rev. B*, 1994, **50**, 17953.
37. J. P. Perdew, K. Burke and M. Ernzerhof, *Phys. Rev. Lett.*, 1996, **77**, 3865.
38. L. F. Wan, B. R. Perdue, C. A. Apblett and D. Prendergast, *Chem. Mater.*, 2015, **27**, 5932.
39. S. Okamoto, T. Ichitsubo, T. Kawaguchi, Y. Kumagai, F. Oba, S. Yagi, K. Shimokawa, N. Goto, T. Doi and E. Matsubara, *Adv. Sci.*, 2015, **2**, 1500072.
40. M. Liu, Z. Rong, R. Malik, P. Canepa, A. Jain, G. Ceder and K. A. Persson, *Energy Environ. Sci.*, 2015, **8**, 964.
41. H. Li, N. L. Okamoto, T. Hatakeyama, Y. Kumagai, F. Oba and T. Ichitsubo, *Adv. Energy Mater.*, 2018, **8**, 1801475.
42. Q. D. Truong, M. Kempaiah Devaraju, P. D. Tran, Y. Gambe, K. Nayuki, Y. Sasaki and I. Honma, *Chem. Mater.*, 2017, **29**, 6245.
43. R. D. Shannon, *Acta Crystallogr., Sect. A*, 1976, **32**, 751.

CHAPTER 12

Aqueous Mg Batteries

MIN DENG[a], DANIEL HÖCHE*[a,b], DARYA SNIHIROVA[a], LINQIAN WANG[a], BAHRAM VAGHEFINAZARI[a], SVIATLANA V. LAMAKA[a] AND MIKHAIL L. ZHELUDKEVICH[a,c]

[a]Helmholtz-Zentrum Geesthacht (HZG), MagIC – Magnesium Innovation Centre, 21502 Geesthacht, Germany; [b]Helmut-Schmidt-University - University of the Federal Armed Forces, Faculty of Mechanical Engineering, Computational Material Design, 22043 Hamburg, Germany; [c]Kiel University, Faculty of Engineering, Institute of Materials Science, 24143 Kiel, Germany
*E-mail: Daniel.Hoeche@hzg.de

12.1 Introduction

Magnesium possesses several appealing qualities for energy storage, including a relatively negative electrode potential (−2.37 V *vs.* SHE) and high volumetric capacity (3832 mA h cm^{-3}).[1–3] Moreover, Mg is relatively low in cost and environmentally benign. All of these merits enable metallic Mg to be a promising anode material for use in non-aqueous and aqueous batteries. Generally, aqueous Mg batteries are not electrochemically rechargeable due to the irreversible formation of $Mg(OH)_2$ caused by the self-discharge reaction of Mg in aqueous solution. Nevertheless, aqueous Mg-based batteries can be mechanically recharged through replacing the anode, which allows these battery systems to have numerous applications, especially in marine devices.

Over the past few decades, different types of aqueous Mg batteries have been designed with various cathode materials to fulfill the requirements of

Energy and Environment Series No. 23
Magnesium Batteries: Research and Applications
Edited by Maximilian Fichtner

Published by the Royal Society of Chemistry, www.rsc.org

different system environments. However, the wider application of aqueous Mg batteries is still limited due to issues concerning the separate parts of the battery cells, *i.e.* the anode, cathode and especially water-based electrolyte. In order to promote the application of aqueous Mg batteries, numerous methods have been adopted to boost the battery performance, including development of novel electrode materials, modification of electrolyte solutions, *etc.* In this chapter, common types of aqueous Mg batteries are summarized. The main issues, such as anode self-corrosion (related to anode impurities), which hinder the wider application of aqueous Mg batteries are discussed. Moreover, some conventional and advanced approaches for boosting the performance of aqueous Mg batteries are covered based on numerous research studies. At last, some outlooks concerning the applications and research directions of aqueous Mg batteries are presented based on the authors' knowledge and experience gained from Mg-based battery development.

12.2 Types of Aqueous Mg Batteries

Various types of aqueous Mg batteries have already been developed and utilized on different occasions, such as dry cell, reserve and mechanically rechargeable types. Here we categorize them based on the different cathode materials and reactants involved in the cathodic reactions, as shown in Figure 12.1 and Table 12.1. Additionally, battery performance boosting related to the cathode material is presented separately, since the cathode characteristics, including the cathode active material, cathode structure and catalysts, exert great influence on the properties of aqueous Mg batteries. The discharge potential of the cathodic reaction and, thus, the discharge voltage of Mg batteries, varies with different cathode active materials. Severe activation polarization and concentration polarization, which are generally related to the sluggish kinetics of the cathodic reaction and limited mass transfer of the cathodic reactants, respectively, appear alongside the discharge process

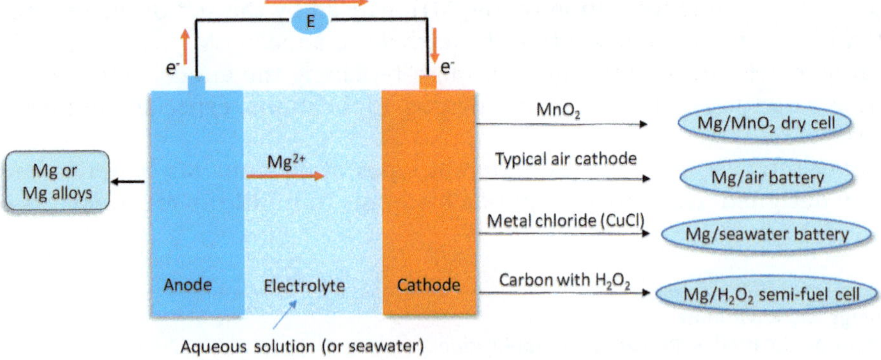

Figure 12.1 Schematic of various primary aqueous Mg-based batteries according to the cathode reactants.

Table 12.1 Characteristics of typical aqueous Mg batteries.

Battery type	Cathode reactant	Electrolyte	Theoretical voltage[a] (V)	Cathode capacity[c] (A h g^{-1})	Theoretical energy density[d] (W h kg^{-1})	Advantages	Application
Mg–MnO_2	MnO_2	Alkaline solution (*e.g.* $Mg(ClO_4)_{2\,aq}$)	2.84	0.308	759	High capacity, higher voltage than Zn–carbon, competitive cost	Military communication equipment, aircraft emergency transmitters
Mg–seawater	AgCl, CuCl, CuI, $PbCl_2$, MnO_2, *etc.*	Seawater	2.59[b]	0.187[b]	447[b]	High energy density, long unactivated shelf-time, safe	Electric torpedoes, marine devices
Mg–H_2O_2	H_2O_2 (fed into catholyte while operating)	Analyte: seawater, Catholyte: seawater, with acid (H_2SO_4)	4.15	1.577	3814	High voltage and power density, environmentally compatible	Unmanned undersea vehicles
Mg–air	O_2 (ambient air)	Neutral saline solution (*e.g.* $NaCl_{aq}$)	3.09	3.350	6800	High energy density, low cost, environmentally benign, safe	Back-up power source, remote military and telecommunication sites, undersea devices

[a]Voltage of a single cell calculated *via* standard electrode potential of each half-reaction.
[b]With AgCl active cathode material.
[c]Calculated for cathode active material only.
[d]Based on anode and cathode active materials only, excluding oxygen from the air and water in electrolyte.

and, thus, might reduce the cell voltage and cathode efficiency. Therefore, advanced cathode active materials such as active carbons with suitable catalysts are always needed to facilitate the cathodic reaction and to decrease the related activation polarization. Besides this, a suitable cathode structure is also important to solve the problems caused by the limited concentration of cathodic reactants due to species transport limitations either from the front or back side, by increasing the cathode/electrolyte interface area for example.

12.2.1 Mg–MnO_2 Dry Cell

The Mg–MnO_2 dry cell uses commercially pure Mg (cp-Mg) or Mg alloys as an anode, manganese dioxide mixed with acetylene black as a cathode and an alkaline solution (*e.g.* magnesium perchlorate with magnesium hydroxide) as electrolyte, including some chromate as an corrosion inhibitor.[4] The discharge reactions of the Mg–MnO_2 dry cell are:

$$\text{Anode: } Mg + 2OH^- \rightarrow Mg(OH)_2 + 2e \quad (12.1)$$

$$\text{Cathode: } 2MnO_2 + H_2O + 2e \rightarrow Mn_2O_3 + 2OH^- \quad (12.2)$$

$$\text{Overall: } Mg + 2MnO_2 + H_2O \rightarrow Mn_2O_3 + Mg(OH)_2 \quad (12.3)$$

When immersed in aqueous solution, Mg suffers from a parasitic reaction with water, which is termed as self-corrosion or self-discharge. The parasitic reaction is often related to local cathodic acting impurities, resulting in the formation of magnesium hydroxide, hydrogen gas and heat. Macroscopically, it can be described as:

$$Mg + H_2O \rightarrow Mg(OH)_2 + H_2\uparrow \quad (12.4)$$

The Mg–MnO_2 dry cell was developed in the early 20th century and it possesses some advantages over the primary Zn–MnO_2 battery, also commonly known as the alkaline battery. Firstly, the theoretical voltage and energy density of the Mg–MnO_2 cell are 2.8 V and 759 W h kg^{-1} (based on the anode and cathode active materials only), respectively, both higher than those of the Zn–MnO_2 system.[5] However, in practice, the voltage is decreased due to a large overvoltage of 0.7–0.9 V. Thus, the Mg–MnO_2 cell can achieve an open-circuit voltage (OCV) of only 1.9–2.0 V, still higher than that of the Zn–MnO_2 battery. Besides this, the Mg–MnO_2 dry cell has a long shelf life in an undischarged condition, even at high temperatures (>>RT), where it is twice that of the Zn system. The excellent storability is attributed to the protective film that forms on the surface of the Mg anode and the strong inhibition against secondary detrimental corrosion caused by chromate inhibitors. Therefore, the Mg–MnO_2 dry cell has been successfully used in military communication equipment and aircraft emergency transmitters.[5]

Traditional MnO_2 powders are normally used to prepare the cathode for Mg dry cells, mixed with acetylene black or carbon black. Basically, among the various crystallographic forms of MnO_2, such as α-, β-, γ- and δ-types, α-MnO_2 is a traditionally attractive material because of its specific properties for industrial applications and molecular sieves. Nevertheless, the morphologies of MnO_2 powders have an impact on their properties as a cathode material. As reported by Srither *et al.*, a cathode prepared with MnO_2 nanoparticles (average diameters of 87 and 46 nm) exhibits a higher discharge capacity (173 and 206 mA h g^{-1}, respectively) and enhanced energy density (299 and 360 W h kg^{-1}, respectively) than a cathode with micron MnO_2 powder, with a discharge capacity of 103 mA h g^{-1} at an energy density of 178 W h kg^{-1}.[6] Besides this, graphene-doped MnO_2 nanoparticles can further improve the performance of Mg–MnO_2 cells compared to a pristine MnO_2 electrode.[7] This is attributed to the higher Mg-ion storage capacity in the increased open space available in the MnO_2–graphene framework. Additionally, some organic nitro compounds have been reported to act as advanced active cathode materials for primary Mg dry cells, such as 2-β-dinitrostyrene.[8] However, as indicated by Renuka, the usage of 2-β-dinitrostyrene as a cathode decreases the voltage of the Mg primary cell in $Mg(ClO_4)_2$ solution (~0.95 V), compared to that of a traditional MnO_2 cathode (~1.5 V), even though a more stable discharge voltage can be obtained.[8]

12.2.2 Mg–Seawater Battery

The Mg–seawater battery uses cp-Mg or Mg alloys as an anode material and seawater from the ocean as an electrolyte solution. The active cathode material varies from silver chloride, cuprous chloride, lead chloride, cuprous iodide, cuprous thiocyanate to manganese dioxide.[9] The current-producing and overall reactions for this seawater battery are as follows (using a Mg/seawater/AgCl system as an example)

$$\text{Anode: Mg} \rightarrow \text{Mg}^{2+} + 2\text{e} \tag{12.5}$$

$$\text{Cathode: AgCl} + \text{e} \rightarrow \text{Ag} + \text{Cl}^{-} \tag{12.6}$$

$$\text{Overall: Mg} + 2\text{AgCl} \rightarrow 2\text{Ag} + \text{MgCl}_2 \tag{12.7}$$

The detrimental side reaction seen in eqn (12.4) between the Mg anode and water also occurs in this battery system.

The water-activated battery was developed in the 1940s to meet the need for an extremely lightweight, small sized and high-powered battery for military applications.[10] During this period, magnesium was used as the anode material in most water-activated batteries due to its attractive electrochemical properties. Normally, the water-activated battery has a long shelf life because it is constructed and stored under dry conditions and the seawater-based electrolyte is separated from the battery system during storage. This battery, usually as a reserve type, is activated at the time of use when the electrolyte is added in by *e.g.* opening a sensor controlled valve. The huge benefit of the water-activated battery is that the electrolyte can be natural seawater from the ocean, decreasing the space and cost for battery storage and transport. The Mg–seawater battery has been successfully used in various applications with different designs and cathode materials.

A Mg/seawater/AgCl battery was first developed as a power source for electric torpedoes. Afterwards, it was adapted as a power source for marine devices, such as sonobuoys, weather balloons, air–sea rescue equipment, emergency lights, *etc.*, due to its high energy density, high power density and long shelf life in an unactivated condition.[11] Silver chloride is a well-known cathode active material for seawater batteries since it offers several advantages compared to other active materials for cathodes.[12–15] Firstly, the high reduction potential of AgCl (0.222 V) contributes towards the high discharge voltage of the Mg–seawater system.

Secondly, silver chloride can be melted, cast into ingots and rolled into sheets of up to 0.08 mm in thickness.[13] It can be used in almost any cell configuration due to its ductility and malleability. Besides this, a base current collector grid is not needed for the usage of an AgCl cathode. However, the high raw material cost hinders its commercial application in a wider range of power sources. Therefore, the request for economical alternatives to AgCl has prompted searches for substitute cathode materials. In this context, some non-silver cathodes have been proposed, such as $PbCl_2$, MnO_2 and some copper compounds, such as CuCl and CuI, despite no alternative being found to achieve discharge properties close to those of AgCl.[14,16–23] Among these non-silver cathodes, CuCl is commercially successful but shows drawbacks of low voltage, inferior energy density, low rate capability and worse resistance to storage under high humidity. CuI was found to have a superior performance and more improvements can be achieved with cathode additives such as sulphur.

The voltage increase of a CuCl electrode caused by the addition of elemental sulphur was found and explained by Amlie *et al.*[24] After sulphur addition, the electrode reaction changes from

$$\mathrm{CuCl + e \rightarrow Cu + Cl^-} \tag{12.8}$$

to

$$\mathrm{CuCl_2 + e \rightarrow CuCl + Cl^-} \tag{12.9}$$

where cupric chloride is generated *via* a chemical reaction between sulphur and cuprous chloride after immersion in an aqueous electrolyte:

$$\mathrm{2CuCl + S + e \rightarrow CuCl_2 + CuS} \tag{12.10}$$

Therefore, the overall reaction can be presented as:

$$CuCl + S + e \rightarrow CuS + Cl^- \qquad (12.11)$$

leading to a theoretical voltage increase of 0.506 V.

The addition of sulphur was reported to enhance the cathode voltage of other copper compounds, such as CuBr, $CuSO_4$, $Cu(HCOO)_2$, $Cu(CH_3COO)_2$ and CuC_2O_4 *via* a similar mechanism.[25] Besides, the type of sulphur allotrope used also has a great influence on the cell voltage of Mg–CuI seawater batteries, as indicated by Renuka when comparing different sulphur additives, *i.e.* cyclohexasulphur (S_6), cyclooctasulphur (S_8) and cyclododecasulphur (S_{12}).[26] Aromatic disulfide, AgCl and Ag_2S as additives for CuI cathodes have been studied, showing significant enhancement in the cell voltage of a Mg seawater system, *e.g.* from 1.08 to 1.6 V after the addition of 10 wt% S and 1.25 wt% Ag_2S into the CuI cathode.[13,27] A voltage increase upon S addition is facilitated by the presence of AgCl and Cu_2S in the CuI cathode. Interesting to know is the fact that the adoption of AgCl as an additive to CuI systems can offer a battery performance close to that supplied by an AgCl active cathode while significantly reducing the silver content.

Beyond these, Mg seawater batteries using hydrogen peroxide or dissolved oxygen in the seawater as cathode reactants have also been developed for different applications. These two types, termed as the Mg–H_2O_2 semi-fuel cell and Mg–air battery, respectively, are covered in Sections 12.2.3 and 12.2.4, respectively.

12.2.3 Mg–H_2O_2 Semi-fuel Cell

The Mg–H_2O_2 semi-fuel cell was developed by the Naval Undersea Warfare Center (NUWC) as a power source for US Navy's low rate, long endurance unmanned undersea vehicles (UUV).[28] This semi-fuel cell uses Mg as an anode and hydrogen peroxide as a cathode reactant. The cathode, serving as the reaction site for hydrogen peroxide reduction, is normally carbon paper with palladium and iridium as a catalyst and is separated from the anode by a conductive ion exchange membrane. This system uses seawater as an anolyte and a catholyte with a solution of seawater, acid and hydrogen peroxide. The cell is activated when the anolyte and catholyte are pumped into the system. In other words, with sufficient active anode material, this system has the ability to produce electrical energy as long as the catholyte is fed to the cathode, being the reason for the naming of it as a "semi-fuel cell". The half-cell reactions and overall reaction are as follows:

$$\text{Anode: } Mg \rightarrow Mg^{2+} + 2e \qquad (12.12)$$

$$\text{Cathode: } H_2O_2 + 2H^+ + 2e \rightarrow 2H_2O \qquad (12.13)$$

$$\text{Overall: } Mg + H_2O_2 + 2H^+ \rightarrow Mg^{2+} + 2H_2O \qquad (12.14)$$

Besides reaction 12.4, secondary decomposition and precipitation reactions can occur:

$$2H_2O_2 \rightarrow H_2O + O_2 \quad (12.15)$$

$$Mg^{2+} + 2OH^- \rightarrow Mg(OH)_2 \quad (12.16)$$

$$Mg^{2+} + CO_3^{2-} \rightarrow MgCO_3 \quad (12.17)$$

The precipitation products of the reactions shown in eqn (12.16) and (12.17) decrease the voltage and energy density of the semi-fuel cell, but this issue can be solved through the addition of reacting acids such as H_2SO_4. The theoretical voltage of the acid Mg–hydrogen peroxide system is 4.14 V. Nevertheless, the practical voltage is only 1.7–1.8 V with some modification and the energy density ranges from 500–520 W h kg^{-1}.[28–30] The working conditions, including anolyte flow rate, catholyte flow rate, hydrogen peroxide concentration, current density and temperature, have a great influence on the discharge performance of this electrochemical system.[30]

In terms of Mg–H_2O_2 semi-fuel cells, the cathode material does not participate in the cathodic reaction. The cathodic reactants are fed in from the outside and then reduced in the overall battery reaction with assistance of the catalyst contained in the cathode, which is needed to accelerate the kinetics of the cathodic reaction. The combination of palladium and iridium has been widely accepted as a catalyst for the Mg–H_2O_2 system, since it shows superior performance compared to other catalysts such as silver and Pd.[28,29,31,32] A combined Pd/Ag catalyst has also been explored by electrodeposition of Pd on Ag coated nickel foam.[33] The Mg–H_2O_2 system with this Pd–Ag catalyzed porous cathode exhibits a maximum power density of 140 mW cm^{-2}, which is higher than for Ag coated Ni foam cathode (110 mW cm^{-2}) under the same operation conditions. Nevertheless, the battery performance of the Mg–H_2O_2 system is still limited by the concentration polarization of H_2O_2 when it is diluted in the flowing catholyte. Therefore, several cathode substrates have been adopted, such as nickel foams, titanium foams, carbon papers, *etc.*, and the cathode structure has been modified to decrease the H_2O_2 concentration effect.[34] Titanium foam shows the best corrosion resistance in 0.5 M H_2SO_4 solution when compared to nickel foam and silver-coated nickel foam. The adoption of a Pd/Ti cathode enhances the power density of the Mg–H_2O_2 system to 110 mW cm^{-2} compared to a Pd–Ag/Ni cathode (80 mW cm^{-2}) and enables the battery system to have good stability during 50 h of testing.[35] Moreover, carbon fibers have been applied to a carbon paper substrate to modify the cathode structure by Bessette *et al.*[36] As reported, after electrodeposition to the prepared carbon fiber cathode, well-formed Pd–Ir catalyst clusters are clearly shown, which are expected to significantly increase the surface area per volume. Hence, a superior cell voltage to current performance in the Mg–H_2O_2 system was achieved.

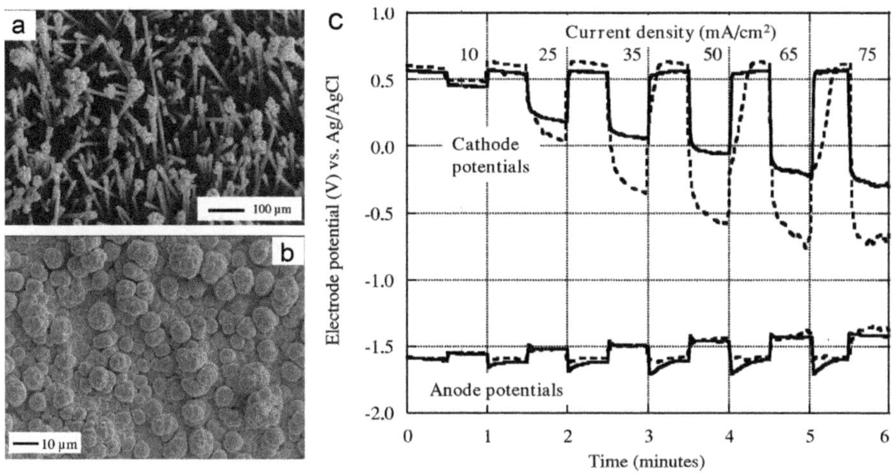

Figure 12.2 Top-view SEM images of (a) a Pd–Ir coated carbon microfiber array cathode, (b) a planar (control) Pd–Ir cathode. (c) The cathode and anode potentials of a Mg–H_2O_2 semi-fuel cell at various current densities: (–) with a carbon microfiber array-based cathode; (– – –) with a planar cathode. Reproduced from ref. 34 with permission from The Electrochemical Society, Copyright 2008.

As a further development, a method called direct charging electrostatic flocking has been used to fabricate 3D, carbon microfiber array (CMA)-based cathodes.[34] These cathodes comprise conductive carbon microfibers (11 μm in diameter, 500 μm in length) that protrude from the titanium substrate (as shown in Figure 12.2a). The CMA-based cathode with Pd–Ir catalyst enables the Mg–H_2O_2 system to have a higher voltage at a high current density (Figure 12.2c), enhanced power density (peak power density increased from 53 to 84 mW cm^{-2}) and equivalent H_2O_2 utilization, compared to planar cathodes with the same catalyst loading.

12.2.4 Mg–Air Battery (Aqueous Type)

The magnesium–air battery uses cp-Mg or a Mg alloy as an anode, an aqueous solution as an electrolyte and oxygen from the air as a cathode reactant. A typical air electrode with a catalyst is adopted as a cathode. The usage of air as a cathode reactant reduces the weight of this battery system and frees up more space for energy storage, contributing to the high energy density of the system. The reactions involved in the Mg–air cell are as follows:

$$\text{Anode: } Mg \rightarrow Mg^{2+} + 2e \quad (12.18)$$

$$\text{Cathode: } O_2 + 2H_2O + 4e \rightarrow 4OH^- \quad (12.19)$$

$$\text{Overall: } 2Mg + O_2 + 2H_2O \rightarrow 2Mg(OH)_2 \quad (12.20)$$

The theoretical voltage of the Mg–air system is 3.1 V and the energy density is 6.8 kW h kg^{-1} (based on the Mg anode only). Unavoidably, the specific voltage and energy density in practice are reduced, because of the parasitic reaction shown in eqn (12.4) again and severe overvoltage induced from different components (IR-drop, *etc.*) of the battery system. Another problem for the Mg–air system is the delayed voltage response to an increased load caused by the surface film of the discharge products, *i.e.* magnesium hydroxide. Nevertheless, the Mg–air batteries have a broad range of applications and good future prospects.

The Mg–air system was initially designed as a reserve type as the electrolyte is separated from the battery system. This battery can be stored for long time and be activated any time after adding electrolyte (water-based) into the system. Therefore, one important application of the Mg–air system is as a back-up power source for electric and solar power in schools or hospitals in case of emergencies. Besides this, the effective energy density of the Mg–air battery can be increased significantly with the addition of corrosion inhibitors to the electrolyte, contributing to its potential as a power source for remote military and telecommunication sites. Furthermore, as mentioned already in Section 12.2.2, a battery system with seawater as the electrolyte, oxygen dissolved in seawater as a cathode reactant and Mg-based material as an anode is also available. This type of battery is expected to have an energy density of 600 W h kg^{-1}, which is higher than that of other undersea power systems such as silver/zinc, Al/H_2O_2 and fuel cells.[37] Unfortunately, due to the limited solubility of oxygen in water, this system can only offer low current density and low power, therefore it is mainly used in long-term low power undersea devices, such as lighthouses, floats, undersea monitoring equipment, UUVs and sea floor borehole broadband seismic observatories.[38–40]

Also, here, the performance of Mg–air batteries is closely associated with the cathode, in which the oxygen reduction reaction (ORR) takes place. The structure of a conventional air cathode is shown in Figure 12.3, indicating four layers, *i.e.* the waterproof breathable layer, the gas diffusion layer (GDL), the catalyst layer and the current collector. The function of the waterproof layer is to prevent electrolyte leakage while enabling oxygen to penetrate from all sides. However, in some cases, this waterproof layer is not necessary as the gas diffusion layer, which is usually made from carbon materials and PTFE, can also offer the same functionality.[41–45] The catalyst layer is composed of effective catalysts to decrease the activation overpotential resulting from the sluggish kinetics of the ORR. Both the GDL and catalysts have a great influence on the performance of Mg–air batteries and much research work has been carried out to develop a superior cathode for a Mg–air system.

In fact, a series of complex electrochemical reactions are involved in the cathode oxygen reduction. In a neutral electrolyte, the related reactions can be expressed as follows, including the reaction shown in eqn (12.19):

$$O_2 + 2H_2O + 2e \rightarrow HO_2^- + OH^- \tag{12.21}$$

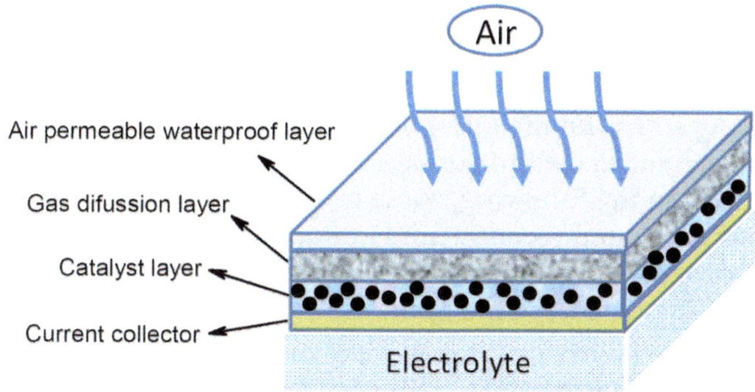

Figure 12.3 Conventional structure of a four-layer air cathode. Adapted from ref. 2 with permission from the Royal Society of Chemistry.

$$HO_2^- + H_2O + 2e \rightarrow 3OH^- \quad (12.22)$$

$$2HO_2^- \rightarrow 2OH^- + O_2 \quad (12.23)$$

Therefore, the oxygen reduction may proceed *via* a four-electron pathway (shown in eqn (12.19)) or a two-electron pathway (the reaction shown in eqn (12.21)) followed by either the reaction shown in eqn (12.22) or (12.23) on the surface of the cathode materials. The direct four-electron ORR pathway is desirable owing to its high efficiency for energy conversion, while the two-electron pathway is unfavorable because of its low efficiency and generated corrosive peroxide species.[1,2] The applied catalytic materials have a great influence on the ORR pathways and mechanisms, which, therefore, affect the battery performance in terms of a constant high power output. Hence, a promising catalyst should have good catalytic ability to favor the ORR *via* a direct four-electron pathway. Note that the ORR pathway varies not only with the catalyst category but also with its structure (surface geometry/morphology) due to its impact on oxygen transport and local surface reaction conditions. Normally most catalysts for other metal–air batteries can also operate as catalysts in Mg–air systems. Platinum shows superior ORR catalytic activity owing to the predominant four-electron pathway of the ORR, but the high price hinders its commercialization in Mg–air batteries. Therefore, the search for economical low-Pt or non-Pt catalysts draws much attention. Pt alloys doped with other transition metals, *e.g.* Fe, Ni and Co, and other modified noble metals such as Pd, Cu and Ag have been proposed as catalysts for the ORR. These catalysts show high catalytic activity and good durability. The adoption of noble metals rather than Pt reduces the cost associated with the cathode materials for Mg–air batteries. Nevertheless, more research has been devoted to explore more economically available catalysts without noble metals, such as carbon-based materials, transition metal oxides (MnO_2) and perovskite-based oxides. Various proposed catalysts for ORR and their modifications have been well summarized by Cheng *et al.* and Zhang *et al.*[1,2] However, most proposed catalysts for ORR were tested and evaluated in alkaline or acidic solutions. Mg–air batteries utilize neutral solutions as the electrolyte to avoid fast self-corrosion in acidic media and high overpotential resulting from the $Mg(OH)_2$ film in alkaline solutions. Suitable catalysts working in neutral electrolytes are required for superior performance of Mg–air batteries. In the following, some novel catalysts are briefly introduced that have been evaluated recently in pH-neutral Mg–air full cells.

Zhao *et al.* proposed a Pd–Sn binary alloy as a catalyst for Mg–air batteries to reduce the cathode cost while maintaining excellent catalytic activity.[46] A cathode was prepared with 24 h heat-treated Pd–Sn as a catalyst and multi-walled carbon nanotubes (MWCNTs) as the support material. The Mg–air battery with this Pd–Sn/MWCNTs-based cathode exhibits a higher power density (more than 112.4 mW cm^{-2}) and increased durability than that with a Pd/C-based catalyst (82 mW cm^{-2}). Besides this, non-metal catalysts as low-cost alternatives have also been explored. Planar iron–polyphthalocyanine (PPcFe) catalysts for Mg–air batteries were fabricated with different PPcFe/carbon ratios and various heat treatment temperatures by Li *et al.*[45] The PPcFe/C-based cathode prepared under optimized conditions (PPcFe/C ratio 1 : 1, 400 °C) was then tested in a Mg–air full cell and compared with a pure Pt/C-based cathode. The results demonstrate that a similar discharge voltage and power density can be obtained with this PPcFe/C-based catalyst in 10 wt% NaCl solution compared with a Pt/C-based catalyst. Furthermore, based on the success of perovskite-based oxides as catalysts for the ORR in alkaline solutions, a mixed-phase mullite ($SmMn_2O_5$) has been prepared and evaluated as a catalyst for Mg–air batteries by Li *et al.*[47] The results show that the mullite-based cathode exhibits a similar catalytic activity to that of a Pt/C-based cathode in 1 M NaCl solution, while better stability is obtained. The enhanced cathode performance is attributed to the catalytic Mn sites and facilitated oxygen adsorption and diffusion by the coexistence of a CeO_2 phase. In addition to the catalysts, the support materials in the catalyst layer also exert a great influence on the cathode performance, especially for MnO_2-based cathodes because of the low electric conductivity of catalytic MnO_2. Support materials, normally based on carbon, must have good electrical conductivity and can offer sufficient reaction sites for the ORR. Modified carbon materials with nanostructures have been explored as catalyst supports for Mg–air batteries, such as graphene, carbon nanofibers (CNFs) and the aforementioned MWCNTs.[41,44,48] Both carbon black and MWCNTs show a catalytic activity for the ORR that is inferior to that of a MnO_2 catalyst. Nevertheless, the catalytic activity of MnO_2 can be significantly enhanced when combining it with carbon materials, especially MWCNTs. The Mg–air battery measurements in 10 wt% NaCl solution demonstrate that the MnO_2/MWCNT-based cathode enables the Mg–air system to have a higher voltage and enhanced power density compared with MnO_2- and MnO_2/C-based cathodes.[41] In detail, the cell voltages for MnO_2/MWCNTs-, MnO_2- and MnO_2/C-based air batteries at a current density of 20 mA cm^{-2} are 1.4, 1.13 and 1.35 V, respectively, while the peak power densities are ~70, 55 and 61 mW cm^{-2}.

Apart from the aforementioned factors, the cathode performance is also related to the porous GDL. Proper pore number and dimensions in the GDL are beneficial for lowering the overpotential resulting from oxygen diffusion limitation, and, hence, enhancing the final battery performance. Therefore, the modification of electronic conduction materials in the GDL, which are generally carbon materials, has attracted attention. Carbon fiber felts (CFF) with a three-dimensional structure and high electronic conductivity have been suggested as alternatives for traditional carbon powders in the GDL.[42,43] As reported, a Mg–air cell with a CFF-based cathode shows higher power density (maximum of about 116 mW cm^{-2}) and good stability during 1100 h of life testing at a current density of 30 mA cm^{-2} in 10 wt% NaCl solution, compared to a traditional carbon powder-based cathode using the same catalysts (maximum of around 83 mW cm^{-2}).[43]

12.2.5 Other Types

Aside from the widely researched systems mentioned above, some new aqueous Mg-based power sources with special design have been proposed recently. Liu *et al.* developed a high-specific-energy magnesium–water battery for full-depth ocean applications with water as a cathode reactant.[49] This battery system uses nickel foam with optimized Pt loading as a cathode material, serving as a location for water reduction (Volmer–Heyrovsky type). The reactions involved in this system are as follows:

$$\text{Anode: } Mg \rightarrow Mg^{2+} + 2e \tag{12.24}$$

$$\text{Cathode: } 2H_2O + 2e \rightarrow H_2 + 2OH^- \tag{12.25}$$

$$\text{Overall: } Mg + 2H_2O \rightarrow Mg(OH)_2 + H_2 \tag{12.26}$$

Instead of dissolved oxygen, the utilization of water as a cathode reactant could overcome the issue of low power density caused by low and non-uniform concentration of oxygen in ocean water. Combined with AZ91 Mg alloy as an anode, this system exhibited a high energy density of 1003 W h kg^{-1} (anode related limitation) and stable discharge for 100 days, as reported. Nevertheless, large amounts of hydrogen are produced during the battery operation and released into the sea, which, consequently, causes considerable energy loss, *e.g.* due to superfluous hydrogen at the very surface and lowered effective surface. Another battery design combining a Mg–water system with a fuel cell was proposed by Hahn *et al.* to take advantage of the produced hydrogen and to increase the energy density.[50] The illustration of this battery system is presented in Figure 12.4. In this system, the Mg–water system is submerged in a seawater electrolyte, supplying power to an outside device and, meanwhile, serves as a hydrogen generator for a fuel cell. The released hydrogen, which is produced both by cathode reaction and self-discharge of

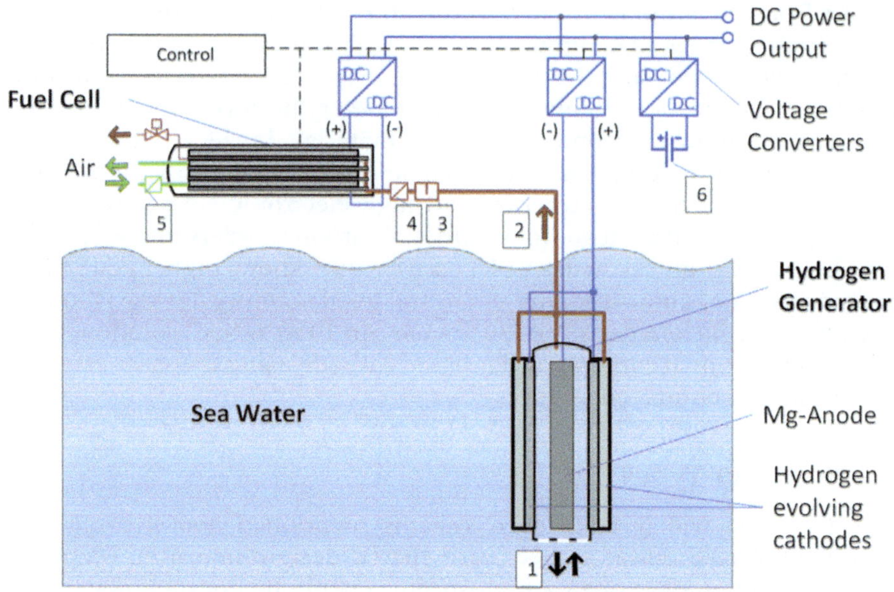

Figure 12.4 Concept of a magnesium hydrogen fuel cell power supply. (1) Bottom opening for sea water exchange and removal of reaction products, (2) hydrogen connection to fuel cell, (3) gas liquid separator, (4) hydrogen gas filter, (5) air filter, (6) secondary battery as an electrical buffer. Reproduced from R. Hahn, J. Mainert, F. Glaw and K.D. Lang, Sea water magnesium fuel cell power supply, *J. Power Sources*, **288**, 26–35., Copyright 2015, with permission from Elsevier.

the Mg anode, is collected and utilized as an anode reactant for a fuel cell above the sea level, which in parallel uses oxygen from the air as a cathode reactant. The anode and cathode reactions involved in this highly attractive fuel cell can be expressed as:

$$\text{Anode: } 2H_2 \rightarrow 4H^+ + 4e \quad (12.27)$$

$$\text{Cathode: } 2O_2 + 4H^+ + 4e \rightarrow 2H_2O \quad (12.28)$$

The overall reaction for the whole battery system can be described as:

$$2Mg + 2H_2O + O_2 \rightarrow 2Mg(OH)_2 \quad (12.29)$$

With a commercial AZ31 rolled sheet as an anode, this battery system can achieve a high energy density of more than 1200 W h kg^{-1} (again limited by the anode) when coupled with the fuel cell. Besides this, because the Mg–water system is open to the ocean, the discharge products (magnesium hydroxide) can be continuously flushed out from the cell lowering the ohmic effects at the surface. The concentration of the electrolyte remains constant, leading to the capability of long-term stable operation of this electrochemical power system. Furthermore, Xie *et al.* reported a new battery operated by direct contact of Mg foil (anode) with foamed polyaniline (PANI, cathode).[51] No additional electrolyte is needed for this battery system. The 3D PANI cathode possesses a through-pore structure with high porosity and strong polarity, contributing to the entrapment of H_2O (humidity) from air. Thus, the involved anode and cathode reactions in this system are the oxidation of Mg and the reduction of water, separately, similar to in the Mg–water system. This proposed battery prototype could eliminate the problems caused by the electrolyte, such as self-discharge of the Mg anode, simplifying the assembling and lowering the cost.

12.3 Current Issues of Aqueous Mg Batteries

Aqueous Mg batteries have been researched and developed for a long time. They have been successfully adopted in several applications, such as the Mg–MnO_2 dry cell in military applications, the Mg–seawater system in marine devices, the Mg–H_2O_2 semi-fuel cell in UUVs and so on, as mentioned above. However, a wider commercial acceptance of the Mg-based batteries has not been realized so far owing to several disadvantages. Firstly, magnesium or magnesium alloys suffer from severe self-corrosion during discharge due to the well-known negative difference effect (NDE) that can be simply described as an increasing tendency of self-corrosion rate with enhanced polarization on Mg anodes mainly related to noble inclusions such as Fe, Cu or Ni within the anode material (typically from bad raw material quality or from inadequate anode material processing). The high self-corrosion rate decreases the utilization efficiency of Mg anodes and the specific energy of the battery system, especially at low operating current densities. Besides this, after partial discharge, the severe self-corrosion of Mg anodes also leads to a poor storage capability and, thus, reduces the service life of the entire battery system. The produced large amounts of hydrogen and the generated heat require proper battery design for gas venting and heat dissipation. Furthermore, the practical voltage is always decreased because of the large overvoltage caused either by discharge product film on the anode surface or by the sluggish kinetics of the cathodic reactions involved. The tightly adherent oxide or hydroxide film on the anode surface would also lead to voltage delay when a higher current density is required for an increasing outside electrical power consumption. Therefore, in order to promote the application of Mg-based batteries, much work has been carried out with the aim of solving these existing problems and boosting the overall performance (U–I characteristics) of aqueous Mg batteries. The main approaches and recent research progress on Mg batteries are presented in the next section.

12.4 Performance Improvement of Aqueous Mg Batteries

Some problems of Mg batteries concerning practical application can be avoided through battery design. For instance, mechanical venting of excess gas is available through a small hole in the seal washer of the Mg–MnO_2 dry cells. The storage time of most Mg-based batteries can be increased when designed as a reserve type. Anode materials can be replaced anytime and taken out of the battery system, so that battery service life can be controlled and extended. Nevertheless, in order to boost the performance of Mg-based batteries (stable U–I characteristics at high utilization efficiency), efforts are normally focused on the main components of the battery system, *i.e.* the anode, the cathode and the electrolyte. Advanced Mg anodes with more negative discharge potential (in a half-cell) and low self-corrosion rate are under development. Suitable cathode materials with catalysts are important for achieving active reaction kinetics. Besides this, battery performance can also be improved through electrolyte modification, *e.g.* the addition of proper additives, leading to electrolyte stabilization and proper electrode/electrolyte interface conditions. The main approaches and progress achieved are summarized in the following section.

12.4.1 Development of Mg Anodes

As the generator of electrons in the electrochemical power system, the Mg anode plays a key role in the battery performance. Therefore, the development of novel Mg-based anodes has attracted considerable attention. Generally, advanced Mg anodes should possess two specific properties. The first one is high utilization efficiency, which can be achieved with its low self-corrosion rate and limited "chunk effect", namely the undermining of the undissolved Mg matrix and the detachment of the secondary phases.[52–54] The second one is at a more negative discharge potential (in a half-cell measurement), which, in principle, can be obtained through negative open circuit potential and low overpotential active dissolution of Mg anodes and easier peeling or even avoidance of the discharge product film formed on the anode surface. In general, the research on Mg anodes is focused on the following three directions: improving the intrinsic properties of pure magnesium (impurity level), adding alloying elements such as Ca and adjusting the microstructure (*e.g. via* extrusion).

12.4.1.1 Improvement of Pure Magnesium

The intrinsic quality of pure Mg has a great influence on its self-corrosion and discharge properties. The corrosion rate of pure magnesium is closely linked to the content of impurities, such as Fe, Ni, Cu, *etc.*[55–58] Pure magnesium of commercial purity grade shows a high corrosion rate and low utilization efficiency in different battery electrolytes as reported, indicating its imperfection as an anode material.[59–61] Improving the intrinsic quality of pure Mg is helpful for improving the discharge properties. High-purity Mg (99.99%) shows a better discharge voltage than some Mg alloys, *e.g.* AM50 and AZ91, as measured in a Mg–air cell under the same test conditions.[62] Besides this, instead of a conventional macroscale structure (either from cast or wrought processing), the adoption of a micro/nanoscale structure to pure Mg can also have positive effects on its discharge properties. Li *et al.* explored the discharge performance of micro/nanoscale structured Mg with various morphologies and shapes when acting as an anode material for Mg–air and Mg–MnO_2 batteries.[63,64] Mg hybrid systems consisting of micro/nano Mg spheres with different diameters can be prepared by vapor deposition processes at different temperatures (as shown in Figure 12.5). All of these prepared materials show a higher discharge voltage than commercial Mg powder when used as an anode for a Mg–MnO_2 cell due to the high electrochemical activity and reduced polarization caused by the increased surface area. Among these materials, sample C prepared at 300 °C shows the best discharge voltage (1.44 V at 25 mA g^{-1} *vs.* the 1.2 V of commercial Mg powder), which is attributed to its higher surface area caused by a mixture of micro- and nanospheres. It is worth noting that, with increasing specific surface area, the Mg anode with a micro/nanostructure unavoidably shows a fast self-corrosion rate. Therefore, a suitable electrolyte, in which the Mg anode exhibits a low self-corrosion rate, is highly desired to ensure high utilization efficiency and energy density. Another way proposed to improve the properties of Mg anode is using porous Mg thin films. Xin *et al.* prepared a series of Mg films with a thickness of 100 nm with various morphologies by magnetron sputtering at different temperatures.[65] Battery measurements indicated that a Mg–air cell with Mg films prepared at 150 °C as an anode and Pt/C as a cathode showed excellent discharge properties in 6 M KOH electrolyte, including a stable discharge voltage (0.8 to 1 V), high open circuit voltage (1.41 V) and large discharge capacity (~820 mA h g^{-1}).

12.4.1.2 Addition of Alloying Elements

Alloying is the most common method used to control the properties of Mg anodes, since numerous alloying elements can be used and they may have quite different impacts on the electrochemical properties of magnesium and the entire cell. Until now, many ternary Mg alloys have been developed as anode materials, such as Mg–Al–Zn, Mg–Al–Pb, Mg–Hg–Ga and Mg–Li–Al alloys aimed at preventing power failure by surface activation *via* the alloying element. Additionally, other alloying elements to these alloys have been explored in order to obtain a better anode performance. Figure 12.6 summarizes some commercially available and newly proposed Mg anodes from the aspects of anodic efficiency and OCP in sodium chloride solution.

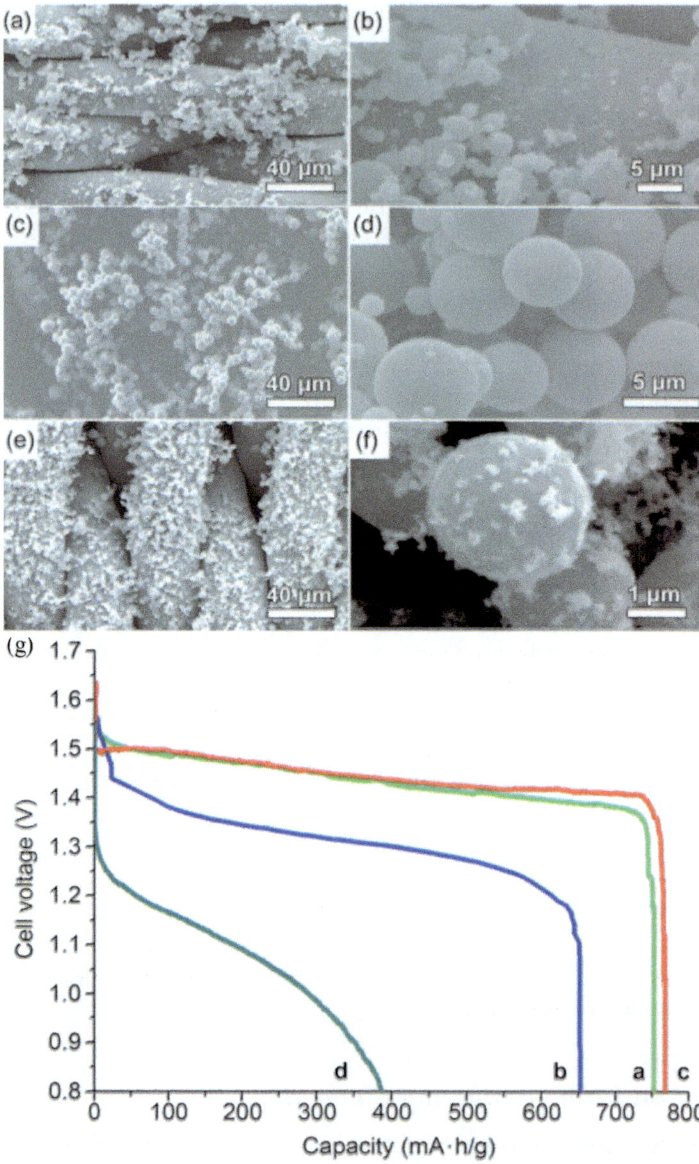

Figure 12.5 SEM images of the vapor-deposited Mg micro/nanoscale materials for different deposition temperatures: (a and b) 240 °C; (c and d) 270 °C; (e and f) 300 °C; (g) Discharge curves of different materials as anodes for a Mg–MnO_2 cell at a current density of 25 mA g^{-1} at a working temperature of 25 °C. a, b, c represents the sample vapors deposited at 240, 270 and 300 °C, respectively. Sample d is made with commercial Mg powder. The utilized electrolyte is a mixture of $Mg(NO_3)_2$ (2.6 M) and $NaNO_2$ (2.6 M). Reproduced from ref. 64 with permission from Springer Nature, Copyright 2009.

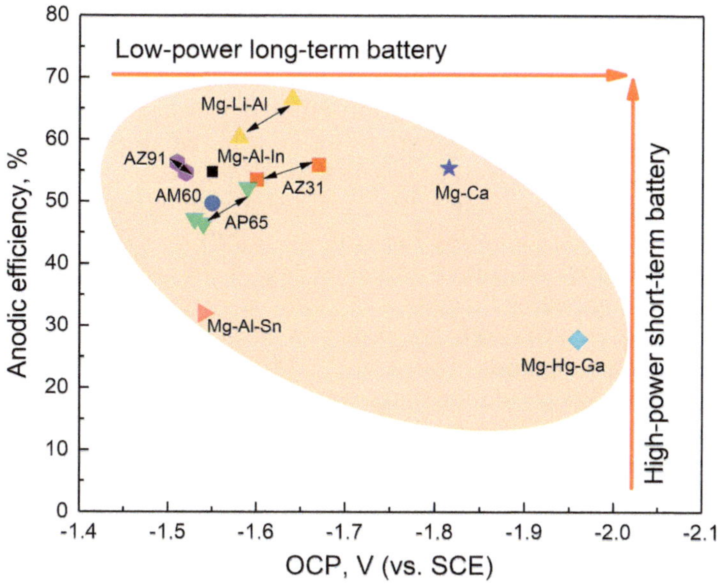

Figure 12.6 Property space for anodic efficiency *versus* the OCP of different Mg alloy anodes. All data are collected from published literatures (ref. 70–81). The anodic efficiencies of all of the anodes shown here were measured at a current density of 10 mA cm^{-2}, excluding Mg–Hg–Ga (efficiency was recorded at 20 mA cm^{-2}). All of the data were determined in 3.5 wt% NaCl solution. The OCP value is calibrated and *versus* SCE, assuming that the potential of a saturated Ag/AgCl electrode is −0.045 V *vs.* SCE electrode.

Mg–Al–Zn alloys, particularly AZ31, AZ61 and AZ91, are the most commonly used anode materials in aqueous Mg batteries, since they are commercially available and have a relatively high corrosion resistance in a Cl^- environment.[66] The good corrosion resistance is due to the reduced anodic kinetics with Al addition below solubility and suitable secondary phase morphology.[66–68] Mg–Al–Zn alloys have been used in practical Mg dry cells, Mg–air batteries, Mg–H_2O_2 semi-fuel cells and have been studied as anodes for Mg–water batteries.[4,28,49,69] They show relatively high utilization efficiency (*e.g.* 82% for AZ31 in 2 M $Mg(ClO_4)_2$ electrolyte at 40 mA cm^{-2}), excellent discharge capacity and contribute towards the high energy density of Mg-based cells.[61] However, Mg-based batteries with Mg–Al–Zn anodes normally exhibit a low cell voltage, which is due to the low open circuit voltage and more protective film caused by Al and Zn accumulation.[68] Therefore, Mg–Al–Zn alloys are mainly used in low power batteries with long service life, for example in harsh environments. Alloying with elemental indium can enhance the discharge properties of the AZ63 alloy as an anode for a Mg–CuCl battery, as reported by Li *et al.*[82] The discharge voltages of the AZ63 alloy at different current densities increase with 0.5 wt% In addition (*e.g.* from 0.45 to 0.6 V at

75 mA cm^{-2}), because the addition of In, as supposed, accelerates the dissolution of the Mg matrix (α-phase) and promotes the self-peeling of discharge products through the dissolution–reprecipitation of indium. Heat treatment was adopted to enhance the utilization efficiency and, eventually, enabled the Mg–Al–Zn–In alloy with excellent overall discharge properties.

Mg–Al–Pb alloy, especially AP65 (Mg–6%Al–5%Pb), was developed by Magnesium Elektron UK to supply a high cell voltage for high-power seawater battery applications.[5] This alloy has a more negative OCV and can supply a higher voltage than AZ alloys and pure Mg in different electrolyte solutions.[61,80,81] The high cell voltage of AP65 alloy is attributed to the promoted peeling or breakdown of discharge products caused by the synergetic effect of Al and Pb (protection and activation).[83,84] The usage of the AP65 alloy as an anode could cause severe sludge, but under some forced-flow discharge conditions the sludge problems can be controlled.[85] This alloy is used in some military applications, for example, as a power source for electric torpedoes. Except for the toxic lead, the disadvantages of Mg–Al–Pb anodes also include the high self-corrosion rate and relatively low utilization efficiency at low current densities, leading to low specific energy and shortened service life of the respective Mg-based batteries.[61,85,86] Therefore, this alloy is normally used for high-power devices with short operating time, like the mentioned electric torpedoes. Some alloying elements to the AP65 alloy, such as Zn, Mn, In, Ce/Y and La, have been studied, aimed at improving its self-corrosion and discharge properties.[80,81,87–91] The addition of 1 wt% In into the AP65 alloy can enhance its discharge potential in 3.5 wt% NaCl solution and increase its utilization efficiency at high current densities (from 82.1 to 87.7% at 180 mA cm^{-2}), but the utilization efficiency at low current density (10 mA cm^{-2}) was decreased from about 46 to 41%, due to a severe chunk effect.[80] Besides this, some work has also been carried out to optimize the composition of the Mg–Al–Pb alloy as an anode material.[83,92] Wen *et al.* suggested that, for an extruded sheet anode material, Mg-6 wt% Al-7 wt% Pb-0.5 wt% Zn alloy is the optimized composition due to its superior overall discharge properties in 3.5 wt% NaCl solution, while Wang *et al.* proposed that Mg-9 wt% Al-2.5 wt% Pb is more suitable because of the better discharge properties with low Pb content.[83,92] Recently, the AP65 alloy has been explored as an anode for Mg–air batteries. With the addition of Ce/Y alloying elements, the AP65 alloy can exhibit a higher discharge voltage and enhanced utilization efficiency than pure Mg and the AZ31 alloy, even at relatively low and moderate current densities (10 or 20 mA cm^{-2}).[81] For instance, the AP65-Ce/Y alloy shows a voltage of 1.28 V and a utilization efficiency of ~60% at 10 mA cm^{-2}, while the AZ31 alloy has a voltage of 1 V with a utilization efficiency of ~56% under the same discharge conditions.

Another Mg alloy developed for high power military applications is the Mg–Hg–Ga alloy, since it can supply a more negative discharge potential than the AP65 and AZ31 alloys.[93,94] The addition of mercury and gallium is well known to accelerate the activation of aluminum anodes.[95–97] An activation mechanism combining amalgamation, amalgam corrosion, liquid metal embrittlement and galvanic corrosion was proposed.[96] Recently, Hg has been adopted in Mg anodes for seawater-activated batteries and a battery system with a Mg–Hg anode can supply a power density of above 150 W h kg^{-1}.[98] As reported by Feng *et al.*, a Mg-6 wt% Hg alloy exhibits extremely negative corrosion potential in 3.5 wt% NaCl solution (around −2.2 V *vs.* SCE), which, in fact, is the most negative corrosion potential reported for any Mg alloy.[66,99] While superior properties in terms of voltage and power density are achieved, Mg–Hg anodes also have drawbacks such as high self-corrosion rate, low utilization efficiency, short service life and increasing sensitivity to environmental benignity. Therefore, gallium has been used to replace mercury in Mg–Hg anodes considering the intrinsic activation effect of Ga on Al anodes. Feng *et al.* have carried out work to investigate the effect of Ga addition and content on the microstructure and respective discharge properties of Mg–Hg anodes as well as the activation mechanism.[98,99] Mg anodes could be activated by the dissolution and subsequent redeposition of Hg and Ga. Complementary, the addition of Ga accelerates the stripping of the corrosion film and the formation of Mg amalgams, which promotes the electrochemical activity. Wang *et al.* studied the effect of the Hg/Ga ratio on the phase constituents and electrochemical performance of Mg–Hg–Ga alloys, indicating that an alloy with a Hg/Ga ratio of 0.68 could be the most suitable anode material for seawater-activated batteries concerning the observed discharge properties (~1.83 V *vs.* SCE in half-cell testing) and moderate self-corrosion rate.[100]

Mg–Li alloys have also been explored as anode materials for aqueous Mg-based batteries, due to the most negative standard electrode potential of lithium (−3.05 V *vs.* SHE). However, the utilization efficiency of Mg–Li alloys as anode material is related to the Li content, since the Li has a significant effect on the self-corrosion rate of the Mg anodes. Mg–Li alloys with a relatively low Li content (<11 wt%) have a poor corrosion resistance resulting from the increased cathodic kinetics or galvanic corrosion between the α-Mg and β-Li phases when the Li content is between 5–11 wt%.[101] Despite the enhanced discharge voltage, the poor corrosion resistance of Mg–Li alloys with relatively low Li content leads to low utilization efficiency (*e.g.* ~36% for Mg-8 wt% Li-3 wt% Al-1 wt% Ce-1 wt% Y alloy at 10 mA cm^{-2}) during discharge, especially at low current densities, compared to the AZ31 alloy, as reported (~54% at 10 mA cm^{-2}).[76,102]

In contrast, with heavy Li alloying (>11 wt%) the corrosion resistance of Mg can be enhanced.[103,104] The crystal structure of the α-Mg matrix changes to bcc after Li addition of above 11 wt%, in contrast to the hcp structure of the pure primary Mg. Taking advantage of this change, the anodic dissolution kinetics of Mg are significantly restricted, which, as proposed, can be attributed to the dynamic development of a carbonate rich surface film.[103,105] As a consequence, only Mg–Li alloys with a high Li content are suitable anode materials for aqueous Mg batteries. A Mg-13 wt% Li alloy as an anode material for Mg batteries was tested by Sivashanmugam *et al.*[106] The results show that the Mg-13 wt% Li alloy has a lower self-corrosion rate than the

AZ31 alloy in various battery electrolytes, for example, 0.0059 mg cm^{-2} min^{-1} for the Mg–Li alloy *vs.* 0.1832 mg cm^{-2} min^{-1} for AZ31 in 2 M $MgCl_2$ solution. Besides this, it shows higher discharge voltages of 0.3–0.4 V at different current densities than the AZ31 alloy in a Mg/CuO full-cell setup. The alloy also possesses high anodic efficiency at relatively low current drains, for example 81% at 8.6 mA cm^{-2}. Recently, combined with other alloying elements, Mg–Li alloys with different Li content have also been proposed as anodes for either Mg–air batteries or Mg semi-full cells.[76,78,107–109] A Mg–air battery with a Mg-14 wt% Li-1 wt% Al-0.1 wt% Ce alloy as an anode shows a higher voltage at different current densities than those based on pure Mg and AZ31 alloy, *e.g.* 1.20, 1.17 and 1.13 V for Mg–Li–Al–Ce, Mg and AZ31, respectively. This alloy also exhibits enhanced anodic efficiency and discharge capacity at all of the tested current densities, supporting its possible application in Mg–air batteries. Another rare earth element, yttrium, has also been investigated as an alloying element for a Mg–Li alloy.[108] The addition of 1.2 wt% Y enhances the corrosion resistance of the Mg-8.5 wt% Li-3.2 wt% Al-1.2 wt% Ce alloy and contributes towards its boosted peak power density (from 91 to 110 mW cm^{-2}) in Mg/H_2O_2 semi-fuel cell measurements. In addition, a Mg-8 wt% Li-3 wt% Al-1 wt% Ce-1 wt% Y alloy was also evaluated as an anode material for a Mg–air battery, indicating good discharge voltage but low utilization efficiency at a relatively low current density (36% at 10 mA cm^{-2} in 3.5 wt% NaCl).[76] Nevertheless, with the addition of 1 wt% Zn, both the voltage and utilization efficiency of this alloy are enhanced, and appear to be higher than those of AZ31 and pure Mg. Zn is mainly distributed in the β-Li phase and thus weakens the galvanic corrosion between the α-Mg and β-Li phases, leading to a relative uniform dissolution of this alloy and a reduced self-corrosion rate.

Indeed, the aforementioned Mg alloys possess some attractive advantages as anode materials. However, they are all associated with specific issues, raising the need for better Mg anode materials in general. Firstly, due to the increasing social attention on environmental issues, new anode materials are needed to replace the widely used toxic Mg–Al–Pb or Mg–Hg–Ga alloys in seawater batteries. Secondly, in terms of a Mg–Li anode, the heavy Li load and need for a fourth or even fifth alloying element strongly reduces the technological tolerances leading to a significant increase in associated costs. In this case, some other low melting point elements (LMPEs), such as Ga, In and Sn, are being considered. Work has been done to explore the possibility of replacing Pb with these LMPEs in Mg–Al anodes. Mg–Al–Sn, Mg–Al–Ga and Mg–Al–In alloys have been tested either in Mg–air batteries or Mg–seawater batteries and compared with Mg–Al–Zn anodes.[71,74,82,110,111] As reported, all of these alloys show a better discharge voltage than Mg–Al–Zn anodes, attributed to the less protective oxide film on the anode surface resulting from these LMPEs. For example, the voltages of Mg-3 wt% Al-1 wt% Ga, Mg-3 wt.%Al-1 wt% Sn and Mg-3 wt.%Al-1 wt% In at 5 mA cm^{-2} are 1.18, 1.21 and 1.31 V respectively, higher than that of the Mg-3 wt.%Al-1 wt% Zn alloy (1.17 V). A dissolution–reprecipitation mechanism was proposed to explain the activation effect of In addition on a Mg anode with the assistance of X-ray diffraction (XRD), X-ray photoelectron spectroscopy (XPS) and energy-dispersive X-ray spectroscopy (EDS). However, these alloys normally have a higher self-corrosion rate than Mg–Al–Zn alloys evaluated either by polarization curves, EIS measurements or by immersion tests. Therefore, the service life of Mg batteries is definitely shortened by adopting these Mg–Al systems alloyed with LMPEs as an anode. Recently, a novel Mg–Ca anode material with both good voltage and low self-corrosion rate was presented.[72] The consideration of Ca addition was based on its more negative electrode potential (−2.87 V *vs.* SHE) than that of magnesium and the grain refinement effect of Ca on Mg metal. The content of Ca was recommended to be lower than 0.5 wt% since the increasing Mg_2Ca secondary phases, accompanied by high Ca content, severely accelerate the self-corrosion rate of Mg–Ca alloys. As presented in Figure 12.7, a Mg–air battery with a Mg-0.1 wt% Ca as an anode has a more stable and higher discharge voltage compared to anodes

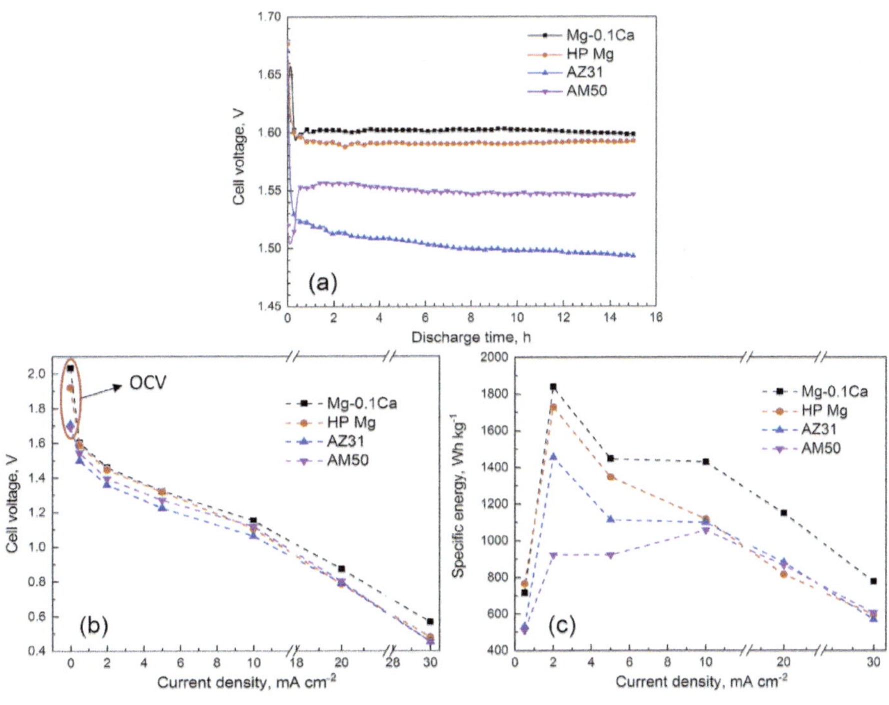

Figure 12.7 Discharge properties of Mg–air batteries with different anodes: (a) discharge curves at a current density of 0.5 mA cm^{-2}; (b) cell voltage at different current densities; (c) specific energy at different current densities. Electrolyte is 3.5 wt% NaCl solution; Cathode is a commercial air cathode with C/MnO_2 catalyst. Reproduced from M. Deng, D. Höche, S.V. Lamaka, D. Snihirova and M.L. Zheludkevich, Mg–Ca binary alloys as anodes for primary Mg–air batteries, *J. Power Sources*, **396**, 109–118., Copyright 2018, with permission from Elsevier.

based on high purity Mg, AZ31 and AM50 alloys. Besides this, a Mg-0.1 wt% Ca anode can also offer enhanced energy density, even higher than that of high purity Mg, which generally has a lower corrosion rate in a technically relevant testing solution such as 3 wt% NaCl than most Mg alloys, as already summarized in some literature reviews.[66,112]

12.4.1.3 Microstructure Tuning

Microstructure, which includes grain size, secondary phase constituent, secondary phase morphology, density of various defects and impurities, *etc.*, also has a significant effect on the self-corrosion and discharge properties of Mg anodes. Therefore, technical processing, such as deformation and subsequent heat treatment, are adopted to adjust the microstructure of Mg anodes and to obtain good overall properties. Note that sheet or rod materials of Mg anodes, which can be obtained through deformation processes, such as rolling or extrusion, are normally required in practical applications due to cell design constraints. Thus, studies concerning the effect of these processes on properties of Mg anodes are quite important as well as iterative processes for optimization. Until now, a number of studies have been carried out to reveal the influence of different processes on the microstructure, self-corrosion and discharge properties of numerous Mg anodes.[60,70,74,79,82,90,91,93,94,113–116]

Heat treatment, which can be applied to both as-cast materials and wrought materials, has a significant impact on the microstructure of Mg anodes, especially the size, quantity and distribution of the secondary phases. Consequently, the OCP of Mg anodes and the reaction kinetics of anode dissolution are affected. As demonstrated by Yan *et al.*, homogenization and aging treatment have a great effect on the self-corrosion and discharge potential of a Mg-4.8 wt% Hg-8wt% Ga alloy anode, resulting from the changing precipitation behavior of the $Mg_{21}Ga_5Hg_3$ and Mg_5Ga_2 secondary phases.[113] The self-corrosion resistance and discharge potential are improved after aging at 473 K for 96 h. This also indicates that anode pretreatment *via* heat aging under some conditions has a negative effect on the overall performance of this Mg–Hg–Ga anode. Therefore, suitable treatment parameters, including temperature and treatment duration, are urgently required. The effect of heat treatment on the discharge properties of Mg–Al–Zn anode with In addition has also been studied by Li *et al.*[82] After microstructural homogenization at 385 °C for 24 h followed by water quenching, the utilization efficiency of an AZ63-0.5 wt% In anode was enhanced (from 67 to 71% at 10 mA cm^{-2}) due to the reduced self-corrosion rate attributed to the dissolution of the secondary phases into the Mg matrix. Besides this, the discharge voltage of a Mg–CuCl seawater-activated battery was increased with this treated alloy as an anode, because of more In dissolved in the α-Mg solid solution after heat treatment, which also promotes the indium-induced dissolution–reprecipitation process.

Deformation processes such as extrusion are normally used to refine materials and to obtain a more uniform microstructure. The induced dislocations, twins, crystallographic textures, various grain orientations, incomplete recrystallization structures and subsequently, the occurrence of high surface energy could also have a remarkable influence on the discharge properties of Mg anodes. Wang *et al.* investigated the effect of hot rolling and subsequent annealing on the discharge properties of a Mg–Al–Pb anode.[114] The results show that hot rolling improves the discharge potential of the AP65 anode, attributed to the induced dislocations and twins, which accelerate the dissolution of the anode. Besides this, the utilization efficiency increased from 78.8% for the as-cast AP65 alloy to 81.5% for the as-rolled one at 180 mA cm^{-2}, because the grains are refined and the strongly cathodic Al–Mn particles are fractured by hot rolling. Subsequent annealing at 150°C for 4 h further improved the overall properties of the hot-rolled AP65 anode material. The more negative discharge potential is due to the accelerated self-peeling of the discharge products, while the increased utilization efficiency can be attributed to the decreased density of dislocations and twins (might be related to a lowered chunk effect). Generally, appropriate heat treatment processes following deformation processes are quite necessary to obtain the desired microstructure. Additionally, the different grain orientations induced by rolling could also have an effect on both the corrosion and discharge performance of Mg anodes, as indicated in the recent publication by Wang *et al.*[117] Extrusion is another widely used process to obtain a specific anode shape and uniform microstructure. In the recent research of Xiong *et al.*, a Mg-6 wt% Al-1 wt% Sn alloy anode for Mg–air batteries was extruded to achieve superior discharge properties.[74] Similarly, both discharge voltage and utilization efficiency are enhanced due to accelerated discharge activity and more uniform anode dissolution. At a discharge current density of 10 mA cm^{-2}, the utilization efficiency increases from 32.2 to 45.2% after extrusion, while the discharge potential in half-cell tests becomes more negative, from −1.605 to −1.670 V (*vs.* SCE).

12.4.2 Electrolyte Modification

The electrolyte is another determining factor for the performance of aqueous Mg-based electrochemical power sources. Firstly, the utilization efficiency of anode materials is closely related to the adopted electrolytes, since the different self-corrosion rates of Mg anodes can be detected in various solutions.[56,61] Secondly, the formation of discharge product films on the anode surface, which has a great influence on the discharge potential of Mg anodes as mentioned, is related to the electrolyte solution. Meanwhile, solution conductivity also varies with different electrolytes. High solution conductivity, which means low solution resistance, is needed to decrease the voltage loss caused by ohmic drop (IR-drop). Therefore, electrolyte modification, even in a water environment, is quite important to gain superior battery performance. Most research concerning electrolytes for Mg-based batteries follows two directions, namely the modification of background electrolytes and the utilization of additives. Note that electrolyte modification may not be feasible for Mg–seawater batteries since the battery system is open and the electrolyte is natural seawater from the ocean.

The induced changes on Mg battery performance by the adjusted characteristics of the background electrolytes, including the soluble salt type, species concentrations, pH, *etc.*, have been deeply investigated over the last few decades.[61,63,86,118–121] Zhao *et al.* tested Mg battery performance in NaCl solutions with different concentrations.[121] Upon increasing the NaCl concentration from 3.5 to 6.5 wt%, the discharge current at a specific voltage increases while the battery life time decreases. This could be attributed to the decreased overpotential resulting from increased solution conductivity and facilitated Mg anode dissolution. Instead of an aggressive NaCl solution, other less-aggressive solutions without Cl^- have been proposed as electrolytes for Mg-based batteries, such as $Mg(ClO)_4$, $MgBr_2$, $Mg(NO_3)_2$, *etc.* The utilization efficiencies of various Mg anodes at different current densities have been demonstrated to be higher in $Mg(ClO_4)_2$ and $MgBr_2$ solutions than in $MgCl_2$ solution, owning to the reduced self-corrosion rate.[61] Besides this, Winther-Jensen *et al.* studied several chloride salts, *i.e.* $MgCl_2$, LiCl, $CaCl_2$ and NaCl, as electrolytes and anode/electrolyte interface modulators for Mg batteries under saturated conditions at pH 11.[118] The results show that, in $MgCl_2$ and LiCl saturated solutions, Mg batteries maintain a cell voltage of above 1.4 V for more than 150 h at 20 μA cm^{-2}, which cannot be achieved in saturated $CaCl_2$ and NaCl solutions. Moreover, a mixture of $MgCl_2$ and LiCl saturated solutions can further enhance the cell voltage and stability. This phenomenon is important for enhancing Mg battery performance, even if the reasons are not yet well explained.

In addition to the modification of the background solution, utilization of electrolyte additives is also an important approach for the previously introduced boosting of Mg battery performance and has already drawn much attention. Most of the prior research on electrolyte additives was focused on corrosion inhibitors. The aim was to improve the anode utilization efficiency through inhibiting the self-corrosion of Mg anodes. Meanwhile, the discharge potential of Mg anodes is also affected as a result of the varied kinetics of anode dissolution. Commonly used inhibitors for aqueous Mg-based batteries include nitrites and chromate compounds.[4,63,64,119,122] However, they are not environmentally friendly. Therefore, more effective and environmentally benign inhibitors for Mg batteries are required. Several new additives for aqueous Mg-based batteries have been evaluated, such as gallium oxide, decyl glucoside, water soluble graphene, sodium phosphate and sodium dodecyl benzenesulfonate.[102,123–125] As reported, decyl glucoside as the electrolyte additive can improve both the cell voltage and the utilization efficiency of Mg–air battery systems.[123] The utilization efficiency of a pure Mg anode at 5 mA cm^{-2} increases from to 86.2% after the addition of 2.5 mM decyl glucoside into the electrolyte, while the voltage increases to 1.36 from 1.15 V in bare electrolyte. The significantly enhanced utilization efficiency can be attributed to the reduced self-corrosion rate resulting from the adsorption of decyl glucoside on the anode surface. The increased cell voltage is due to the enhanced open circuit voltage, which indicates the cathodic inhibition effect of decyl glucoside on local acting cathodes within the anode material. Recently, Lamaka *et al.* reported work representing results of the systematic screening of magnesium corrosion inhibitors.[126] In this work, the inhibition effect of 151 individual compounds on various Mg alloys and three grades of pure Mg was evaluated through hydrogen evolution tests in 0.5 wt% NaCl solution. The results indicate a number of new inhibitors that show higher inhibition efficiency on Mg corrosion than chromate. This work shines some light on the selection of effective electrolyte additives for aqueous Mg-based batteries. Nevertheless, more work is still needed to find suitable additives for fully assembled Mg-based batteries taking into consideration both utilization efficiency and cell voltage.

Recently, Höche *et al.* proposed a new concept to simultaneously improve the anode efficiency and discharge voltage *via* the addition of Mg^{2+} and Fe^{3+} complexing agents.[59] This idea was based on their prior research on iron complexing agents as effective corrosion inhibitors for magnesium.[55,126,127] As proposed, iron re-deposition on Mg surface accelerates the self-corrosion of impurities containing Mg.[55] Therefore, as illustrated in Figure 12.8a, preventing the re-plating of iron with strong iron complexing agents becomes a feasible method by which to suppress the self-corrosion of magnesium.[126,127] The principle for battery performance boosting based on this mechanism is presented in Figure 12.8b. Electrolyte additives as strong Fe^{3+} and moderate Mg^{2+} complexing agents were added into the battery system. The self-corrosion of Mg anodes is expected to be suppressed by preventing Fe-re-deposition and impurity activation, while the cell voltage is supposed to be enhanced through inhibiting the formation of a $Mg(OH)_2$ film and activation of the surface *via* Mg complexing. An excellent organic additive, namely salicylate, was found in this work. With the addition of 0.05 M salicylate, the utilization efficiency of a cp-Mg (220 ppm Fe) in 0.5 wt% NaCl solution was enhanced from 13.5 to 27.2%, while the discharge potential became more negative, from −1.47 V (*vs.* Ag/AgCl) to −1.61 V (*vs.* Ag/AgCl). Note that the anode material used in this work was cp-Mg that shows a high self-corrosion rate and inferior discharge performance in electrolyte solutions. Nevertheless, it indicates a new direction for choosing novel electrolyte additives for aqueous Mg-based batteries. Thus, research on complexing agents as electrolyte additives for superior Mg alloy anodes deserves more attention.

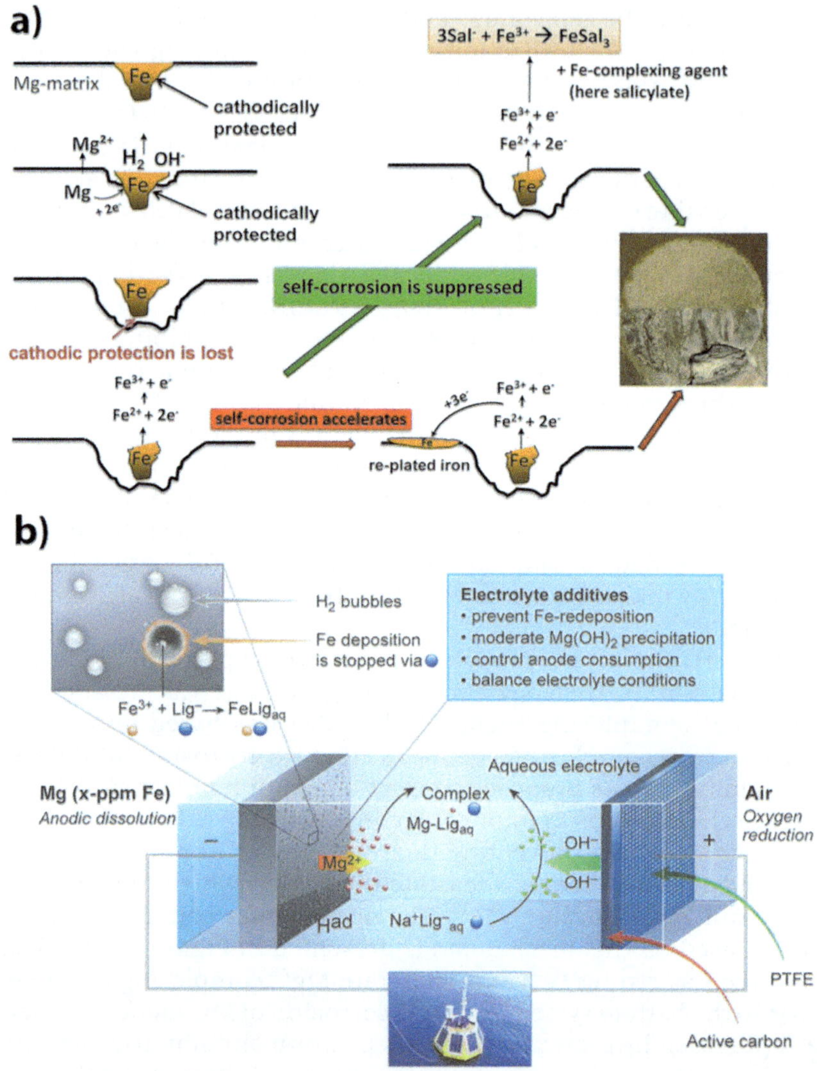

Figure 12.8 (a) Iron re-deposition mechanism on Mg corrosion and the working principles of additives (here salicylate, Sal) as a corrosion inhibitor *via* preventing Fe-re-plating. (b) Working principles of complexing agents as additives for the performance boosting of Mg-based batteries. Main aspects of the additive effect are indicated and emphasize the scope of the discovery. Reproduced from ref. 59, https://doi.org/10.1038/s41598-018-25789-8, under the terms of the CC BY 4.0 licence, http://creativecommons.org/licenses/by/4.0/.

12.5 Outlook

In this chapter aqueous Mg-based batteries were introduced with respect to battery types, involved reactions, hindrances to practical applications and also research progress. For sure, Mg-based batteries with superior theoretical battery performance are promising future power sources, which fulfill the social requirements in terms of economic feasibility and clean energy. The various battery types supported by different cathode reactants enable Mg-based electrochemical systems to be used for applications in different fields, such as marine devices, military equipment, backup power, *etc.* Nevertheless, the practical applications and commercialization are still hindered because of the imperfect battery performance, such as low/unstable cell voltage and insufficient specific energy density. Therefore, in order to widen the application of aqueous Mg-based batteries, more work should be done on improving all of the different components of the battery system, *i.e.* the anode, cathode, electrolyte and the cell design itself.

Novel Mg anodes with a low self-corrosion rate and more negative discharge potential are required. Mg alloys with non-toxic alloying elements should be developed to replace the usage of high-quality but environmentally harmful alloys, such as Mg–Hg–Ga and Mg–Al–Pb. Meanwhile, except for good electrochemical performance, Mg anodes with simplified composition, such as binary or at most ternary alloys, are needed to reduce the cost associated with the anode preparation. The focus needs to be shifted from Mg–Al- and Mg–Li-based alloys to other novel Mg alloys, such as the Mg–Ca system.

The cathode, including the cathode structure, cathode reactants and catalysts, also plays a key role in the performance of aqueous Mg-based systems. Firstly, the cathode structure needs to be optimized, supporting enough mechanical strength and reducing battery inner resistance. Secondly, effective and low-cost cathode reactants are important, especially for Mg–seawater systems. Non-silver alternatives with a discharge performance close to that of AgCl are still required. Besides this, concerning cathodes for Mg–air and Mg–H_2O_2 systems, highly effective catalysts need to be discovered, considering not only high catalytic activity but also low cost. Modified MnO_2 and perovskite-based oxides deserve more attention as catalysts for Mg–air systems. In addition, the importance of substrate materials for cathode catalysts should also be noted. Adoption of a porous nanostructure is a feasible approach, since the significantly increased surface area can reduce polarization induced by mass transfer and gas diffusion processes.

Electrolyte modification, especially if they are water based, as an effective way of boosting battery performance, draws increasing attention nowadays. Non-aggressive solutions can be used as electrolytes to ensure relatively high anode efficiency, but the effect of these solutions on cell voltage and power density should be considered thoroughly. Electrolyte additives provide another possible and very easy to handle alternative. Except for corrosion inhibitors, other new additives that can simultaneously enhance the anode efficiency and cell voltage need to be developed, such as the newly proposed dual acting complexing agents.

Acknowledgement

Min Deng and Linqian Wang are grateful for the award of fellowships and funding from the China Scholarship Council No. 201606370031 and No. 201706370183.

References

1. F. Cheng and J. Chen, *Chem. Soc. Rev.*, 2012, **41**, 2172.
2. T. Zhang, Z. Tao and J. Chen, *Mater. Horiz.*, 2014, **1**, 196.
3. M. A. Rahman, X. Wang and C. Wen, *J. Electrochem. Soc.*, 2013, **160**, A1759.
4. R. C. Kirk and A. B. Fry, *J. Electrochem. Soc.*, 1948, **94**, 277.
5. D. Linden and T. B. Reddy, *Handbook of Batteries*, McGraw-Hill, New York, 2002.
6. S. R. Srither, M. Selvam, S. Arunmetha, R. Yuvakkumar, K. Saminathan and V. Rajendran, *Sci. Adv. Mater.*, 2013, **5**, 1372.
7. K. Narthana, R. Naveen, K. Saminathan, P. Siva, M. Selvam and P. Saha, *Int. J. Res. Adv. Eng.*, 2016, **2**, 1.
8. R. Renuka, *J. Power Sources*, 2000, **87**, 4.
9. N. G. Wang, R. C. Wang, C. Q. Peng, C. W. Hu, Y. Feng and B. Peng, *Trans. Nonferrous Met. Soc. China*, 2014, **24**, 2427.
10. J. B. Mullen and P. L. Howard, *Trans. Electrochem. Soc.*, 1946, **90**, 529.
11. R. Balasubramanian, A. Veluchamy, N. Venkatakrishnan and R. Gangadharan, *J. Power Sources*, 1995, **56**, 197.
12. J. B. Mullen and P. L. Howard, *J. Electrochem. Soc.*, 1946, **90**, 529.
13. R. Renuka, *J. Appl. Electrochem.*, 1997, **27**, 1394.
14. F. Sammoura, K. B. Lee and L. Lin, *Sens. Actuators, A*, 2004, **111**, 79.
15. K. V. Rao, *Def. Sci. J.*, 2001, **51**, 161.
16. J. R. Coleman, *J. Appl. Electrochem.*, 1971, **1**, 65.
17. L. J. Burant, *US Pat.*, 4,021,597, 1977.
18. I. C. Blake, *US Pat.*, 2,491,640, 1949.
19. J. B. Ockerman, *US Pat.*, 3,433,678, 1967.
20. L. E. Pucher, *J. Electrochem. Soc.*, 1952, **99**, 203C.
21. J. M. Dines, *US Pat.*, 2,716,671, 1955.
22. W. K. McCarter, J. F. Flood and J. T. Fischer, *US Pat.*, 5,395,707, 1995.
23. M. Hiroi, *J. Appl. Electrochem.*, 1985, **15**, 201.
24. R. F. Amlie, H. N. Honer and P. Ruetschi, *J. Electrochem. Soc.*, 1965, **112**, 1073.
25. M. Hiroi, *J. Appl. Electrochem.*, 1985, **16**, 431.
26. R. Renuka, *Mater. Chem. Phys.*, 1999, **59**, 42.
27. R. Renuka, *J. Appl. Electrochem.*, 1999, **29**, 271.
28. M. G. Medeiros, R. R. Bessette, C. M. Deschenes, C. J. Patrissi, L. G. Carreiro, S. P. Tucker and D. W. Atwater, *J. Power Sources*, 2004, **136**, 226.
29. M. G. Medeiros and E. G. Dow, *J. Power Sources*, 1999, **80**, 78.
30. M. G. Medeiros, R. R. Bessette, C. M. Deschenes and D. W. Atwater, *J. Power Sources*, 2001, **96**, 236.
31. M. G. Medeiros, E. G. Dow and R. R. bessette, *US Pat.*, 6,465,124B1, 2002.
32. M. G. Medeiros, E. G. Dow, R. R. Bessette and J. M. Cichon, *US Pat.*, 6,740,220 B1, 2004.
33. W. Yang, S. Yang, W. Sun, G. Sun and Q. Xin, *Electrochim. Acta*, 2006, **52**, 9.

34. C. J. Patrissi, R. R. Bessette, Y. K. Kin and C. R. Schumacher, *J. Electrochem. Soc.*, 2008, **155**, B558.
35. C. Shu, E. Wang, L. Jiang, Q. Tang and G. Sun, *J. Power Sources*, 2012, **208**, 159.
36. R. R. Bessette, M. G. Medeiros, C. J. Patrissi, C. M. Deschenes and C. N. LaFratta, *J. Power Sources*, 2001, **96**, 240.
37. Ø. Hasvold and N. Størkersen, *J. Power Sources*, 2001, **96**, 252.
38. W. S. D. Wilcock and P. C. Kauffman, *J. Power Sources*, 1997, **66**, 71.
39. Ø. Hasvold, T. Lian, E. Haakaas, N. Størkersen, O. Perelman and S. Cordier, *J. Power Sources*, 2004, **136**, 232.
40. M. Shinohara, E. Araki, M. Mochizuki, T. Kanazawa and K. Suyehiro, *J. Power Sources*, 2009, **187**, 253.
41. P. Yue, Z. Li, S. Wang and Y. Wang, *Int. J. Hydrogen Energy*, 2015, **40**, 6809.
42. Y. Xue, H. Miao, S. Sun, Q. Wang, S. Li and Z. Liu, *J. Power Sources*, 2015, **297**, 202.
43. C. Shu, E. Wang, L. Jiang and G. Sun, *Int. J. Hydrogen Energy*, 2013, **38**, 5885.
44. M. Jiang, H. He, C. Huang, B. Liu, W.-J. Yi and Z.-S. Chao, *Electrochim. Acta*, 2016, **219**, 492.
45. Z. Li, J. Yang, G. Xu and S. Wang, *J. Power Sources*, 2013, **242**, 157.
46. C. Zhao, Y. Jin, W. Du, C. Ji and X. Du, *J. Electroanal. Chem.*, 2018, **826**, 217.
47. Y. Li, X. Zhang, H.-B. Li, H. D. Yoo, X. Chi, Q. An, J. Liu, M. Yu, W. Wang and Y. Yao, *Nano Energy*, 2016, **27**, 8.
48. W. Yang, S. Yang, J. Guo, G. Sun and Q. Xin, *Carbon*, 2007, **45**, 397.
49. Q. Liu, Z. Yan, E. Wang, S. Wang and G. Sun, *Int. J. Hydrogen Energy*, 2017, **42**, 23045.
50. R. Hahn, J. Mainert, F. Glaw and K. D. Lang, *J. Power Sources*, 2015, **288**, 26.
51. P. Xie, M. Z. Rong and M. Q. Zhang, *Angew. Chem., Int. Ed.*, 2016, **55**, 1805.
52. G. A. Marsh and E. Schaschl, *J. Electrochem. Soc.*, 1960, **107**, 960.
53. M. Andrei, F. d. Gabriele, P. L. Bonora and D. Scantlebury, *Mater. Corros.*, 2003, **54**, 5.
54. M. E. Straumanis and B. K. Bhatia, *J. Electrochem. Soc.*, 1963, **110**, 357.
55. D. Hoche, C. Blawert, S. V. Lamaka, N. Scharnagl, C. Mendis and M. L. Zheludkevich, *Phys. Chem. Chem. Phys.*, 2016, **18**, 1279.
56. G. L. Song and A. Atrens, *Adv. Eng. Mater.*, 1999, **1**, 11.
57. J. D. Hanawalt, C. E. Nelson and J. A. Peloubet, *Trans. AIME*, 1942, **147**, 273.
58. R. E. Mcnulty and J. D. Hanawalt, *J. Electrochem. Soc.*, 1942, **81**, 423.
59. D. Höche, S. V. Lamaka, B. Vaghefinazari, T. Braun, R. P. Petrauskas, M. Fichtner and M. L. Zheludkevich, *Sci. Rep.*, 2018, **8**, 7578.
60. B. Campillo, C. Rodriguez, J. Genesca, J. Juarez-Islas, O. Flores and L. Martinez, *J. Mater. Eng. Perform.*, 1997, **6**, 449.

61. R. Udhayan, N. Muniyandi and P. B. Mathur, *Br. Corros. J.*, 1992, **27**, 68.
62. Y. D. Milusheva, R. I. Boukoureshtlieva, S. M. Hristov and A. R. Kaisheva, *Bulg. Chem. Commun.*, 2011, **43**, 42.
63. W. Li, C. Li, C. Zhou, H. Ma and J. Chen, *Angew. Chem.*, 2006, **118**, 6155.
64. C. Li, F. Cheng, W. Ji, Z. Tao and J. Chen, *Nano Res.*, 2009, **2**, 713.
65. G. Xin, X. Wang, C. Wang, J. Zheng and X. Li, *Dalton Trans.*, 2013, **42**, 16693.
66. K. Gusieva, C. H. J. Davies, J. R. Scully and N. Birbilis, *Int. Mater. Rev.*, 2014, **60**, 169.
67. M.-C. Zhao, M. Liu, G. Song and A. Atrens, *Corros. Sci.*, 2008, **50**, 1939.
68. A. Pardo, M. C. Merino, A. E. Coy, F. Viejo, R. Arrabal and S. Feliú, *Electrochim. Acta*, 2008, **53**, 7890.
69. R. P. Hamlen, E. C. Jerabek, J. C. Ruzzo and E. G. Siwek, *J. Electrochem. Soc.*, 1969, **116**, 1588.
70. M. Yuasa, X. Huang, K. Suzuki, M. Mabuchi and Y. Chino, *Mater. Trans.*, 2014, **55**, 1202.
71. X. Li, H. Lu, S. Yuan, J. Bai, J. Wang, Y. Cao and Q. Hong, *J. Electrochem. Soc.*, 2017, **164**, A3131.
72. M. Deng, D. Höche, S. V. Lamaka, D. Snihirova and M. L. Zheludkevich, *J. Power Sources*, 2018, **396**, 109.
73. M. Yuasa, X. Huang, K. Suzuki, M. Mabuchi and Y. Chino, *J. Power Sources*, 2015, **297**, 449.
74. H. Xiong, K. Yu, X. Yin, Y. Dai, Y. Yan and H. Zhu, *J. Alloys Compd.*, 2017, **708**, 652.
75. J. L. Wu, R. C. Wang, Y. Feng and C. Q. Peng, *J. Alloys Compd.*, 2018, **765**, 736.
76. N. Wang, R. Wang, Y. Feng, W. Xiong, J. Zhang and M. Deng, *Corros. Sci.*, 2016, **112**, 13.
77. X. Liu, S. Liu and J. Xue, *J. Power Sources*, 2018, **396**, 667.
78. Y. Ma, N. Li, D. Li, M. Zhang and X. Huang, *J. Power Sources*, 2011, **196**, 2346.
79. N. Wang, R. Wang, C. Peng and Y. Feng, *Corros. Sci.*, 2014, **81**, 85.
80. N. Wang, R. Wang, C. Peng, B. Peng, Y. Feng and C. Hu, *Electrochim. Acta*, 2014, **149**, 193.
81. Y. Feng, W. Xiong, J. Zhang, R. Wang and N. Wang, *J. Mater. Chem. A*, 2016, **4**, 8658.
82. J. Li, B. Zhang, Q. Wei, N. Wang and B. Hou, *Electrochim. Acta*, 2017, **238**, 156.
83. L. Q. Wang, R. C. Wang, Y. Feng, M. Deng and N. G. Wang, *J. Electrochem. Soc.*, 2017, **164**, A438.
84. N.-g. Wang, R.-c. Wang, C.-q. Peng, Y. Feng and X.-y. Zhang, *Trans. Nonferrous Met. Soc. China*, 2010, **20**, 1403.
85. J. F. King and S. L. Bradshaw, *US Pat.*, 4,332,864, 1982.
86. R. Udhayan and D. P. Bahtt, *J. Power Sources*, 1996, **63**, 103.
87. N.-g. Wang, R.-c. Wang, C.-q. Peng, Y. Feng and X.-y. Zhang, *Trans. Nonferrous Met. Soc. China*, 2010, **20**, 1936.

88. B. Chen, R.-c. Wang, C.-q. Peng, Y. Feng and N.-g. Wang, *Trans. Nonferrous Met. Soc. China*, 2014, **24**, 423.
89. H.-x. Jin, R.-c. Wang, C.-q. Peng, K. Shi and Y. Feng, *J. Cent. South Univ.*, 2012, **19**, 2086.
90. Y. Shi, C. Peng, Y. Feng, R. Wang and N. Wang, *Mater. Des.*, 2017, **124**, 24.
91. Y. Shi, C. Peng, Y. Feng, R. Wang and N. Wang, *J. Alloys Compd.*, 2017, **721**, 392.
92. L. Wen, K. Yu, H. Xiong, Y. Dai, S. Yang, X. Qiao, F. Teng and S. Fan, *Electrochim. Acta*, 2016, **194**, 40.
93. K. Yu, X. Tan, Y. Hu, F. Chen and S. Li, *Corros. Sci.*, 2011, **53**, 2035.
94. J. Zhao, K. Yu, Y. Hu, S. Li, X. Tan, F. Chen and Z. Yu, *Electrochim. Acta*, 2011, **56**, 8224.
95. A. H. Al-Saffar, V. Ashworth, W. A. Grant and R. P. Procter, *Corros. Sci.*, 1978, **18**, 687.
96. J. B. Bessone, *Corros. Sci.*, 2006, **48**, 4243.
97. D. O. Flamini, S. B. Saidman and J. B. Bessone, *Corros. Sci.*, 2006, **48**, 1413.
98. Y. Feng, R. Wang, K. Yu, C. Peng, J. Zhang and C. Zhang, *J. Alloys Compd.*, 2009, **473**, 215.
99. Y. Feng, R.-c. Wang, K. Yu, C.-Q. Peng and W.-x. Li, *Trans. Nonferrous Met. Soc. China*, 2007, **17**, 1363.
100. L. Wang, P. Li and L. He, *IOP Conf. Ser.: Mater. Sci. Eng.*, 2017, **229**, 012008.
101. Y. Song, D. Shan, R. Chen and E.-H. Han, *Corros. Sci.*, 2009, **51**, 1087.
102. D. Cao, L. Wu, Y. Sun, G. Wang and Y. Lv, *J. Power Sources*, 2008, **177**, 624.
103. W. Xu, N. Birbilis, G. Sha, Y. Wang, J. E. Daniels, Y. Xiao and M. Ferry, *Nat. Mater.*, 2015, **14**, 1229.
104. M. Sahoo, *J. Mater. Sci.*, 1982, **17**, 3564.
105. L. Hou, M. Raveggi, X. B. Chen, W. Xu, K. J. Laws, Y. Wei, M. Ferry and N. Birbilis, *J. Electrochem. Soc.*, 2016, **163**, C324.
106. A. Sivashanmugam, T. P. Kumar, N. G. Renganathan and S. Gopukumar, *J. Appl. Electrochem.*, 2004, **34**, 1135.
107. Y.-Z. Lv, Y.-Z. Jin, Z.-B. Wang, M. Liu, Y.-F. Li, L. Wang and D.-X. Cao, *Ionics*, 2014, **21**, 429.
108. Y. Lv, Y. Xu and D. Cao, *J. Power Sources*, 2011, **196**, 8809.
109. D. Cao, L. Wu, G. Wang and Y. Lv, *J. Power Sources*, 2008, **183**, 799.
110. S. Yuan, H. Lu, Z. Sun, L. Fan, X. Zhu and W. Zhang, *J. Electrochem. Soc.*, 2016, **163**, A1181.
111. T. Zheng, Y. Hu, Y. Zhang, S. Yang and F. Pan, *Mater. Des.*, 2018, **137**, 245.
112. F. Cao, Z. Shi, G.-L. Song, M. Liu and A. Atrens, *Corros. Sci.*, 2013, **76**, 60.
113. F. Yan, W. Ri-chu, P. Chao-qun, Q. Ke, W. Nai-guang, Z. Chun and Z. Jia-pei, *Corros. Sci.*, 2010, **52**, 3474.
114. N. Wang, R. Wang, C. Peng, Y. Feng and B. Chen, *Corros. Sci.*, 2012, **64**, 17.
115. Y. Feng, R.-c. Wang and C.-q. Peng, *Intermetallics*, 2013, **33**, 120.

116. G. Huang, Y. Zhao, Y. Wang, H. Zhang and F. Pan, *Mater. Lett.*, 2013, **113**, 46.
117. N. Wang, Y. Mu, W. Xiong, J. Zhang, Q. Li and Z. Shi, *Corros. Sci.*, 2018, **144**, 107.
118. B. Winther-Jensen, M. Gaadingwe, D. R. Macfarlane and M. Forsyth, *Electrochim. Acta*, 2008, **53**, 5881.
119. S. Sathyanarayana and N. Munichandraiah, *J. Appl. Electrochem.*, 1981, **11**, 33.
120. J. L. Robinson and P. F. King, *J. Electrochem. Soc.*, 1961, **108**, 36.
121. H. Y. Zhao, P. Bian and D. Y. Ju, *J. Environ. Sci.*, 2009, **21**, S88.
122. R. Renuka and S. Ramamurthy, *J. Power Sources*, 2000, **87**, 144.
123. M. A. Deyab, *J. Power Sources*, 2016, **325**, 98.
124. M. M. Dinesh, K. Saminathan, M. Selvam, S. R. Srither and V. Rajendran, *J. Power Sources*, 2015, **276**, 32.
125. Y. Li, J. Ma, G. Wang, F. Ren, Y. Zhu and Y. Song, *J. Electrochem. Soc.*, 2018, **165**, A1713.
126. S. V. Lamaka, B. Vaghefinazari, D. Mei, R. P. Petrauskas, D. Höche and M. L. Zheludkevich, *Corros. Sci.*, 2017, **128**, 224.
127. S. V. Lamaka, D. Höche, R. P. Petrauskas, C. Blawert and M. L. Zheludkevich, *Electrochem. Commun.*, 2016, **62**, 5.

CHAPTER 13

Life Cycle Analysis of a Magnesium–Sulfur Battery

CLAUDIA TOMASINI MONTENEGRO[a], JENS F. PETERS[a], ZHIRONG ZHAO-KARGER[a], CHRISTOPHER WOLTER[c] AND MARCEL WEIL*[a,b]

[a]Helmholtz Institute Ulm for Electrochemical Energy Storage (HIU), Ulm, Germany; [b]Institute for Technology Assessment and System Analysis (ITAS), Karlsruhe Institute of Technology (KIT), Karlsruhe, Germany; [c]Custom Cells Itzehoe GmbH, Itzehoe, Germany

*E-mail: marcel.weil@kit.edu

13.1 Introduction

More than ever in our globalized society, humankind faces great challenges to decrease its impact on the environment. This has triggered the definition of quantifiable goals in the context of a sustainable energy transition. Regarding climate change, the 2016 Paris Agreement defined the goals to limit the increase in the global average temperature to below 2 °C above pre-industrial levels.[1]

In order to keep track of the 2 °C scenario, the role of energy storage systems for mobile and stationary applications is crucial. Among the numerous different storage technologies, lithium-ion batteries (LIBs) are experiencing by far the highest growth rates in the market.[2] Their versatility and high performance (*e.g.* high energy density, high power rate capability, life cycling and cost flexibility to scale to large capacity sectors) make

Energy and Environment Series No. 23
Magnesium Batteries: Research and Applications
Edited by Maximilian Fichtner

Published by the Royal Society of Chemistry, www.rsc.org

them the technology of choice for both stationary and mobile applications. However, their evolution faces challenges associated with both performance and cost. These limitations include a maximum achievable energy intensity of around 250–300 W h kg^{-1}, cost reduction limits of 100–150 US$/kW h,[3] resource availability[4] and associated supply risks, plus safety concerns associated with thermal runway and short circuiting due to dendrite formation.

Consequently, research on other battery chemistries is being conducted to identify new technologies with high energy densities, increased technical performance, and which are based on more abundant materials.[5] Magnesium rechargeable batteries (MRBs) are among the most promising candidates for substituting LIBs in this regard.[6] Their potential advantages in comparison to existing LIBs are their higher volumetric energy density (3837 mA h cm^{-3} for Mg *vs.* 2062 mA h cm^{-3} for lithium), low cost due to the use of cheap and abundant materials, and reduction in safety hazards associated with the absence of dendrite formation.[7]

While promising in theory, current MRBs are still in an early stage of research and face severe drawbacks, including the lack of stable and compatible electrolytes and high capacity cathodes.[6] Besides solving these technical obstacles, a prospective analysis of the potential environmental impacts associated with MRBs is required in order to identify the main drivers for their environmental impacts, which in turn can guide future MRB development. Currently, no analysis of the environmental impacts associated with MRBs is available. Thus, the aim of this study is to provide an environmental assessment of these batteries using a prospective life cycle assessment approach (LCA). Nonetheless, as pointed out by Sullivan *et al.* in 2012[8] and Peters *et al.* in 2018,[9] the analysis of the life cycle environmental impacts associated with different batteries (namely lithium-ion, nickel–cadmium, lead–acid and sodium–sulfur batteries) is hampered by two factors. Firstly, the stage of the research and development (R&D), and secondly, a lack of information about the assembly or the manufacturing process, which, due to either confidentiality or competitive reasons are usually not released. For this reason, the modelled battery is based on a Mg–S cell tested on a pilot scale by Wagner *et al.* in 2018[10] as part of the Mag–S project funded by the German Federal Ministry of Education (BMBF) for the "Development and manufacturing of rechargeable magnesium batteries".

The results of this prospective LCA serve to identify the processes that greatly contribute towards the carbon footprint and other environmental impacts of the analyzed Mg–S battery, and to understand its potential environmental performance in comparison to existing LIBs.

13.1.1 Status of the MRB

Detailed explanations of the stage of development of MRBs were presented by Mohtadi and Mizuno in 2014,[11] Yoo *et al.* in 2013,[12] and Park *et al.* in 2015.[13] In this section, a brief overview is given to justify the prospective approach considered for the estimation of the environmental impacts of the Mg–S battery within this chapter.

Progress in the field of MRBs towards a commercial level has been hampered primarily by the difficulties in finding a suitable electrolyte compatible with high energy cathodes.[7,11] In this regard, there have been two breakthroughs in the R&D of MRBs. The first was registered in 1990, when Gregory *et al.*[14] developed ethereal solutions with organoborate or organoaluminate anions.[12] However, these electrolytes have low voltages as a limitation and are chemically aggressive.[13] Thus, further research led to the second breakthrough, which came in 2000.[15] The outcome was the development of the Mg organoaluminate complex, or dichloro complex (DCC), which was the basis for the first Mg prototype developed by Aurbach *et al.* in 2000.[12,15] In spite of the stable electrochemical window, these electrolytes still face some challenges that have so far hampered the evolution of this Mg battery prototype to a commercial level.

One of the major challenges is corrosion due to the potential interaction of chloride with steel and aluminum, required as current collectors in the cell.[11] Another drawback is the volatility and chemical tendency of the electrolytes to form peroxides of tetrahydrofuran, which is the preferred solvent in DDC electrolytes. Despite these hurdles, the potential higher energy density of the MRBs triggered further research in companies (*i.e.* Sony, LG, Toyota, and Pellion Technologies) and research groups around the world.[12] Current research in the field is directed primarily towards the identification of high capacity redox compound cathodes or Mg ion insertion cathodes. Due to the natural abundance, low cost and low toxicity of the materials, the Mg–S couple is considered to be a promising candidate.[7,16] Additionally, Mg–S batteries exhibit a high theoretical specific energy of 1722 W h kg^{-1}, well above that of Li_2S/graphite (908 W h kg^{-1}).[7] Nevertheless, as with other MRB candidates, the Mg–S cell faces challenges associated with the electrolyte and the cathode, including loss of sulfur in the cathode, capacity loss and overpotential during charge.[7] In order to overcome these challenges, in 2017, Zhao-Karger *et al.*[17,18] established a robust synthesis route for an alkoxyborate electrolyte[18] ($Mg[B(hfip)_4]_2 \cdot 3DME$), hereafter referred to as $Mg[B(hfip)_4]_2$. Due to the unique features of $Mg[B(hfip)_4]_2$, which include non-corrosiveness, a scalable production process, air and hydrolysis stability, low cost and chemical compatibility with electrode materials, was the electrolyte of choice for the Mg–S pouch cell prototype developed by Wagner *et al.* in 2018.[10]

13.2 LCA Method

The estimation of the potential environmental impacts is conducted using a prospective LCA approach, following the ISO standards 14040/44.[19,20] Unlike other environmental assessment tools (*e.g.* environmental impact assessment (EIA)), LCA considers the whole life cycle of the MRB (Mg–S battery), thus providing a holistic picture and avoiding environmental burden shifting.[21] In consequence, LCA is widely applied for the estimation of the potential environmental impacts of batteries, including lithium-ion, lead-acid (PbA), nickel–cadmium (NiCd), nickel metal hydride, sodium–sulfur and sodium–ion batteries.[8,9,22]

13.2.1 Goal and Scope

The goal of this study is the identification and quantification of the environmental impacts associated with the manufacturing of a hypothetical Mg–S battery.[10] The environmental impacts are estimated based on the provided battery capacity. In consequence, the functional unit, *i.e.* the basis of comparison for the following LCA, is defined as 1 W h of storage capacity provided by the battery. This also allows identification of the main hotspots associated with manufacturing a Mg–S battery, while providing a first indicative environmental comparison of the Mg–S battery with a LIB, which is useful to deduce future design recommendations.

13.2.2 System and System Boundaries

As depicted in Figure 13.1, the product system considered for the analysis of the Mg–S battery comprises seven stages: (i) cathode production, (ii) anode production, (iii) separator production, (iv) electrolyte synthesis, (v) Mg–S cell assembly, (vi) battery management system (BMS) and casing manufacture, and (vii) manufacture of the Mg–S battery. The system boundaries include the production and assembly of the Mg–S battery from the extraction of the raw material to the factory gate (cradle-to-gate approach). This is helpful for identifying environmental hotspots in the battery production process, and is frequently used for the LCA of LIBs.[23,24] Use phase and end-of-life stage are disregarded in this approach, which might be considered a significant limitation when it comes to comparing different battery types. However, due to the early stage of development of MRBs, this can be considered reasonable since including use and end-of-life stages for such an early stage technology would be based on assumptions with high uncertainties.

13.2.3 Data Sources and Assumptions

The life cycle inventory (LCI) of the Mg–S battery system under analysis is developed following a hybrid approach. Primary data is used for modelling both cell components and assembly of the Mg–S cell.[10] For the electrode components, this includes mixing the active materials with binders and additives to cast them onto the metal foil of the cathode. Nevertheless, due to the lack of data, information about the manufacture of LIBs is used as a reference for the design of the Mg–S battery. Hence, it has been assumed that on a commercial level the manufacture process of the Mg–S cell set-up is similar to that of LIBs, as previously undertaken in the assessment of sodium batteries

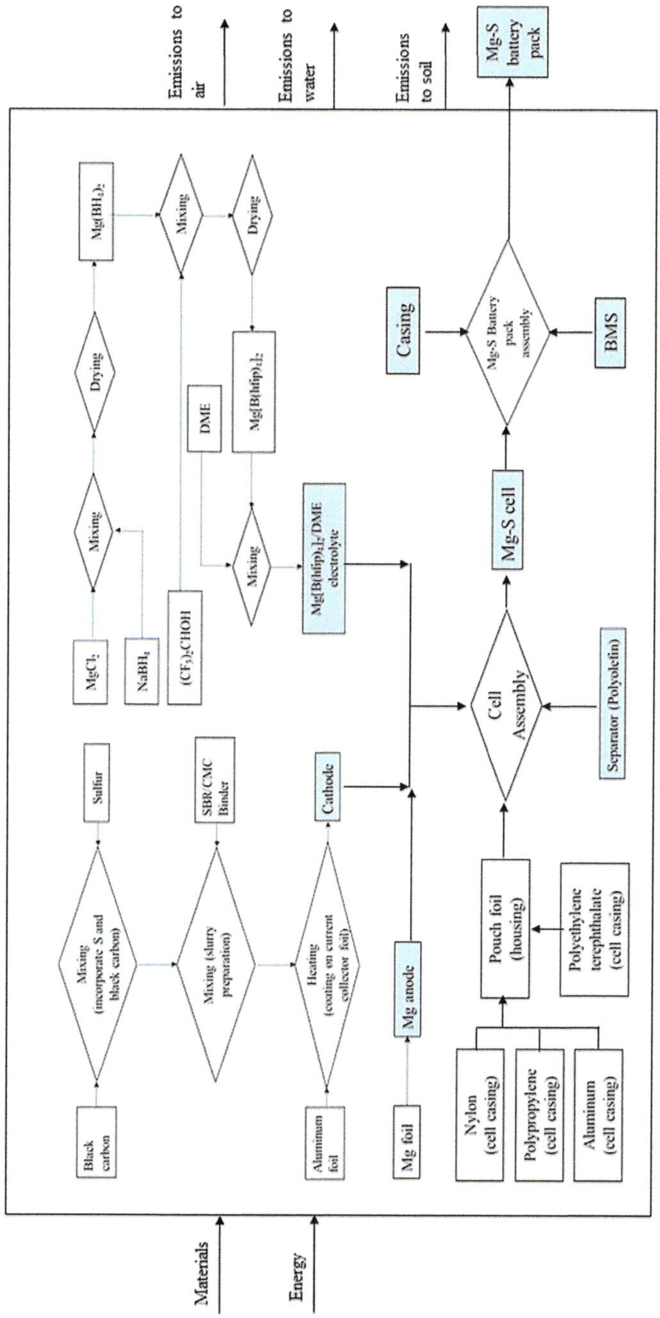

Figure 13.1 System boundaries of the Mg–S battery.

by Peters *et al.* in 2016.[22] While it is still unclear what a commercial production line for MRBs would look like, this approach allows a comprehensive comparison between the Mg–S battery and LIB on a life cycle basis, with a focus on the battery components and materials.[22]

13.2.4 Battery Cell Layout

13.2.4.1 Prototype Pouch Cell

The composition of the Mg–S cell is given in Table 13.1. The energy density of this prototype cell reaches 57 W h kg^{-1}.[10] This prototype cell is a monolayer cell on a pilot scale, and the cell layout is therefore not optimized for achieving maximum energy densities (as can be observed by the high mass share of the pouch cell housing). However, due to lack of information about possible commercial MRB cell layouts, this composition is used as a basis for the following assessments, assuming that the Mg–S battery is mounted in an automotive battery pack.

13.2.5 Data for Mg–S Battery Production and Assembly

Since there is no data for the large-scale production of Mg–S cells or Mg–S battery packs, the BMS, pack casing and energy demand for cell manufacturing are assumed to be similar to those of LIBs. The inventory data has been derived from the unified LCI of LIBs developed by Peters *et al.* in 2018,[9] assuming that the weight composition of the battery pack includes 80% of cells, 5.5% BMS and 14.5% of pack casing. With the Mg–S battery prototype cells previously defined, the whole battery pack would thus reach 45.6 W h kg^{-1} on a battery pack level. Energy required for the battery manufacture is sourced from the unified criteria developed by Peters *et al.* in 2018.[9] A summary of the information sources considered for the development of the LCI is provided in Table 13.2.

Table 13.1 Composition of the Mg–S cell.

Item	Function	Material	Amount (mg)	Share (%)
Anode	Active material	Mg foil	427	6.36
Cathode	Active material	Sulfur	421	6.28
	Conductive carbon	Black carbon	5	0.075
	Binder	SBR/CMC	5	0.075
	Collector foil	Al foil	88	1.32
Separator	Separator	Polyolefin	700	10.43
Electrolyte	Electrolyte	Mg fluorinated alkoxyborate-based	2060	30.72
Housing	Pouch container	Al composite	3000	44.74
Total	Pouch cell		6706	100

Table 13.2 Information source for the LCI of the Mg–S cell.

Input	Information source
Mg–S cell composition	Wagner *et al.*, 2018[10]
Electrolyte production path	Zhao-Karger *et al.*, 2017[17] and Zhao-Karger *et al.*, 2018[18]
Mg–S battery pack (BMS and casing)	Peters *et al.*, 2016[22]
Energy required for Mg–S battery manufacture	Peters *et al.*, 2018[9]

13.2.5.1 Life Cycle Inventory

As in previous studies, where Mg foil was used as an anode material,[25–27] the anode of the Mg–S cell set up defined by Wagner *et al.* in 2018[10] is prepared with magnesium foil of 100 μm thickness. Most of the magnesium (86%) of the 930 kilotons worldwide primary magnesium production originates from China, where it is mainly produced by the Pidgeon process (a variant of the thermal-reduction of MgO using coal as a primary fuel[28]). Depending on the country, other processes are alternatively used; either different variants of the thermal-reduction process (*i.e.* Magnetherm or Bolzano), or an electrolytic process with magnesium chloride ($MgCl_2$).[29] The LCIs of both magnesium production pathways are sourced from the ecoinvent database v3.4,[30] which is based on information provided by the International Magnesium Association (IMA) and reflects the market conditions previously explained.

Since the process of Mg foil formation is recent, there is no information for the corresponding LCI.[31,32] Therefore, the corresponding material and energy requirements of Mg foil formation are assumed to be similar to Al foil formation processes, despite the inherent differences associated with the hot rolling steps due to the different crystallization behavior of Mg.[33]

Since the inventory for the cathode is not available in the ecoinvent database, it is modelled assuming a slurry-casting method for the electrode coating process. In this method, the composite cathode is prepared by mixing a slurry comprising 50 wt% sulfur, 40 wt% carbon black and 10 wt% carboxy–methyl–cellulose/styrene–butadiene rubber (CMC/SBR) binder.[10] This mixture is cast on the aluminum current collector foil.[10,34] For proprietary reasons, the cathode manufacturing process is not disclosed by the project partner. Thus, electricity and heat requirements for both the slurry and casting are derived from a study by Notter *et al.* in 2010.[23] Furthermore, information for the LCI of the sulfur, electricity, heat, chemical plant, aluminum foil and carbon black are taken from ecoinvent version 3.4.[35] The LCI of the CMR/SBR binder stems from a study by Peters *et al.* in 2016.[22]

For the $Mg[B(hfip)_4]_2$ electrolyte, the LCI is built considering the synthesis route detailed by Zhao-Karger *et al.* in 2018,[18] according to Figure 13.1. Information about the energy and infrastructure requirements for the industrial production of $Mg[B(hfip)_4]_2$ is derived from the study by Sutter *et al.* in 2007,[36] considering a stoichiometric allocation. This LCI comprises three data sets that correspond to the production of $Mg[B(hfip)_4]_2$, and its precursors, namely magnesium borohydride ($Mg(BH_4)_2$) and magnesium chloride ($MgCl_2$).

For the latter, inventory data are derived from information about the Australian Magnesium process,[29] which requires anhydrous magnesium chloride ($MgCl_2$) for the production of magnesium in an electrolytic cell. According to this source, the synthesis route for $MgCl_2$ is based on the concentration of brine produced from the leaching between hydrochloric acid and magnesite. For proprietary reasons, neither the reaction nor the material requirements are disclosed for the dehydration process. However, information about the energy requirements for the dehydration process and the $MgCl_2$ brine production were provided by Johnson and Sullivan in 2014.[37] Thus, the LCI for the production of anhydrous $MgCl_2$ is derived from the stoichiometric allocation of the previously defined reaction, based on the LCI given by Johnson and Sullivan.[37] Regarding the production of magnesium borohydride ($Mg(BH_4)_2$), it is based on the synthesis route developed by Zhao-Karger *et al.* in 2017.[17] Since there is no information on the infrastructure and energy requirements for the production process, the approach used by Sutter *et al.* in 2007[36] is followed for the development of the LCI for $Mg(BH_4)_2$.

13.2.6 Results of the Environmental Impacts Associated with a Mg–S Battery

The environmental impact assessment of the production of the MRB (Mg–S battery) is conducted based on the CML 2001 impact assessment method,[38] using Open LCA version 1.6.3.[39] The environmental impacts associated with the Mg–S battery are given for 10 impact categories: global warming potential (GWP), abiotic depletion potential (ADP), acidification potential (AP), eutrophication potential (EP), freshwater aquatic ecotoxicity potential (FAETP), marine water aquatic ecotoxicity potential (MAETP), human toxicity potential (HTP), terrestrial ecotoxicity potential (TETP), stratospheric ozone depletion potential (ODP) and photochemical oxidation potential (POCP).

A breakdown of the potential environmental impacts of the Mg–S battery pack to the component level is provided in Figure 13.2. Electricity requirements for manufacturing the battery cells, the BMS and the pouch cell packaging are identified as the main drivers for environmental impacts in the majority of the assessed impact categories. The contributions of the individual battery components towards the environmental impacts within each category are discussed in the following section.

13.2.6.1 Abiotic Depletion Potential

The abiotic depletion potential (ADP), *i.e.* resource depletion, describes the decrease in the availability of natural resources, which includes fossil energy resources and minerals (Guinée *et al.*, 2001[40]). It is measured in terms of equivalents of antimony (Sb-eq.), which is the reference substance. The estimated ADP amounts to 2.45×10^{-3} kg antimony eq. W h^{-1}. As shown in Figure

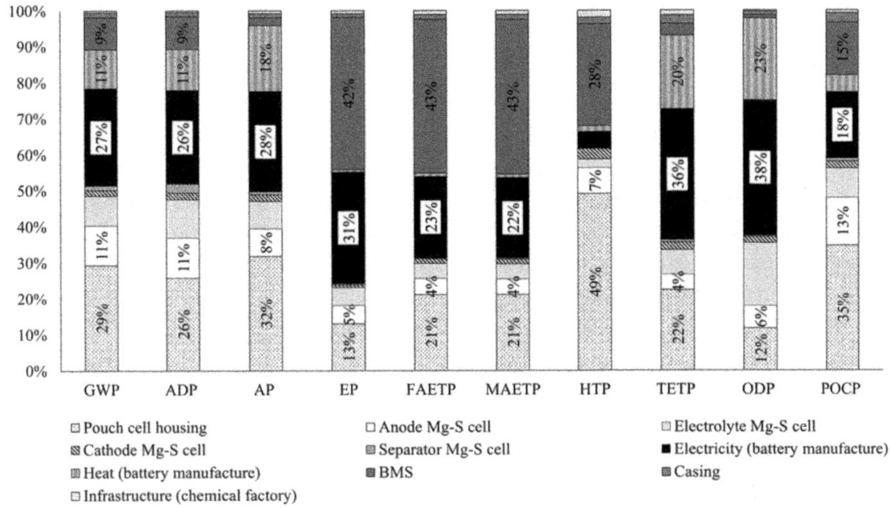

Figure 13.2 Environmental profile of the assessed Mg–S battery.

13.2, the Mg–S cell (excluding manufacturing energy) contributes almost 52% towards this impact category. Apart from that, the electricity and heat requirements for battery manufacturing (37%) are also highly relevant to the total ADP impact.

When looking at the Mg–S cell alone, the major share of ADP (1.27×10^{-3} kg Sb eq. W h^{-1} in total) is caused by the pouch cell housing (50%), anode (21%, 2.72×10^{-4} kg Sb eq. W h^{-1}) and electrolyte (21%, 2.72×10^{-4} kg Sb eq. W h^{-1}). The contribution of the Al and Mg foils alone makes up 67 and 99% of the ADP associated with the pouch cell housing and the anode, respectively. This is basically due to the high demand of electricity for their manufacturing and corresponding consumption of resources. Regarding the electrolyte, salt manufacturing is the dominating process (72%) in its ADP.

13.2.6.2 Global Warming Potential

In the context of climate change, the evaluation of the global warming potential (GWP) is an important measure of the contribution of greenhouse gas emissions to radiative forcing (Guinée *et al.*, 2001[40]). The estimated GWP of the Mg–S battery amounts to 3.23×10^{-1} kg CO_2 eq. W h^{-1}. The contribution of the Mg–S cell alone makes up 51% of the estimated GWP, mainly as a result of the pouch cell housing, anode and electrolyte, which contribute 57, 21, and 16%, respectively. While the main drivers for the GWP estimated for the pouch cell housing are Al foil (76%), nylon (6%) and polypropylene (5%), the GWP associated with the anode is dominated by the Mg foil (99%). Regarding the GWP estimated for the electrolyte, manufacture of $Mg[B(hfip)_4]_2$ salt is the main cause, contributing 84% to its GWP. The electricity and heat requirements for manufacturing the battery cell are identified as the second and third drivers, accounting for 27 and 11% of the total GWP of the Mg–S battery. According to a study by Peters *et al.* in 2018,[9] the dry room conditions are the main driver for the electricity demand during cell manufacturing. Consequently, due to the high air and hydrolysis stability of $Mg[B(hfip)_4]_2$ in comparison with LIB electrolytes,[17] the dry room conditions in the cell manufacturing process might be less extreme, which would lower the GWP. Nevertheless, due to the lack of information about the manufacturing process of the battery, and due to the sensitivity of fresh Mg metal surfaces to humid air this possible reduction is not included in the analysis.

13.2.6.3 *Acidification Potential*

The acidifying effect of anthropogenic chemicals is quantified and aggregated as SO_2 eq. in the acidification potential (AP) (Guinée *et al.*, 2001[40]). The main contributor to the total AP of 1.63×10^{-3} kg SO_2 eq. W h^{-1} are the Mg–S cell (50%), and electricity and heat (46%) required during battery manufacture. The pouch cell housing (69%), the anode (16%) and the electrolyte (15%) are the main drivers for the AP of the Mg–S cell. The APs of the housing and anode are dominated by Al foil and Mg foil, which respectively contribute towards 76 and 99% of their respective APs. The AP of 1.22×10^{-4} kg SO_2eq. W h^{-1} estimated for the electrolyte is mainly due to the salt manufacture, responsible for 87% of the total AP.

13.2.6.4 *Eutrophication Potential*

The potential of nutrient enrichment in sea, lakes and watercourses is quantified by the eutrophication potential (EP) (Guinée *et al.*, 2001[40]). Both the BMS (42%) and electricity requirements for battery manufacturing (31%) are identified as the dominant contributors towards the total EP, which amounts to 9.61×10^{-4} kg PO_4eq. W h^{-1}. For the BMS, this is mainly due to the wires and cables that contribute approximately 94% towards the total of 4.07×10^{-4} kg PO_4 eq. W h^{-1} (being copper mining associated with high EP impacts). An additional 24% of the total EP is caused by the Mg–S cell itself (2.32×10^{-4} kg PO_4eq. W h^{-1}), mainly due to the pouch cell housing (53%), the electrolyte (21%) and the anode (20%).

13.2.6.5 *Eco Toxicity Potentials*

The potential effects of toxic substances on aquatic and terrestrial ecosystems is described by the freshwater, marine aquatic, and terrestrial ecotoxicity potentials (FAETP, MAETP and TETP; Guinée *et al.*, 2001).[40] The total FAETP, MAETP and TETP (0.233, 8.30 and 7.73×10^{-5} kg 1,4DCBeq. W h^{-1}, respectively) of the Mg–S battery are shown in Figure 13.2. The main drivers for these impacts are the BMS and the electricity required for cell manufacture, but also the Mg–S cell itself. Regarding the cell, the pouch cell housing, the electrolyte and the anode are important contributors.

13.2.6.6 *Human Toxicity Potential*

The potential damage to human health is accounted for in the human toxicity potential (HTP) (Guinée *et al.*, 2001 [40]). The Mg–S cell (62%) and electricity requirements for cell manufacturing (28%) dominate the estimated HTP (3.25×10^{-1} kg 1,4DCBeq. W h^{-1}). Within the battery cell, the pouch cell housing (79%), anode (12%) and cathode (5%) are the main drivers of human toxicity impacts.

13.2.6.7 *Stratospheric Ozone Depletion Potential*

Ozone depletion potential (ODP) describes the effect of anthropogenic emissions in the reduction of the ozone layer (Guinée *et al.*, 2001 [40]). The total ODP amounts to 2.45×10^{-8} kg CFC-11eq. W h^{-1}, mainly due to the Mg–S cell itself (37%), electricity (36%) and heat requirements (12.5%) for the cell manufacturing process. The estimated ODP associated with the Mg–S cell is mainly due to the electrolyte (46%), the pouch cell housing (32%) and the anode (16%). The salt manufacture is the main contributor to the ODP associated with the electrolyte. The Al foil required for the current collector mainly contributes towards the ODP associated with the pouch cell housing. With regard to the ODP of the anode, it is mainly produced by the Mg foil required for the anode.

13.2.6.8 *Photochemical Ozone Creation Potential*

The photochemical ozone creation potential (POCP) describes the contribution to ozone formation produced by the interaction of air pollutants and sunlight (Guinée *et al.*, 2001 [40]). For the Mg–S battery, the total POCP amounts to 9.13×10^{-5} kg ethylene eq. W h^{-1}. The Mg–S cell (59%), electricity required for the cell manufacture (18%) and BMS (15%) are identified as the main contributors to the total POCP. While the Mg–S cell contribution to this impact category is due to the manufacture of the pouch packaging, anode, and the electrolyte, contributing 35, 13 and 8%, respectively. Al foil is the main driver of the POCP (85%) associated with the pouch cell housing. Regarding the anode, the Mg foil production (98%) is identified as the main cause of POCP, while the salt manufacture (68%) is the main cause of the POCP of the electrolyte.

13.2.7 Sensitivity Analysis

A series of "what if" cases is considered to evaluate the effect of varying assumptions for parameters that have been identified as relevant for the outcome of the study. These "what if" cases include the variation of different materials for manufacturing the pouch cell, the gravimetric energy density and the electricity mix used for manufacturing the cell and the battery.

Table 13.3 Influence of the increase in the energy density on a W h basis.

Impact category	Original MRB	MRB-500
GWP (kg CO_2 eq. W h^{-1})	3.23×10^{-1}	3.85×10^{-2}
ADP (kg antimony eq. W h^{-1})	2.45×10^{-3}	3.06×10^{-4}
AP (kg SO_2 eq. W h^{-1})	1.63×10^{-3}	1.77×10^{-4}
EP (kg PO_4 eq. W h^{-1})	9.61×10^{-4}	1.12E-0
FAETP (kg 1,4 DCB eq. W h^{-1})	2.33×10^{-1}	2.53×10^{-2}
MAETP (kg 1,4 DCB eq. W h^{-1})	8.30×10^{-1}	9.01×10^{-2}
HTP (kg 1,4 DCB eq. W h^{-1})	3.25×10^{-1}	3.02×10^{-2}
TETP (kg 1,4 DCB eq. W h^{-1})	8.42×10^{-5}	9.20×10^{-6}
ODP (kg CFC-11 eq. W h^{-1})	2.46×10^{-8}	3.05×10^{-9}
POCP (kg ethylene eq. W h^{-1})	9.13×10^{-5}	1.10×10^{-5}

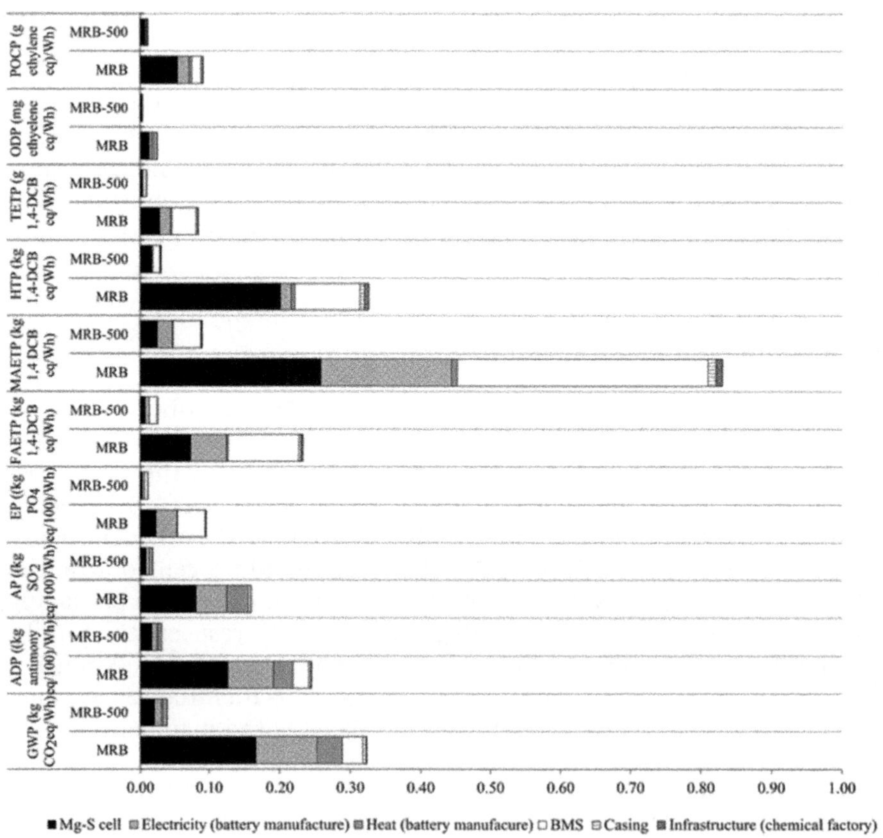

Figure 13.3 Effect of an increase in the energy density on the estimated environmental impacts.

13.2.7.1 Influence of the Mg–S Battery Energy Density

The initially considered MRB (Mg–S battery) prototype shows a comparably low energy density of 57 W h kg^{-1}, which can be considered far from competitive with existing LIBs, but also for future MRBs. Therefore, a hypothetical energy density of 500 W h kg^{-1} is assumed (MRB-500), which has been assumed as an objective value for developing future MRBs.[12] The results of the environmental assessment of both systems are given in Table 13.3 and are further broken down into battery components in Figure 13.3.

The impacts per W h are reduced proportionally between the initial MRB, with 45.6 W h kg^{-1}, and the battery with a hypothetical 500 W h kg^{-1}. This highlights the relevance of a high energy density for the environmental performance of a battery; in other words, a higher energy density requires the production of less battery mass per W h of storage capacity, especially for the pouch and housing, and thus has lower environmental impacts. Therefore, from an environmental perspective, improving the performance of future MRBs is identified as a key factor.

13.2.7.2 Influence of the Electricity Mix Considered for Mg–S Cell and Battery Manufacture

As previously discussed, electricity requirements are identified as one of the principal drivers for the estimated environmental impacts associated with the Mg–S battery. In view of this finding, the influence of different electricity mixes used for battery manufacturing on the total environmental impacts is evaluated. While for the base case, a European electricity mix is assumed, the influence of using an electricity mix of either high carbon intensity (*e.g.* a Chinese electricity mix), or with a high share of renewable energy sources and thus of low carbon intensity (*e.g.* a Swiss electricity mix) is assessed in the following section.

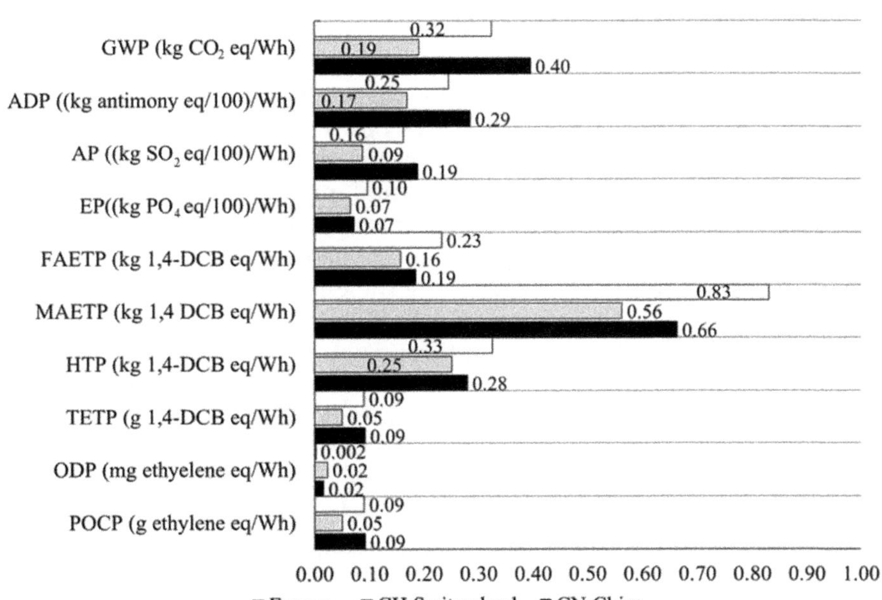

Figure 13.4 Effect of the electricity mix on selected environmental impacts.

As can be seen in Figure 13.4, due to the high share of coal power in the Chinese electricity mix (76%),[41,42] the estimated GWP, ADP, AP, TETP and POCP are increased significantly compared to the base case, while using the Swiss electricity mix reduces them. However, due to the emissions from waste management, the highest values for toxicity impacts (HTP, FAETP, and MAETP) and EP are obtained with the European electricity mix, followed by the Chinese and Swiss electricity mixes.

In a climate change context, the results show that the carbon content of the electricity mix used for cell and pack manufacture strongly determines the carbon footprint of the Mg–S battery, but also other impacts, including AP, TETP, ADP and POCP. Associated toxicity impacts (*e.g.* FAETP, MAETP and HTP) are related to the waste management of the residues from lignite mining driven by the electricity mix. These results can be explained because the European electricity mix relies more heavily on lignite, which exhibits less toxicity impact from lignite spill than the Chinese electricity mix that is more strongly hard-coal oriented.

13.2.7.3 Influence of the Packaging Material of the Pouch Cell

The analysis in Section 13.2.6 revealed a high contribution of the pouch cell package to the total environmental impacts of the Mg–S battery. According to a study by Wagner *et al.* in 2018,[10] the high mass of the pouch is a result of its very early stage of development. If the share of the pouch housing remains as a specific requirement of MRB cells, it would be relevant for this type of battery to find a pouch material that minimizes the corresponding environmental impacts, while still providing the function of guaranteeing integrity of the cell. Based on the information provided in the 2018 study by Wagner *et al.*,[10] the material of the pouch cell is Al and plastic composite, but polypropylene (PE) and polyvinyl chloride (PVC) are common alternative materials for pouch cells.[43] The results for the Mg–S battery when assuming these (hypothetical) alternative pouch materials for the battery cell are provided in Table 13.4, while Figure 13.4 displays the contribution of the individual components towards the total environmental impacts.

Although hypothetical, the results of this analysis point out that the substitution of the aluminum composite cell pouch material (See Figure 13.5) shows significant potential for reducing the environmental impacts of the Mg–S battery. This is the case especially for the impact categories where the battery cells show high contributions, such as GWP, ADP, ODP or POCP. Therefore, impact reductions of up to 40% can be achieved by the use of an alternative material. In categories dominated by other components or by the manufacture energy demand of the Mg–S battery, such as the EP or toxicity categories, the potential for reducing impacts by using an alternative pouch material is naturally lower. A reduction in the environmental impacts is achieved in all assessed impact categories for both the PE pouch and the PVC pouch, while PE shows generally better results than PVC. However, this simple sensitivity analysis does not consider the properties of the alternative pouch materials at all. Simple PE or PVC pouches might show

Table 13.4 Influence of different materials for the pouch cell on the environmental impacts.

Impact category	Cell with Al-composite	Cell with polypropylene (PE)	Cell with polyvinyl chloride (PVC)
GWP (kg CO_2 eq. W h^{-1})	3.23×10^{-1}	2.54×10^{-1}	2.76×10^{-1}
ADP ((kg antimony eq. 100) W h^{-1})	2.45×10^{-1}	2.16×10^{-1}	2.15×10^{-1}
AP ((kg SO_2 eq./100)W h^{-1})	1.63×10^{-1}	1.20×10^{-1}	1.36×10^{-1}
EP ((kg PO_4 eq./100)W h^{-1})	9.61×10^{-2}	8.64×10^{-1}	8.81×10^{-1}
FAETP (kg 1,4 DCB eq. W h^{-1})	2.33×10^{-1}	1.89×10^{-1}	1.92×10^{-1}
MAETP (kg 1,4 DCB eq. W h^{-1})	8.30×10^{-1}	6.74×10^{-1}	6.88×10^{-1}
HTP (kg 1,4 DCB eq. W h^{-1})	3.25×10^{-1}	1.71×10^{-1}	1.84×10^{-1}
TETP (g 1,4 DCB eq. W h^{-1})	8.42×10^{-2}	6.86×10^{-2}	8.66×10^{-2}
ODP (mg CFC-11 eq. W h^{-1})	2.46×10^{-2}	2.92×10^{-2}	2.30×10^{-2}
POCP (g ethylene eq. W h^{-1})	9.13×10^{-2}	6.56×10^{-2}	7.46×10^{-2}

inferior resistance or durability than the composite material, and eventually require an even higher mass share of the cell pouch for assuring the same function. In fact, commercial LIB pouch cells exclusively use aluminum composite, driven by the superior properties of this material. However, for MRBs the requirements might vary (as can be seen by the high pouch mass share required by the first Mg–S cell), which is why different pouch packages can be considered. Despite the shortcomings of this simplified analysis; it helps to point out the relevance of the aluminum content in the cell pouch for the overall environmental performance of the considered Mg–S battery prototype cell.

13.2.7.4 Comparison with LIBs

The development of new battery chemistries like those of MRBs aims to overcome the limitations of existing LIBs in terms of performance, safety and environmental impacts. Since MRBs need to compete with LIBs also in terms of environmental performance, the environmental impacts of the Mg–S battery is compared with that of a common LIB. The LIB model used for this purpose is an NMC-C type battery based on the study by Ellingsen *et al.* in 2014.[24] As previously mentioned, the Mg–S battery prototype pouch cell shows an extremely high mass share for the cell pouch, significantly reducing the energy density. Although no information about whether this is a specific requirement of MRB cells in general or just a result of lacking an optimized prototype cell, for the following comparison equal mass shares are assumed for the pouch cells. The average pouch cell mass share considered for both the LIB and the MRB is 3% according to a study by Peters *et al.* in 2018.[9] The mass shares of the remaining cell components are scaled linearly, as shown in Table 13.5. Consequently, there is an increase in the energy density of the Mg–S cell to 100.5 W h kg^{-1} on a cell level. In addition to the analysis on a cell level, the comparison is also done at the battery pack level, assuming hypothetical automotive battery packs with identical mass shares of pack components for the two cell chemistries (also shown in Table 13.5). The gravimetric

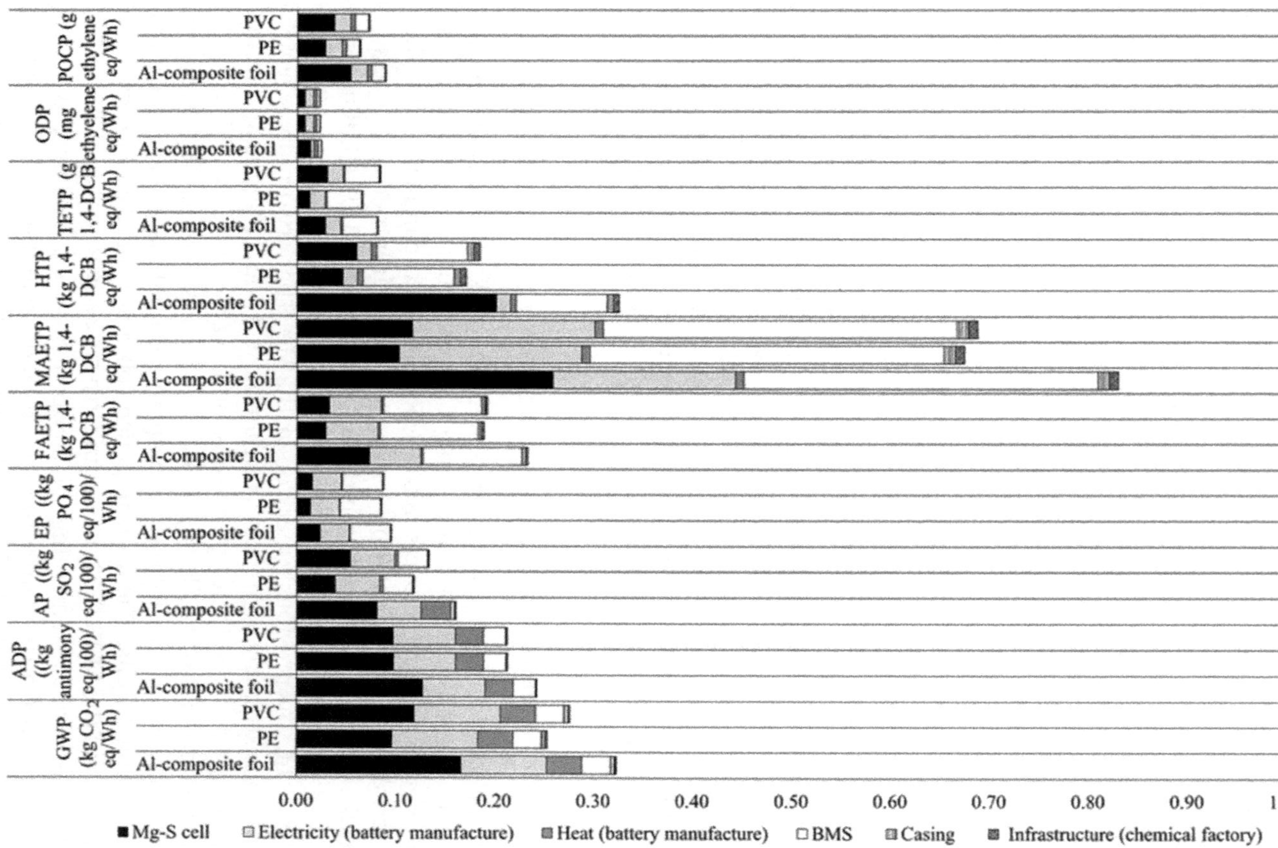

Figure 13.5 Influence of pouch cell material variation on the environmental impacts.

Table 13.5 Optimized composition of the Mg–S cell based on the NMC pouch cell.

	Composition of the cell	
Item	NMC pouch cell share	Actual Mg–S pouch cell share
Anode	37%	11%
Cathode	42%	13%
Electrolyte	16%	55%
Housing	3.0%	3.0%
Separator	2.2%	18%
	Composition of the battery	
Item	NMC battery share	Mg–S battery share
BMS	5.5%	5.5%
Casing	14.5%	14.5%
Cell	80%	80%

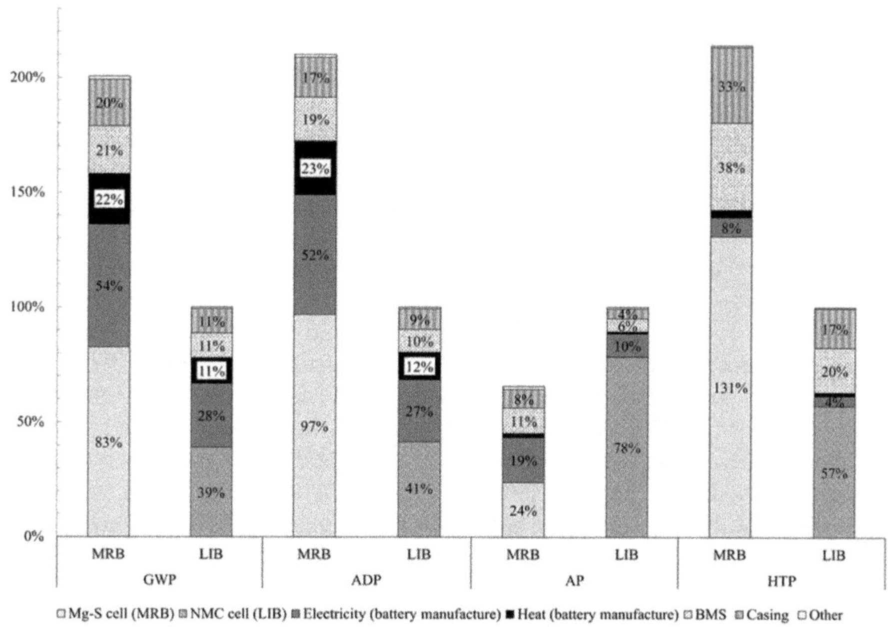

Figure 13.6 Contribution analysis of the LIB battery and the Mg–S battery.

energy density of the optimized Mg–S battery is 80.4 W h kg^{-1}, while the corresponding value for the NMC battery is 155.7 W h kg^{-1}.[19]

The results of the comparison per W h of storage capacity are provided in Figure 13.6. For the selected environmental impacts, these are dominated by cell manufacture (cell components and manufacturing energy). Compared to the NMC-C battery, the Mg–S battery has higher impacts in terms of global warming (GWP), resource depletion (ADP) and acidification (AP). Except for GWP, the main contributors to the potential impacts associated with Mg–S battery production are the electrolyte salt $Mg[B(hfip_4)_2]$, followed by the energy demand for cell manufacturing. In contrast, for the NMC-C battery, impacts are driven mainly by the cathode active material (nickel cobalt manganese hydroxide manufacture). The latter is also responsible for the major share of GWP of the NMC cell, while for the Mg–S battery, the main contribution comes from the manufacture of the Mg foil required for the anode. A key for the higher impacts associated with the MRB is its comparably low energy density. This can be observed by the contribution from the manufacturing energy, which is assumed to be identical per amount (mass) of battery produced. However, due to the lower energy density of this first Mg–S battery, its contribution per W h of storage capacity is significantly higher, since a higher mass of battery would need to be produced for the same capacity.

13.3 Conclusions

Magnesium rechargeable batteries (MRBs) are considered an environmentally friendly technology for future energy storage systems. Due to the lack of information about the environmental impacts associated with this technology, the present assessment is a first attempt towards the identification and quantification of the potential environmental hotspots of MRBs. This study provides a preliminary comparison between a Mg–S battery and a LIB to understand similarities and differences in their environmental profiles. However, it should be noted that due to the preliminary stage of the development of the Mg–S battery, high uncertainties are associated with the model. Therefore, the outcome of this analysis should be considered purely on an indicative level.

A key driver for environmental impacts is the manufacture of the pouch cell packaging, and within this, the contained aluminum. It is identified as the main hotspot for global warming potential (GWP), abiotic depletion (ADP), human toxicity (HTP), acidification (AP) and photochemical oxidation potential (POCP). However, the mass share of the modelled prototype pouch package is roughly 50% of the total battery cell mass, which is considered far from realistic. When comparing this value with corresponding values from existing LIBs, it can be observed that the actual mass share of the pouch cell is roughly 3%. Thus, the mass composition of the Mg–S battery as provided for the prototype pouch cell seems neither optimized, nor realistic for a future battery system.

For the packaging of the prototype pouch cell, an aluminum composite material is used, which could be replaced by PVC or PE foil. In this case and with the high cell pouch mass share of the prototype cell, the GWP of the battery could be reduced from 3.23×10^{-1} (kg CO_2 eq. W h^{-1}) to 2.76×10^{-1} (kg CO_2 eq. W h^{-1}) and 2.54×10^{-1} (kg CO_2 eq. W h^{-1}). Thus, if the high mass of the pouch package is a requirement of MRB cells, the future development of Mg–S cells could evaluate PE and PVC foil for pouch cell manufacture.

The main contributor of both toxicity (freshwater FAETP and marine aquatic MAETP) and eutrophication potential (EP) is identified as the battery management system (BMS) (and not the battery cells), which is attributed to the emissions from sulfidic tailing associated with the manufacturing of the wiring board of the BMS. Also, it is interesting to note that due to the electricity requirements, electrolyte manufacture is identified as the main cause of the estimated stratospheric ozone depletion potential (ODP). In addition to aluminum and cable manufacture, another hotspot for all of the estimated impacts is the electricity mix considered for the battery and cell manufacture.

For all of the assessed impact categories, the electricity demand for cell manufacturing contributes significant shares. Due to the lack of more detailed information, the same energy demand as for LIB cell manufacturing is assumed in the present assessment. However, MRB electrolytes are less hygroscopic than those used commonly for LIB, thus potentially reducing energy demand for manufacturing significantly (the dry room required due to the hygroscopic properties of the electrolyte is now one of the main drivers of electricity demand for LIB cell manufacturing). Here, MRBs offer significant potential for further reducing environmental impacts. Apart from that, the energy source used for battery production is of great importance for the environmental profile of the battery. The use of a relatively low carbon intensive energy mix (from Switzerland) is beneficial, whereas the use of a predominantly coal-based electricity mix (from China) is disadvantageous regarding the environmental impacts.

Regarding the comparison between the LIB and the MRB, it is interesting to note that for both hypothetical battery pack configurations, the battery cells are the main cause of potential environmental impacts, firstly due to the required materials, but also due to the energy requirements for the battery manufacturing process. While the latter has been assumed to be identical to that of the LIB, MRB electrolytes are potentially less hygroscopic and therefore offer significant improvement potential. The use of an electricity mix based on renewables would also decrease impacts further, though this applies identically to both MRBs and LIBs. The key to further reducing the environmental impacts of the MRB is the optimization of the cell layout, thus achieving higher energy densities. In the current configuration, the prototype cells contain high shares of electrolyte and separator, which were required from a design perspective to reach the energy density of the Mg–S cell (57 W h kg^{-1}). Reducing these, and increasing the share of the active materials would increase energy densities significantly, and would potentially change the results (the comparison is done per W h, for higher energy densities, with less cell mass required, thus reducing the associated impacts). In this sense, achieving higher energy densities by improving and optimizing the battery cell layout can be pointed out as a relevant factor for reducing the environmental impacts associated with future MRBs. However, according to the current state of the MRB laboratory prototype cell, this improvement in battery energy density is hindered by the small share of active materials because of the high amount of the electrolyte and the separator. Consequently, the composition of the cell is identified as a relevant optimization factor for reducing the environmental impacts of MRBs.

In summary, it has to be pointed out that the present life cycle assessment of MRB is at a preliminary stage and therefore associated with significant uncertainties, which have to be taken into account when interpreting the results. However, this kind of information is crucial to identify and quantify the potential environmental impacts of these batteries in a prospective way in order to understand their role within the sustainability context compared to other energy storage systems. Furthermore, if MRBs were produced on a commercial level, then this information is also required for the development of policies for the disposal of these kind of batteries. Additionally, the outcome of this study can be used to optimize future Mg–S systems on a materials, cell and battery level. Moreover, future LCA studies could be useful to answer the question at hand about the environmental impacts associated with the recyclability of Mg–S batteries.

References

1. IPCC, *Climate Change 2014: Synthesis Report. Contribution of Working Groups I, II and III to the Fifth Assessment Report of the Intergovernmental Panel on Climate Change*, Intergovernmental Panel on Climate Change Geneva, Switzerland, 2014.
2. J.-M. Tarascon and M. Armand, in *Materials for Sustainable Energy: A Collection of Peer-Reviewed Research and Review Articles from Nature Publishing Group*, World Scientific, 2011, p. 171.
3. P. T. Moseley and J. Garche, *Electrochemical Energy Storage for Renewable Sources and Grid Balancing*, Newnes, 2014.
4. M. Weil, Z. Saskia and J. Peters, in *Behaviour of Lithium-Ion Batteries in Electric Vehicles. Battery Health, Performance, Safety and Cost*, ed. B. Liaw, Springer 2017.
5. C. Vaalma, D. Buchholz, M. Weil and S. Passerini, *Nat. Rev. Mater.*, 2018, **3**, 18013.
6. D. Aurbach, Y. Gofer, Z. Lu, A. Schechter, O. Chusid, H. Gizbar, Y. Cohen, V. Ashkenazi, M. Moshkovich and R. Turgeman, *J. Power Sources*, 2001, **97**, 28.
7. Z. Zhao-Karger and M. Fichtner, *MRS Commun.*, 2017, **7**, 770.
8. J. L. Sullivan and L. Gaines, *Energy Convers. Manage.*, 2012, **58**, 134.
9. J. F. Peters and M. Weil, *J. Cleaner Prod.*, 2018, **171**, 704.
10. N. Wagner, M. Fichtner, J. Häcker, Z. Zhao-Karger, H.-G. Bremes, C. Wolter, F. Kampmann, R. Richter and T. Danner, *2nd International Symposium on Magnesium Batteries*, Ulm, Germany, 2018.
11. R. Mohtadi and F. Mizuno, *Beilstein J. Nanotechnol.*, 2014, **5**, 1291.
12. H. D. Yoo, I. Shterenberg, Y. Gofer, G. Gershinsky, N. Pour and D. Aurbach, *Energy Environ. Sci.*, 2013, **6**, 2265.
13. J. G. Kim, M. S. Park, Y. J. Kim, N. S. Choi and J. S. Kim, *Isr. J. Chem.*, 2015, **55**, 570.
14. T. D. Gregory, R. J. Hoffman and R. C. Winterton, *J. Electrochem. Soc.*, 1990, **137**, 775.
15. D. Aurbach, Z. Lu, A. Schechter, Y. Gofer, H. Gizbar, R. Turgeman, Y. Cohen, M. Moshkovich and E. Levi, *Nature*, 2000, **407**, 724.
16. J. Zhu, J. Zou, H. Cheng, Y. Gu and Z. Lu, *Green Energy Environ.*, 2018, DOI: 10.1016/j.gee.2018.07.001.

17. Z. Zhao-Karger, M. E. G. Bardaji, O. Fuhr and M. Fichtner, *J. Mater. Chem. A*, 2017, **5**, 10815.
18. Z. Zhao-Karger, R. Liu, W. Dai, Z. Li, T. Diemant, B. Vinayan, C. Bonatto Minella, X. Yu, A. Manthiram and R. J. Behm, *ACS Energy Lett.*, 2018, **3**, 2005.
19. ISO 14040, *Environmental Management–Life Cycle Assessment–Principles and Framework*, 2006.
20. ISO 14044, *Environmental Management–Life Cycle Assessment–Requirements and Guidelines*, 2006.
21. A. Azapagic, *Chem. Eng. J.*, 1999, **73**, 1.
22. J. Peters, D. Buchholz, S. Passerini and M. Weil, *Energy Environ. Sci.*, 2016, **9**, 1744.
23. D. A. Notter, M. Gauch, R. Widmer, P. Wager, A. Stamp, R. Zah and H.-J. Althaus, *Environ. Sci. Technol.*, 2010, **44**, 6550.
24. L. A. W. Ellingsen, G. Majeau-Bettez, B. Singh, A. K. Srivastava, L. O. Valøen and A. H. Strømman, *J. Ind. Ecol.*, 2014, **18**, 113.
25. H. S. Kim, T. S. Arthur, G. D. Allred, J. Zajicek, J. G. Newman, A. E. Rodnyansky, A. G. Oliver, W. C. Boggess and J. Muldoon, *Nat. Commun.*, 2011, **2**, 427.
26. T. Gao, M. Noked, A. J. Pearse, E. Gillette, X. Fan, Y. Zhu, C. Luo, L. Suo, M. A. Schroeder and K. Xu, *J. Am. Chem. Soc.*, 2015, **137**, 12388.
27. X. Yu and A. Manthiram, *ACS Energy Lett.*, 2016, **1**, 431.
28. Mineral Commodity Summaries Online Multimedia 09/10/18.
29. S. Mathaudhu, A. Luo, N. Neelameggham, E. Nyberg and W. Sillekens, *Essential Readings in Magnesium Technology*, Springer, 2016.
30. E. Moreno Ruiz, New data in version 3.1, Changes from v2.2, https://www.ecoinvent.org/files/140516_moreno_new_data.pdf, Ecoinvent Centre, 2016, accesed June, 2018.
31. F. Zarandi and S. Yue, in *Magnesium Alloys-Design, Processing and Properties*, InTech, 2011.
32. Passoni *et al.*, *US Pat.*, US 2012/0227455A, 2012.
33. M. Hakamada, T. Furuta, Y. Chino, Y. Chen, H. Kusuda and M. Mabuchi, *Energy*, 2007, **32**, 1352.
34. M. Hagen, D. Hanselmann, K. Ahlbrecht, R. Maça, D. Gerber and J. Tübke, *Adv. Energy Mater.*, 2015, **5**, 1401986.
35. B. P. Weidema, C. Bauer, R. Hischier, C. Mutel, T. Nemecek, J. Reinhard, C. Vadenbo and G. Wernet, *Overview and methodology: Data quality guideline for the ecoinvent database version 3*, The Ecoinvent Centre, St. Gallen, 2013.
36. J. Sutter, *Life cycle inventories of highly pure chemicals*, Ecoinvent report No. 19, Swiss Center for Life Cycle Inventories, Dübendorf, 2007.
37. M. C. Johnson and J. L. Sullivan, *Lightweight Materials for Automotive Application: An Assessment of Material Production Data for Magnesium and Carbon Fiber*, ANL/ESD-14/7107574, Argonne National Lab. (ANL), United States, 2014.
38. J. Guinée, *Int. J. Life Cycle Assess.*, 2001, **6**, 255.
39. GreenDelta, OpenLCA v1.6.3, 2018, https://openlca.org.

40. J. Guinée, M. Gorrée, R. Heijungs, G. Huppes, R. Kleijn, A. de Koning, L. van Oers, A. W. Sleeswijk, S. Suh and H. U. de Haes, *Spatial Planning and Environment (VROM) and Centre of Environmental Science (CML)*, The Netherlands, 2001.
41. K. Treyer and C. Bauer, *Int. J. Life Cycle Assess.*, 2016, **21**, 1236.
42. K. Treyer and C. Bauer, *Int. J. Life Cycle Assess.*, 2016, **21**, 1255.
43. M. Herrmann, *Packaging Materials-a review, AIP Conference Proceedings*, 2014, vol. 1597, No. 1, p. 121.

Subject Index